AF775904

Consumers and food

Understanding and shaping consumer behaviour

It is widely recognised that agriculture is a significant contributor to global warming and climate change. Agriculture needs to reduce its environmental impact and adapt to current climate change whilst still feeding a growing population, i.e. become more 'climate-smart'. Burleigh Dodds Science Publishing is playing its part in achieving this by bringing together key research on making the production of the world's most important crops and livestock products more sustainable.

Based on extensive research, our publications specifically target the challenge of climate-smart agriculture. In this way we are using 'smart publishing' to help achieve climate-smart agriculture.

Burleigh Dodds Science Publishing is an independent and innovative publisher delivering high quality customer-focused agricultural science content in both print and online formats for the academic and research communities. Our aim is to build a foundation of knowledge on which researchers can build to meet the challenge of climate-smart agriculture.

For more information about Burleigh Dodds Science Publishing simply call us on +44 (0) 1223 839365, email info@bdspublishing.com or alternatively please visit our website at www.bdspublishing.com.

Related titles:

Understanding the behaviour and improving the welfare of chickens
Print (ISBN 978-1-78676-422-5); Online (ISBN 978-1-78676-424-9, 978-1-78676-425-6)

Understanding the behaviour and improving the welfare of pigs
Print (ISBN 978-1-78676-443-0); Online (ISBN 978-1-78676-444-7, 978-1-78676-445-4)

Understanding the behaviour and improving the welfare of dairy cattle
Print (ISBN 978-1-78676-459-1); Online (ISBN 978-1-78676-461-4, 978-1-78676-462-1)

Improving the nutritional and nutraceutical properties of wheat and other cereals
Print (ISBN 978-1-78676-479-9); Online (ISBN 978-1-78676-481-2, 978-1-78676-482-9)

Developing smart agri-food supply chains: Using technology to improve safety and quality
Print (ISBN 978-1-78676-749-3); Online (ISBN 978-1-78676-751-6, 978-1-78676-752-3)

Understanding and improving the functional and nutritional properties of milk
Print (ISBN 978-1-78676-819-3); Online (ISBN 978-1-78676-821-6, 978-1-78676-822-3)

Energy-smart farming: Efficiency, renewable energy and sustainability
Print (ISBN 978-1-78676-835-3); Online (ISBN 978-1-78676-837-7, 978-1-78676-838-4)

Understanding and optimising the nutraceutical properties of fruit and vegetables
Print (ISBN 978-1-78676-850-6); Online (ISBN 978-1-78676-852-0, 978-1-78676-853-7)

Chapters are available individually from our online bookshop:
https://shop.bdspublishing.com

climate•SMART•publishing
IN AGRICULTURAL SCIENCE

BURLEIGH DODDS SERIES IN AGRICULTURAL SCIENCE
NUMBER 144

Consumers and food

Understanding and shaping consumer behaviour

Edited by Professor Marian Garcia Martinez, The University of Kent, UK

burleigh dodds
SCIENCE PUBLISHING

Published by Burleigh Dodds Science Publishing Limited
82 High Street, Sawston, Cambridge CB22 3HJ, UK
www.bdspublishing.com

Burleigh Dodds Science Publishing, 1518 Walnut Street, Suite 900, Philadelphia, PA 19102-3406, USA

First published 2023 by Burleigh Dodds Science Publishing Limited
© Burleigh Dodds Science Publishing, 2024. All rights reserved.

This book contains information obtained from authentic and highly regarded sources. Reprinted material is quoted with permission and sources are indicated. Reasonable efforts have been made to publish reliable data and information but the authors and the publisher cannot assume responsibility for the validity of all materials. Neither the authors nor the publisher, nor anyone else associated with this publication shall be liable for any loss, damage or liability directly or indirectly caused or alleged to be caused by this book.

No part of this publication may be reproduced, stored in a retrieval system or transmitted in any form or by any means electronic, mechanical, photocopying, recording or otherwise without the prior written permission of the publisher.

The consent of Burleigh Dodds Science Publishing Limited does not extend to copying for general distribution, for promotion, for creating new works, or for resale. Specific permission must be obtained in writing from Burleigh Dodds Science Publishing Limited for such copying.

Permissions may be sought directly from Burleigh Dodds Science Publishing at the above address. Alternatively, please email: info@bdspublishing.com or telephone (+44) (0) 1223 839365.

Trademark notice: Product or corporate names may be trademarks or registered trademarks and are used only for identification and explanation, without intent to infringe.

Notice
No responsibility is assumed by the publisher for any injury and/or damage to persons or property as a matter of product liability, negligence or otherwise, or from any use or operation of any methods, products, instructions or ideas contained in the material herein.

Library of Congress Control Number: 2023940750

British Library Cataloguing in Publication Data
A catalogue record for this book is available from the British Library

ISBN 978-1-80146-354-6 (print)
ISBN 978-1-80146-356-0 (PDF)
ISBN 978-1-80146-355-3 (ePub)
ISSN 2059-6936 (print)
ISSN 2059-6944 (online)

DOI: 10.19103/AS.2023.0129

Typeset by Deanta Global Publishing Services, Dublin, Ireland

Contents

© Burleigh Dodds Science Publishing Limited, 2024. All rights reserved.

© Burleigh Dodds Science Publishing Limited, 2024. All rights reserved.

Part 2 Product attributes

© Burleigh Dodds Science Publishing Limited, 2024. All rights reserved.

© Burleigh Dodds Science Publishing Limited, 2024. All rights reserved.

Series list

© Burleigh Dodds Science Publishing Limited, 2024. All rights reserved.

© Burleigh Dodds Science Publishing Limited, 2024. All rights reserved.

© Burleigh Dodds Science Publishing Limited, 2024. All rights reserved.

© Burleigh Dodds Science Publishing Limited, 2024. All rights reserved.

© Burleigh Dodds Science Publishing Limited, 2024. All rights reserved.

© Burleigh Dodds Science Publishing Limited, 2024. All rights reserved.

© Burleigh Dodds Science Publishing Limited, 2024. All rights reserved.

© Burleigh Dodds Science Publishing Limited, 2024. All rights reserved.

© Burleigh Dodds Science Publishing Limited, 2024. All rights reserved.

© Burleigh Dodds Science Publishing Limited, 2024. All rights reserved.

Introduction

In 2021, the Behavioural Insights Team, established by the UK Government in 2010 to better understand consumer behaviour in key areas of public policy such as diet and health, produced a report: *Net Zero: Principles for Successful Behaviour Change Initiatives* which explored how theories of consumer behaviour could help in programmes to boost public engagement in achieving net zero, particularly in areas such as changing food purchasing habits (Londakova et al., 2021). The report was almost immediately withdrawn, presumably to avoid accusations that the government might be interfering with freedom of choice in the sensitive area of citizens' decisions about diet and food (see the discussion in Reid, 2023 pp.187-190). As the Report noted:

'Though everyone has a degree of agency in changing their behaviour, and well-crafted messages from government can certainly be influential, behaviour is simply too profoundly driven by factors in the environment rather than in hearts and minds. As it stands, low-carbon behaviours are often more costly, less convenient, less enjoyable, and rarely the default choice'

These themes have been echoed in more recent publications by the Behavioural Insights Team, now an independent social purpose company (Park et al., 2023).

These reports, and the reactions to them, highlight both how politically sensitive this subject can be as well as how challenging it is to understand the complex processes underlying food purchasing behaviour. In particular, such research highlights the considerable gap between intention and actual purchasing behaviour, particularly where ethical and environmental issues are concerned. This volume reviews what we know about changing food purchasing behaviours amongst consumers so that farmers, food manufacturers, retailers and policymakers can better meet and engage with customer needs and expectations.

The chapters are split into two Parts: Part 1 chapters focus on understanding consumer attitudes and patterns behaviour. Chapters draw specific attention to areas such as the use of duality models to process information regarding food and nutrition and the neuropsychology of consumer food choice. A chapter on the geographical patterns of food-purchasing behaviour in sub-Saharan Africa is also included, which is then followed by a chapter that reviews sustainable food consumption attitudes and behaviour, focusing specifically on generational cohort differences. The final chapter of Part 1 examines the behavioural changes in relation to sustainable food consumption. Part 2 chapters address product attributes that can affect consumers' perceptions and attitudes, such as nutrition labels and health claims in food and preference for

© Burleigh Dodds Science Publishing Limited, 2024. All rights reserved.

locally-sourced food products. Chapters also review the effects of consumer perception of food safety and quality in purchase decisions, attitudes towards organic foods and also consumer attitudes to environmental sustainability issues in agricultural and food production.

Part 1 Understanding consumer attitudes and patterns of behaviour

The first Part of the book opens with a chapter that focuses on using duality models to understand how consumers process information about food and nutrition. Chapter 1 begins by discussing the elaboration likelihood model, arguably the most prominent dual-mode model in nutrition research. This is then followed by a discussion of the contribution of dual-system models as well as duality models' potential to further our understanding of nutrition information processing, focusing on nutrition message and labelling effectiveness. It also describes contingency effects in nutrition information processing as well. A section on how duality models can guide future nutrition research is also provided.

Chapter 2 examines the neuropsychology of consumer food choice. The chapter first describes the continuum of consumer choice, focusing on consumer choice and consumer behaviour, rewards of consumption as well as the importance of valuation and its sources in food consumption. This is followed by an overview of the continuum of food consumption, which focuses on areas such as normal eating, overeating, compulsive eating, binge eating and addictive eating. A section on treatments of extreme food consumption is also included, highlighting both pharmacological and non-pharmacological forms of treatment forms. The chapter concludes by emphasising the importance of understanding the neurophysiological dimensions of food consumption.

The next chapter draws attention to the geographical patterns of food-purchasing behaviour in sub-Saharan Africa (SSA). Chapter 3 begins by providing an overview of the multiplicity of factors that are shaping the food marketing and purchasing behaviours of consumers in SSA and the implications they carry for policy, strategy and future research in the field. The chapter then moves on to discuss some of the theoretical perspectives that have guided empirical investigations into food marketing processes in SSA. This is followed by a review of some major shifts in consumer preferences and purchase behaviours. The chapter also examines changes in the food distribution infrastructure, highlighting the roles that modern food retail firms are playing in the process. Finally, the chapter explores the implications of these changes for policy, strategy and future research.

Chapter 4 explores generational cohorts' attitudes towards and purchasing behaviour relating to different dimensions of sustainable food

© Burleigh Dodds Science Publishing Limited, 2024. All rights reserved.

consumption (SFC). It begins by providing a general definition of SFC. The chapter then moves on to identify the relevant generational cohorts, such as the GI, 'Baby Boomer', Generation X, Generation Y (also known as Millennials) and Generation Z. This then provides a context for the rest of the chapter, as it goes on to examine current research on SFC attitudes with these generational cohorts in mind. A case study on SFC and generational cohorts in Greece is also included, which provides an overview of the methodology that was used by the authors to compile the data required to create the case study and also displays the results of the study. The case study also includes an overview of the policy implications and potential limitations.The final chapter of Part 1 draws attention to creating an infrastructure for cultured meat. Chapter 4 begins by highlighting current gaps in existing infrastructure, focusing on aspects such as the talent gap and funding gaps, where lack of experience and money within the cultured meat industry can affect cultured meat's overall impact in society. To combat these issues, the chapter moves on to review possible avenues for remedy, such as the importance of working towards a more collaborative ecosystem as well as the importance of improving community engagement and education regarding cultured meat.

The final chapter of Part 1 focuses on behavioural change towards sustainable food consumption. Chapter 5 highlights the factors that affect this transition considering the multidimensionality of food, such as environmental, social, economic, health and ethical effects. It considers consumer perceptions, attitudes and behaviours towards sustainable food consumption. Consumer values, perceptions and intentions affect consumers' food choices and motives, thereby affecting sustainable food consumption behaviours. In this context, the chapter provides knowledge and information about the heterogeneity and different segmentation of consumers, which is essential for the characterisation of sustainable food products and related diet compositions.

Part 2 Product attributes

The first chapter of Part 2 focuses on understanding consumer perceptions and attitudes towards nutrition labels and health claims in food. Chapter 6 first provides an overview of nutritional labelling and health claims, which provides a basis for the next section's discussion on developing a systemic review of nutrition labels and health claims research. The chapter then moves on to review several key research areas, such as consumers' understanding, perceptions and preferences for nutrition labels and health claims, how these labels and claims can influence health choices as well as their impact on consumers' attitudes and behaviours. It also examines label design of nutrition labels and health claims and how it is crucial to develop more market-oriented nutrition labels and health claims.

© Burleigh Dodds Science Publishing Limited, 2024. All rights reserved.

Chapter 7 reviews the literature regarding the conceptualisation of local food as well as consumer motivations and barriers, to purchase local food products, drawing largely on evidence from Europe and North America. This is complemented with a case study of cooperative food labelling in southern Hungary, which illustrates the market opportunities and frequently encountered obstacles faced by local food initiatives. The chapter identifies, as a critical impediment, a lack of practitioner and scholarly consensus regarding what constitutes a local food, with consequent disagreement regarding the socio-economic and environmental outcomes of local food systems. In turn, consumers often struggle to comprehend the benefits of local foods, hindering informed decision making. The chapter concludes with an agenda for future research.

The next chapter analyses the effects of consumer perception of food safety and quality in food purchase decisions. Chapter 8 covers a wide range of topics that explain consumer perceptions of food safety and quality and how they relate to subsequent purchase decisions. It begins with a discussion of the numerous factors that consumers consider as they make decisions about what food to purchase. It examines consumers' intent to purchase specific types of food and how intent is correlated with purchase behaviour per se. Finally, the recent Covid-19 pandemic is known to have affected consumer perceptions of safety in general. This chapter ends with a discussion of how the pandemic has affected consumer perceptions of food safety and, ultimately, their purchase behaviour. Recognising the broad scope of this topic, the discussion in this chapter focuses on recent studies of purchase behaviour by North American and European consumers.

The focus of Chapter 9 is understanding consumer attitudes to organic food by using benchmark profiling and deviation analysis. The chapter begins by providing an overview of how organic food consumption has grown in recent years and how attitudes towards this type of food have changed as a consequence of this. It performs a profile deviation analysis on a cross-sectional sample of 1011 Australian organic food consumers. The chapter uses a combination of configuration theory and sustainable consumption research to understand and explain the consumer performance implications of a misalignment with a benchmark organic consumer profile. Sections on theoretical framework and hypothesis development, method and data and results and discussion are included to support the research and theory that was compiled by the authors.

The final chapter of the book discusses understanding consumer attitudes to environmental sustainability issues in agricultural and food production. Chapter 10 first provides an overview of the different product attributes that exist within food which help consumers to obtain the information they need regarding specific products. This is followed by an overview of the

© Burleigh Dodds Science Publishing Limited, 2024. All rights reserved.

environmental attributes that can exist within food products. Moving on from this, the chapter goes on to highlight how 'ecolabelling' is crucial to consumers when it comes to verifying certain environmental attributes in food products. A section providing an overview of various studies that have assessed consumer preferences in relation to their 'willingness-to-pay' is also included, focusing specifically on cross-country studies, individual European and North American countries as well as countries in Asia.

References

Londakova, K. K., Park, T., Reynolds, J. and Wodak, S. 2021. *Net Zero: Principles for Successful Behaviour Change Initiatives*. BEIS Research Paper No. 2021/063. Department for Business, Energy and Industrial Strategy (UK Government), London, UK.

Park, T., Londakova, K., Brennan, I., Schien, A., Reynolds, J., Whincup, E., Chan, E., Pelenur, M. and Halpern, D. 2023. How to build a Net Zero society: Using behavioural insights to decarbonise home energy, transport, food and material consumption. Behavioural Insights Team (https://www.bi.team/wp-content/uploads/2023/01/How-to-build-a-Net-Zero-society_Jan-2023.pdf).

Ward, N. 2023. *Net Zero, Food and Farming: Climate Change and the UK Agri-Food System*. Earthscan: Routledge, Oxford, UK.

© Burleigh Dodds Science Publishing Limited, 2024. All rights reserved.

Part 1

Understanding consumer attitudes and patterns of behaviour

Chapter 1

Using duality models to understand how consumers process information about food and nutrition

Steffen Jahn, Martin Luther University Halle-Wittenberg, Germany; and Yasemin Boztuğ, Georg-August University Göttingen, Germany

1 Introduction

In recent decades, psychological research has postulated that human thinking operates in a way in which information is processed either intuitively or systematically. Multiple terms have been used for this dualism, including automatic/controlled, heuristic/systematic, intuitive/analytic, or impulsive/reflective (Evans, 2008). A number of theories and models exist to describe this duality (Strack and Deutsch, 2015). However, it is possible to draw a distinction between two main approaches (Kruglanski and Orehek, 2007):

- Dual-mode models; and
- Dual-system models.

Dual-mode models are information focused: they focus on different types of information and how they are processed. Dual-system models are process focused: they do not view information processing as differing based on the type of information but, rather, which system is 'in charge' of processing information (Kahneman, 2011; Kruglanski and Orehek, 2007).

http://dx.doi.org/10.19103/AS.2023.0129.03
© Burleigh Dodds Science Publishing Limited, 2024. All rights reserved.

This chapter discusses these models, especially Petty and Cacioppo's (1986) elaboration likelihood model (ELM) and Kahneman's (2011) dual-system model, and how they inform our understanding of food-related information processing. We begin with the elaboration likelihood model (Petty and Cacioppo, 1986), which is arguably the dual-mode model best known to nutrition researchers. We follow with a discussion of the contribution of dual-system models and duality models to our understanding of nutrition information processing. We conclude this chapter with a discussion of implications for future research.

2 Dual-mode models

The ELM was introduced by Cacioppo and Petty (1984) in an attempt to unify disparate streams of attitude research. Its core tenet is that there are multiple routes to message persuasion and attitude change, with two basic routes (Petty and Cacioppo, 1986):

- The central route; and
- The peripheral route.

The central route involves thoughtful consideration of message arguments (i.e. issue-relevant information), whereas the peripheral route involves comparatively little thought about issue-relevant information. Instead, simple rules of thumb (heuristics) or simple association processes can lead to attitude change. The two routes to persuasion are assumed to coexist in inverse relation to each other. A high elaboration likelihood is associated with central processing, while a low elaboration likelihood is associated with more peripheral thinking (Kruglanski and Orehek, 2007).

People following the central route are supposed to think carefully with high motivation about message arguments, draw inferences about the merits of an argument, and explore messages with associations from memory (Petty and Cacioppo, 1986). Motivation for careful thought can stem from the perceived personal relevance of the message or a person's general preference for thinking (the need for cognition); the ability to think can also depend on the person being distracted or how frequently the message is repeated (Petty and Cacioppo, 1986). In peripheral processing, by contrast, issue-unrelated information such as the attractiveness of the source of a message can have a substantial influence on attitudes. The peripheral route is used when people either lack the motivation or ability to think carefully (Petty and Cacioppo, 1986).

The primary domain of the ELM is the persuasiveness of messages and the consequences of message-elicited attitude change. While both the central and peripheral routes can elicit attitude change, attitudes resulting from more

© Burleigh Dodds Science Publishing Limited, 2024. All rights reserved.

effortful thinking better predict behavior and influence behavior over a longer period of time (Petty et al., 2009).

3 Dual-system models

Dual-mode models describe two routes to judgments in a specific domain. Dual-system models (e.g. Kahneman, 2011; Strack and Deutsch, 2004) go further by invoking two mental faculties or systems, which interactively determine social judgment and behavior (Deutsch and Strack, 2006). Dual-system models can be said to represent the state-of-the-art in cognitive psychology (Kruglanski and Orehek, 2007).

The core tenet of dual-system models is that social cognition and behavior is a function of a reflective system and an impulsive system, each operating according to different kinds of representation and processes (Deutsch and Strack, 2006). To characterize the duality of information processing, the terms System 1 and System 2 have become popular at the beginning of the twenty-first century (Evans, 2003; Kahneman, 2003, 2011; Stanovich and West, 2000).

System 1 refers to the mental faculty characterized as 'automatic, largely unconscious, and relatively undemanding of computational capacity' (Stanovich and West, 2000, p. 658). It allows for automatic, fast thinking in which simple heuristics (mental short cuts) are used to make decisions (Kahneman, 2011). A prominent example is the representativeness heuristic (which relies on similarity with a previous experience) being (unconsciously) used when making a probability judgment (Kahneman and Tversky, 1972). However, if this automatic, fast procedure does not produce a clear decision, System 2 is activated.

System 2 refers to the mental faculty responsible for comprehensive or systematic thinking in which a decision emerges from a comparison of multiple features. Decisions made using System 2 are ultimately justifiable, both to oneself and to others. While heuristics can also be used to simplify choice – such as eliminating lunch options based on experiences of similar, previous occasions – decision-makers are consciously aware that they are using a short cut in contrast to the rapid, instinctive processing that characterizes judgment heuristics (Dhar and Gorlin, 2013).

Kahneman (2011) describes System 2 as being 'The Lazy Controller'. He argues that, by default, System 1 is in charge of most information processing. System 2 merely acts as a controller and intervenes only when there is doubt about the suitability of System 1's decision. This is different to the ELM which describes the automatic process as the peripheral route. In dual-system models, the main approach to information processing involves heuristics rather than elaboration.

© Burleigh Dodds Science Publishing Limited, 2024. All rights reserved.

4 Duality models in nutrition research

To gain insight into how duality models are used in nutrition research, we examined papers published in two influential nutrition journals that regularly cover nutrition information processing topics:

- Appetite; and
- Food Quality and Preference.

Both journals have high impact in the field (2022 Impact Factors for *Appetite* and *Food Quality and Preference* were 5.4 and 5.3, respectively). We used 'elaboration likelihood' and 'dual-system' as search terms in Google Scholar to identify papers referencing dual-mode and dual-system models, respectively. We also tracked all papers published in these journals that cite Petty and Cacioppo (1986) or Kahneman (2011). Because the Petty and Cacioppo (1986) article is so closely intertwined with the ELM, the search term 'elaboration likelihood' was effective in returning all relevant articles. Since more publications are associated with dual-system models, such as Kahneman and Frederick (2002) and Strack and Deutsch (2004), tracking citations of Kahneman (2011) returned more articles than the search term 'dual system'.

More than 400 relevant articles have been published in both journals in the last decade (2013–22). Analyzing each journal separately and across two 10-year spans (2003–12 and 2013–22), we find that, in the last decade, duality models have been used more frequently in research on nutrition information processing (see Fig. 1). Between 2003 and 2012, the elaboration likelihood model was more popular than dual-system models but this has changed since.

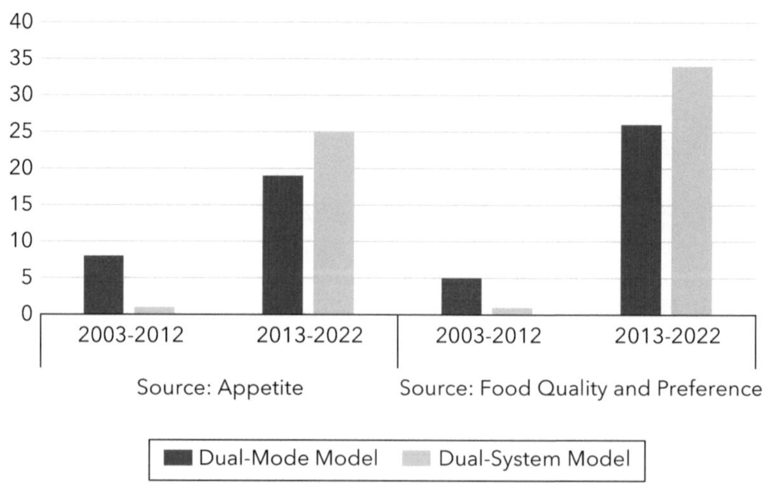

Figure 1 Number of articles using duality models in two nutrition journals.

© Burleigh Dodds Science Publishing Limited, 2024. All rights reserved.

Although the main ideas of dual-system models were popularized in the early 2000s (e.g. Evans, 2003; Kahneman, 2003; Kahneman and Frederick, 2002; Strack and Deutsch, 2004), the huge success of Kahneman's best-selling book, *Thinking, Fast and Slow* (2011), arguably provided the boost for increasing the use of dual-system models in the last 10 years. We also found that articles tend to stick with one class of models, with only one paper – Wilson et al.'s (2016) extensive review of healthful nudges – citing both Kahneman (2011) and Petty and Cacioppo (1986).

In terms of content, popular topics among the identified articles include

- nutrition message framing (Aschemann-Witzel and Grunert, 2015; Churchill et al., 2014);
- the use of health claims or nutrition labels (Hung et al., 2017; Ma and Zhuang, 2021; Temple 2020); and
- contingency effects in nutrition information processing, such as goals and time constraints (van Herpen and van Trijp, 2011).

Other work looked at topics such as information salience (Wilson et al., 2016) and psychological mindsets (Rucker and He, 2016). As these three themes seem representative of nutrition information processing studies published in other nutrition journals as well, we will briefly summarize key findings related to each. This will help us better understand how duality models of nutrition information processing can contribute to the ongoing conversation around healthy and sustainable eating.

4.1 Nutrition message effectiveness

Petty et al. (2009) provide an extensive review of how the ELM can be used to guide health message design. A key implication is to resist the temptation to inject peripheral cues into communications because it might only have a short-term impact (Petty et al., 2009). The goal of health communication should rather be to form 'highly elaborated, accessible, well-integrated and confidently held attitudes' to elicit sustained behavioral change (Petty et al., 2009, p. 21).

Attempts at inducing elaboration should then address motivation and the ability to carefully process the information. Motivation increases if a health message is personally relevant, which can be accomplished by tailoring messages to different groups of people (i.e. the message's target groups). Message tailoring also offers the opportunity to include tailored arguments. For example, Churchill et al. (2014) found that 'individuals low in self-efficacy, who typically have the poorest health behaviours, may be more likely to make healthier choices if they receive messages including an "every year" [versus

© Burleigh Dodds Science Publishing Limited, 2024. All rights reserved.

"every day"] temporal frame' (p. 134). Nour et al. (2018) found that young adults prefer messages that use language they would speak, including 'slang' terms.

Given that argument quality impacts attitude formation in the central route (Petty and Cacioppo, 1986), argument tailoring is a powerful way to ensure the respective target group receives messages they find both relevant and compelling (Kreuter et al., 2000). In a nutrition context, message and argument tailoring could involve highlighting a food's superior healthfulness to consumers with a pronounced health (or functional eating) goal as well as highlighting the food's low-energy content to appeal to consumers with an active dieting (or symbolic eating) goal, respectively (Schuch et al., 2015).

Another perspective on motivation-offer psychological mindsets – such as promotion versus prevention, power versus powerlessness, and fixed versus growth mindsets –have been shown to help or hurt with portion control (Rucker and He, 2016). For example, low-power mindsets can be associated with larger portion size, promotion mindsets with successful initiation of behavior change (e.g. eating less), and prevention mindsets with keeping off lost weight in the long term (Rucker and He, 2016).

If different subsets of the population are known to have particular concerns about certain foods, arguments can be developed to address these particular concerns. For example, firms selling genetically modified (GM) foods can communicate a man-made cue along with a statement of intent to lower moral opposition to GM-labeled products (Hingston and Noseworthy, 2018). Even when less stakes are involved, such as food fortification with vitamin D, consumers have varying levels of concern depending on the category (e.g. sausage vs. milk; Jahn et al., 2019). As a consequence of proactively addressing concerns, the message will become even more effective (Petty et al., 2009).

In addition to motivation, the audience's ability to process the message is increased by the use of language that is easily understood and permits self-pacing (Petty et al., 2009). Among Danish consumers, for example, 'soft' information frames have been found to be more effective than scientifically phrased frames (Aschemann-Witzel and Grunert, 2015). An interesting option in this regard is to use infographics, which have been shown to result in the elaboration of more message-relevant thoughts and more favorable attitudes as compared to presenting the same message as plain text (Lee and Lee, 2022). At least for European consumers, however, motivation outweighs the ability in explaining health-claim use (Hung et al., 2017).

4.2 Nutrition label effectiveness

Nutritional labeling can take different forms (Campos et al., 2011; Garretson and Burton, 2000; Hersey et al., 2013; Williams, 2005):

© Burleigh Dodds Science Publishing Limited, 2024. All rights reserved.

- The provision of detailed information in the form of the nutritional table on the back of the pack - so-called back-of-pack or BOP labels;
- The abbreviated representation of the nutritional table on the front of the pack - so-called front-of-pack or FOP labels - or at the shelf location; and
- The use of nutritional and health claims on the front of the pack - so-called nutrition claims (e.g. 'low fat') and health claims (e.g. 'walnuts improve the elasticity of blood vessels').

While food manufacturers are obliged to provide a standardized BOP label, the use of a label on the front of packaging is largely voluntary. Numerous variants of such FOP labels exist, ranging from multi-color traffic light systems (including the NutriScore system) to numerical reference point-based identifiers (e.g. guideline daily amount (GDA) labels) to seal-like symbols linked to specific criteria, such as the green keyhole symbol common in Scandinavia, the 'USDA Organic' label or the EU quality mark 'Protected Designation of Origin'. Illustrations of the different label variants can be found in Hersey et al. (2013) and Temple (2020).

Citing consumers' preference for simplified information processing, it is often concluded that nutrition labels should be simple (Andrews et al., 2011; Grunert et al., 2012). In support of this claim, Pettigrew and colleagues (2021, p. 3569) argue that '(s)implicity in the form of the provision of a summary indicator of product healthiness also appears to be a FoPL [front-of-pack label] design element worthy of consideration'. Yet research exists that indicates the superiority of nutrition labels that provide more detailed information (Werle et al., 2022). Research also indicates that nutrition and health claims can create cognitive biases on product evaluation and purchase, but front-of-package labels can reduce these biases (Jahn et al., 2023; Talati et al., 2017).

Based on a dual-system view, Sanjari et al. (2017) offer an integrative framework that can account for those seemingly contradicting findings. In a nutshell, Sanjari et al. (2017) argue that instead of trying to identify the silver bullet, or one-size-fits-all solution, different label formats have varying effectiveness given the situation in which a decision is being made. If a situation facilitates System 1 information processing, reductive and evaluative nutrition labels are claimed to be superior (Sanjari et al., 2017). If a situation facilitates System 2 processing, however, detailed and nuanced label formats should allow consumers to make better decisions (Sanjari et al., 2017).

In support of this claim, Li et al. (2022) found that, under time pressure, consumers prefer traffic light labels. In a review of 45 eye-tracking studies on nutrition information processing, Ma and Zhuang (2021) found further evidence of System 1/System 2 processing but concluded that detailed examination is still needed.

© Burleigh Dodds Science Publishing Limited, 2024. All rights reserved.

4.3 Contingency effects in nutrition information processing

Contingency effects refer to specific situations that facilitate intuitive or elaborate processing of the same stimulus. The ELM, for example, states that consumers' motivation and ability to elaborate message arguments determine whether the central or peripheral route is being used (Petty and Cacioppo, 1986). We have mentioned that Kahneman (2011) argues for intuitive (System 1) thinking to be the default, with System 2 being hesitant to override conclusions drawn from System 1. A general tendency to employ System 2 more frequently can be found among people who exert higher self-control and analytic intelligence (Kahneman, 2011; Stanovich and West, 2000).

In the context of decision-making, Dhar and Gorlin (2013) specify that the activation of systems is determined by characteristics of a decision situation. If a choice option is superior in all its characteristics to one or more alternatives, it can be called a dominant option. If a dominant option is present, the choice decision is made by System 1. The situation is similar for asymmetrically dominant options, as known from the attraction effect. If none of the available options stands out, System 2 will control the information processing (Dhar and Gorlin, 2013). System 2 is also called upon when decision-makers are under particular pressure to justify themselves or their attention is drawn to reference information.

Inspired by Dhar and Gorlin's (2013) approach to link dual-system models with choice situation characteristics and based on a review of 59 nutrition-information studies, Sanjari et al. (2017) identified contextual and personal characteristics that determine whether System 1 or System 2 will guide nutrition information processing. Similarities between the approaches exist for some variables, such as health motivation and fatigue (see Sanjari et al., 2017). Health motivation would be an example of motivation, whereas fatigue is an example of (lacking) ability to elaborate message arguments. From a dual-system perspective, health motivation is a personal characteristic, while fatigue is a contextual variable. Consequently, the ELM and dual-system model would make similar predictions, such that high health motivation increases the likelihood of message elaboration through the central route or System 2 processing, while fatigue is associated with the peripheral route or System 1 processing.

However, time pressure would qualify as a contextual variable (Sanjari et al., 2017) but its consequences might be motivational rather than ability related. This means that consumers have the ability to comprehensively process the information but might choose to decide quickly (Dhar and Gorlin, 2013). Conversely, nutrition knowledge is a personal characteristic describing a consumer's ability rather than motivation (Sanjari et al., 2017). The four contingency variables are summarized in Table 1. Although differences

© Burleigh Dodds Science Publishing Limited, 2024. All rights reserved.

Table 1 Situational factors influencing the processing of nutrition information

Time pressure
Consumers under time pressure or other forms of stress process nutrition information heuristically, only focusing on a subset of the information or ignoring it completely. The form of representation is also decisive: colors are partially considered, numerical information less so.
Fatigue
Factors such as fatigue, hunger, an increasing number of choices, or multitasking increase cognitive load and thus hinder the ability to process information comprehensively. As a result, consumers often resort to heuristics.
Health motivation
Health motivation increases the personal relevance of nutrition information, especially in relation to nutrients relevant to health. Consumers with a high health motivation make cognitive efforts to actively seek and process detailed nutrition information. Pleasure-seeking consumers, on the other hand, may ignore this information or pay attention to selected nutrients as taste indicators. For example, these consumers derive a poorer taste perception from low fat or sugar content.
Nutrition knowledge
Consumers with moderate-to-high nutrition knowledge are able to correctly process numerical information and classify specific information (e.g. 'sugar content: 16 g/100 g'). However, there is a risk that some consumers will use their knowledge advantage to develop supposedly 'intelligent' heuristics that lead to suboptimal decisions in some situations. This can be expressed, for example, in the fact that only a subset of nutrients is considered or characteristics such as 'natural ingredients' are considered as substitutes for 'healthy food'. Consumers with little knowledge tend to look at a subset of nutrients. Which nutrients comprise these subsets, however, is externally influenced: when a particular nutrient is stigmatized in public discourse, the likelihood increases that this nutrient will be considered to reject the food, thereby neglecting all other nutrients.

between dual-mode and dual-system models exist on how these variables are classified, each approach nonetheless acknowledges their relevance for (nutrition) information processing.

Knowing about nutrition labeling formats and how contextual factors and personal characteristics affect processing makes it possible to identify different situations where consumers process nutrition information in a certain way. For example, a health-conscious consumer with an intermediate knowledge of nutrition could make a meal decision rested and without time pressure. It can be assumed that this decision-making situation deviates from a situation in which motivation is predominantly pleasure oriented, nutritional knowledge is low, and time pressure and exhaustion are high. To preserve economy, the situations outlined are characterized by congruent context or personal characteristics. For the sake of simplicity, one can assume a context is either

- adverse (e.g. high time pressure and/or severe consumer exhaustion); or
- favorable (e.g. no time pressure and/or rested consumers).

© Burleigh Dodds Science Publishing Limited, 2024. All rights reserved.

Situations in which rested consumers are under time pressure or exhausted shoppers have enough time are not considered. For these situations, however, it can be assumed that there is a tendency towards an adverse context for information processing.

Sanjari et al. (2017) discuss four prototypical situations for processing nutrition information:

- Ignoring;
- Cursory viewing;
- Skimming; and
- Extensive viewing

In situations characterized by adverse contextual factors and where there is a lack of consumer interest and knowledge about healthy eating, nutrition information is processed by System 1. In many cases, this leads to numeric information being almost completely ignored (Situation 1). At best, additives such as 'organic' or 'low-fat' or product and packaging designs that suggest health (e.g. the use of brown instead of refined sugar or packaging with a lot of green color) are used. Since the products mentioned do not necessarily represent healthier alternatives, there is a risk in these particular situations that consumers will be misled. In addition, nutrition information plays a minor role in decision-making compared to other product attributes.

On the other hand, if the context factors are favorable, parts of the nutrition information can be given greater consideration, even in the absence of health knowledge or interest (Situation 2). However, due to the lack of motivation or knowledge, consumers do not process the information in depth but use simple heuristics that can be used, for example, to identify particularly healthy or particularly unhealthy products. They therefore react in particular to the existence of seal-like symbols and consider the color schemes used in traffic light systems.

The presence of sufficient nutritional knowledge and/or the motivation to eat healthily leads to an activation of System 2. However, if the contextual factors are adverse, heuristics are more likely (albeit more sophisticated) to be used and the nutrition information is not fully processed (Situation 3). In this situation, consumers typically limit themselves to assessing individual nutrients, such as salt, proteins, or sugar. The restriction enables consumers to decide in less time, which is why this type of information processing can be maintained even under time pressure and stressful situations.

Once consumers have both the knowledge and the time, they are more likely to process nutrition information comprehensively (Situation 4). This includes that numerical information is also considered to a greater extent and several products are compared with one another. Since consumers want to carry out analyses that

© Burleigh Dodds Science Publishing Limited, 2024. All rights reserved.

are as precise and detailed as possible in this situation, simple identifiers (such as seal-like symbols or pure color codes) are unsuitable. Monochrome GDA labels, on the other hand, already offer a lot of information that consumers are looking for. If this information is not sufficient, the BOP information can be viewed.

5 How duality models can guide future nutrition research

We have mentioned that a large number of nutrition studies interested in food-related decision-making have focused on selected aspects, such as

- the role of objective nutrition knowledge in nutrition label processing (Andrews et al., 2021);
- label fluency (Gomez et al., 2017); and
- time constraints of decision-makers (Van Herpen and Van Trijp, 2011).

Full utilization of duality models appears to be an exception though. In this section, we aim to highlight a few areas where duality models of nutrition information processing can advance our understanding of the best ways to promote healthier behaviors and identify future research opportunities. The guiding question is, How Can We Get Consumers to Consider More Relevant Information?

A promising way of motivating consumers to use detailed nutrition information is to situationally trigger mindfulness. Mindfulness can be trained, for example, in a 2-week course (Mrazek et al., 2013) or through meditation (Katterman et al., 2014). In healthcare, 'mindfulness-based stress reduction' and 'mindfulness-based cognition therapy' are used (O'Reilly et al., 2014). Repeated practice of mindfulness can facilitate System 2 processing and associated self-regulation because it improves a person's ability to control and sustain attention to relevant stimuli (Ostafin, 2015). If, as a result, consumers' attention is increased situationally or over the long term, this will have a positive effect on the use of nutrition information. From an ELM perspective, mindfulness implies attitude formation through the central route, which then has a better chance to change behavior (Petty and Cacioppo, 1986). Increased mindfulness can also lead to paying more attention to physiological signals and eating less accordingly (Van de Veer et al., 2016). In addition, mindfulness is associated with more sustainable food choices (Werner et al., 2020). Against this background, future research can clarify on the mindfulness's role in overcoming System 1-System 2 conflicts, including food neophobia and willingness to pay for plant-based foods (Jahn et al., 2021).

Whereas mindfulness can be a valuable resource to improve decision-making through systematic thinking, an obstacle to such an attempt is habit (Verplanken et al., 1997). This behavior also applies to habitual purchases (Hoyer, 1984; Wood and Neal, 2009) and means that for regularly purchased

© Burleigh Dodds Science Publishing Limited, 2024. All rights reserved.

foods, the provision of nutrition information is rarely accompanied by the use of that information. This phenomenon can be traced back to mental rest (as in 'The Lazy System 2'; Kahneman, 2011): as long as System 2 sees no need for action – after all, there is only a small risk of making a bad purchase with habitual purchases (Henderson et al., 2021) – it will not intervene and information processing will therefore be guided by System 1. While breaking habits is a difficult task (Gardner et al., 2021; Wood and Neal, 2009), Henderson and colleagues (2021) discuss proactive rewards as a means to disrupt thinking minimization. Future research may explore this issue further, especially its applicability to nutrition thinking minimization.

Mindfulness and habit describe two consumer characteristics that affect nutrition information use. Food manufacturers and restaurants can also influence to what extent consumers use relevant information. Restaurateurs, for example, can present nutrition information at a time when their guests are less stressed and pressed for time. For example, this is more likely to be the case with orders at the table or in advance at the workplace than with orders in a cafeteria or canteen. VanEpps et al. (2016) tested consumer responses to calorie information in a realistic out-of-home dining context. The authors produced different versions of menus which, among other things, indicated the calorie content of a dish as well as a color (green, yellow, and red). Consumers were more likely to choose lower-calorie dishes compared to a control group where the menu contained no nutrition information (apart from ingredients that must be declared). This means that, when the contextual factors of food choices are favorable, the likelihood of considering nutrition information increases, regardless of the level of nutrition knowledge.

However, when the context factors are adverse, the information should be easy to grasp and lead to clear conclusions without much effort. At the same time, however, care must be taken to ensure that no false conclusions can be drawn from the information that cannot withstand a comprehensive examination using a large amount of information. The danger arises, for example, when products are labeled as 'low-fat' but contain high amounts of sugar (Jahn et al., 2023). More research is necessary to understand the alignment of 'System 1 friendly' information provision with the prevention of misleading or ambiguous claims.

In order to take the information format into account, it is advantageous if uniform standards are established. Such standards may also apply to evaluation criteria. If the nutrition information refers to different criteria (e.g. different portion sizes or different definitions of regionality and sustainability), consumers will be misled as they rarely take the different evaluation criteria into account when making their decision (Elshiewy et al., 2016). By contrast, if uniform standards are established, consumers only have to 'decipher' the nutrition label format once and then devote few cognitive resources to

© Burleigh Dodds Science Publishing Limited, 2024. All rights reserved.

interpreting it. If this succeeds, consumers will be given a decision-making aid that can lead to an increased choice of healthy and sustainable food.

Another solution is to provide detailed information separately to those consumers who really seek it (e.g. due to their motivation). An app that provides detailed information on request is conceivable here. Another advantage of an app is that this additional information is not only considered at the moment of decision-making but can also be viewed at a later point in time. If the later point in time is characterized by less stress, the detailed information is processed more extensively and can guide future decision-making. Research into digital tools to support nutrition information processing is only beginning. Designing digital displays and mobile apps in a way that is responsive to the user's current information processing mode seems like a fascinating opportunity for nutrition researchers and practitioners.

6 Conclusion

Duality models teach us that it is unwise to expect people to systematically evaluate all food-related information. Rather, they inform us about the situations in which consumers make intuitive choices, some of which might be detrimental to one's own goals. While the academic literature on food and nutrition information processing makes increasing reference to duality models, more theorizing and empirical testing is needed to realize the full potential of duality models in this regard.

We have outlined the main tenets of two types of duality models, dual-mode models and dual-system models. While all duality models highlight differences between intuitive and systematic thinking, dual-mode models are interested in which type of information is effective in facilitation of attitude change (Kruglanski and Orehek, 2007). Dual-system models, in contrast, are concerned with the processes that lead to intuitive versus systematic thinking (Kruglanski and Orehek, 2007). The differences between dual-mode and dual-system models have profound implications for policy recommendations. For example, dual-mode models such as the elaboration likelihood model recommend that health messages refrain from incorporating peripheral cues in order to facilitate robust attitude change (Petty et al., 2009). From a dual-systems perspective, however, one could argue that such health messages might not be processed in many if not most situations because people are exhausted or do not have the knowledge to think through all arguments (Sanjari et al., 2017). We have discussed mindfulness and situation-based messaging as ways to overcome this conflict.

Concluding, this chapter aimed at providing an overview of duality models of the ways consumers process information about food and nutrition. We hope it stimulates broader use of these theories to derive and test specific hypotheses

© Burleigh Dodds Science Publishing Limited, 2024. All rights reserved.

about health message and nutrition label design as well as nutrition information processing more generally.

7 Where to look for further information

A profound introduction to dual-system models is D. Kahneman, Thinking, Fast and Slow (see Kahneman 2011 in the References for full details). Kahneman is professor emeritus at Princeton University and recipient of the Nobel Prize in Economics. The book is readable, wise and full of self-help value. Readers interested in learning more about flaws in human judgment may also consider Noise, a book coauthored by D. Kahneman, O. Sibony and C. R. Sunstein. Noise has first been published in 2021 at William Collins (ISBN: 978-0-00-848016-5).

The U.S. Food & Drug Administration (FDA) as well as the UK National Health Service (NHS) provides information on how to understand and use food labels:

- https://www.fda.gov/food/new-nutrition-facts-label/how-understand-and-use-nutrition-facts-label.
- https://www.nhs.uk/live-well/eat-well/food-guidelines-and-food-labels/how-to-read-food-labels/.

The FDA also provides an interactive Nutrition Facts Label (including a fun facts quiz):

- https://www.accessdata.fda.gov/scripts/InteractiveNutritionFactsLabel/#intro.

8 References

Andrews, J. C., Burton, S. and Kees, J. (2011). Is simpler always better? Consumer evaluations of front-of package nutrition symbols?, *Journal of Public Policy and Marketing* 30(2), 175–190.

Andrews, J. C., Netemeyer, R. G., Burton, S. and Kees, J. (2021). What consumers actually know: the role of objective nutrition knowledge in processing stop sign and traffic light front-of-pack nutrition labels, *Journal of Business Research* 128, 140–155.

Aschemann-Witzel, J. and Grunert, K. G. (2015). 'Influence of 'soft' versus 'scientific' health information framing and contradictory information on consumers' health inferences and attitudes towards a food supplement, *Food Quality and Preference* 42, 90–99.

Cacioppo, J. T. and Petty, R. E. (1984). The elaboration likelihood model of persuasion, *Advances of in Consumer Research* 11, 673–675.

Campos, S., Doxey, J. and Hammond, D. (2011). Nutrition labels on pre-packaged foods: A systematic review, *Public Health Nutrition* 14(8), 1496–1506.

Churchill, S., Good, A. and Pavey, L. (2014). Promoting the avoidance of high-calorie snacks. The role of temporal message framing and eating self-efficacy, *Appetite* 80, 131–136.

© Burleigh Dodds Science Publishing Limited, 2024. All rights reserved.

Deutsch, R. and Strack, F. (2006). TARGET ARTICLE: Duality models in social psychology: from dual processes to interacting systems, *Psychological Inquiry* 17(3), 166–172.

Dhar, R. and Gorlin, M. (2013). A dual-system framework to understand preference construction processes in choice, *Journal of Consumer Psychology* 23(4), 528–542.

Elshiewy, O., Jahn, S. and Boztuğ, Y. (2016). Seduced by the label: how the recommended serving size on nutrition labels affects sales volume, *Journal of the Association for Consumer Research* 1(1), 104–114.

Evans, J. St. B. T. (2003). In two minds: Dual-process accounts of reasoning, *Trends in Cognitive Sciences* 7(10), 454–459.

Evans, J. St. B. T. (2008). Dual-processing accounts of reasoning, judgment, and social cognition, *Annual Review of Psychology* 59(1), 255–278.

Gardner, B., Richards, R., Lally, P., Rebar, A., Thwaite, T. and Beeken, R. J. (2021). Breaking habits or breaking habitual behaviours? Old habits as a neglected factor in weight loss maintenance, *Appetite* 162, 105183.

Garretson, J. A. and Burton, S. (2000). Effects of nutrition facts panel values, nutrition claims, and health claims on consumer attitudes, perceptions of disease-related risks, and trust, *Journal of Public Policy and Marketing* 19(2), 213–227.

Gomez, P., Werle, C. O. C. and Corneille, O. (2017). The pitfall of nutrition facts label fluency: Easier-to-process nutrition information enhances purchase intentions for unhealthy food products, *Marketing Letters* 28(1), 15–27.

Grunert, K. G., Bolton, L. E. and Raats, M. M. (2012). Processing and acting on nutrition labeling on food: the State of Knowledge and New Directions for Transformative Consumer Research. In: Mick, D. G., Pettigrew, S., Pechmann, C. and Ozanne, J. L. (Eds). *Transformative Consumer Research for Personal and Collective Well-Being*, New York: Routledge Academic, pp. 333–352.

Henderson, C. M., Steinhoff, L., Harmeling, C. M. and Palmatier, R. W. (2021). Customer inertia marketing, *Journal of the Academy of Marketing Science* 49(2), 350–373.

Hersey, J. C., Wohlgenant, K. C., Arsenault, J. E., Kosa, K. M. and Muth, M. K. (2013). Effects of front-of-package and shelf nutrition labeling systems on consumers, *Nutrition Reviews* 71(1), 1–14.

Hingston, S. T. and Noseworthy, T. J. (2018). Why consumers don't see the benefits of genetically modified foods, and what marketers can do about it, *Journal of Marketing* 82(5), 125–140.

Hoyer, W. D. (1984). An examination of consumer decision making for a common repeat purchase product, *Journal of Consumer Research* 11(3), 822–829.

Hung, Y., Grunert, K. G., Hoefkens, C., Hieke, S. and Verbeke, W. (2017). 'Motivation outweighs ability in explaining European consumers' use of health claims', *Food Quality and Preference* 58, 34–44.

Jahn, S., Elshiewy, O., Döring, T. and Boztuğ, Y. (2023). 'Truthful yet misleading: Consumer response to 'low fat' food with high sugar content, *Food Quality and Preference* 109, 104900.

Jahn, S., Furchheim, P. and Strässner, A. M. (2021). Plant-based meat alternatives: Motivational adoption barriers and solutions, *Sustainability* 13(23), 13271.

Jahn, S., Tsalis, G. and Lähteenmäki, L. (2019). How attitude towards food fortification can lead to purchase intention, *Appetite* 133, 370–377.

Kahneman, D. (2003). A perspective on judgment and choice: mapping bounded rationality, *American Psychologist* 58(9), 697–720.

Kahneman, D. (2011). *Thinking, Fast and Slow*, NY: Farrar, Straus and Giroux.

© Burleigh Dodds Science Publishing Limited, 2024. All rights reserved.

Kahneman, D. and Frederick, S. (2002). Representativeness revisited: psychology for behavioral economics'. In: Gilovich, T., Griffin, D. W. and Kahneman, D. (Eds). *Heuristics and Biases: the Psychology of Intuitive Judgment*, Cambridge, UK and New York: Cambridge University Press, pp. 49–81.

Kahneman, D. and Tversky, A. (1972). Subjective probability: A judgment of representativeness, *Cognitive Psychology* 3(3), 430–454.

Katterman, S. N., Kleinman, B. M., Hood, M. M., Nackers, L. M. and Corsica, J. A. (2014). Mindfulness meditation as an intervention for binge eating, emotional eating, and weight loss: A systematic review, *Eating Behaviors* 15(2), 197–204.

Kreuter, M. W., Farrell, D., Olevitch, L. and Brennan, L. (2000). *Tailoring Health Messages: Customizing Communication with Computer Technology*, New Jersey: Lawrence Erlbaum Associates.

Kruglanski, A. W. and Orehek, E. (2007). Partitioning the domain of social inference: dual mode and systems models and their alternatives, *Annual Review of Psychology* 58(1), 291–316.

Lee, N. and Lee, S. (2022). Visualizing science: the impact of infographics on free recall, elaboration, and attitude change for genetically modified foods news, *Public Understanding of Science* 31(2), 168–178.

Li, X., Wang, Q. and Zhang, C. Q. (2022). Do individuals use nutrition labels on food packages to make healthy choices? Testing the dual-process Model in two laboratory-based experiments, *Nutrients* 14(18), 3732.

Ma, G. and Zhuang, X. (2021). Nutrition label processing in the past 10 years: contributions from eye tracking approach, *Appetite* 156, 104859.

Mrazek, M. D., Franklin, M. S., Phillips, D. T., Baird, B. and Schooler, J. W. (2013). Mindfulness training improves working memory capacity and GRE performance while reducing mind wandering, *Psychological Science* 24(5), 776–781.

Nour, M. M., Rouf, A. S. and Allman-Farinelli, M. (2018). Exploring young adult perspectives on the use of gamification and social media in a smartphone platform for improving vegetable intake, *Appetite* 120, 547–556.

O'Reilly, G. A., Cook, L., Spruijt-Metz, D. and Black, D. S. (2014). Mindfulness-based interventions for obesity-related eating behaviours: A literature review, *Obesity Reviews* 15(6), 453–461.

Ostafin, B. D. (2015). Taming the wild elephant: Mindfulness and its role in overcoming automatic mental processes. In: Ostafin, B. D., Robinson, M. D. and Meier, B. P. (Eds). *Handbook of Mindfulness and Self-Regulation*, New York: Springer, pp. 47–63.

Pettigrew, S., Dana, L. M., Talati, Z., Tian, M. and Praveen, D. (2021). The role of colour and summary indicators in influencing front-of-pack food label effectiveness across seven countries, *Public Health Nutrition* 24(11), 3566–3570.

Petty, R. E., Barden, J. and Wheeler, S. C. (2009). The Elaboration Likelihood Model of persuasion: developing health promotions for sustained behavioral change. In: Diclemente, R. J., Crosby, R. A. and Kegler, M. C. (Eds). *Emerging Theories in Health Promotion Practice and Research*, San Francisco, CA: Jossey-Bass, pp. 185–214.

Petty, R. E. and Cacioppo, J. T. (1986). The Elaboration Likelihood Model of persuasion. In: Berkowitz, L. (Ed.). *Advances in Experimental Social Psychology* (vol. 19), New York: Academic Press, pp. 123–205.

Rucker, D. D. and He, S. (2016). Psychological mindsets affect consumption: how different mindsets help (hurt) portion control, *Appetite* 103, 425–431.

© Burleigh Dodds Science Publishing Limited, 2024. All rights reserved.

Sanjari, S. S., Jahn, S. and Boztuğ, Y. (2017). Dual-process theory and consumer response to front-of-package nutrition label formats, *Nutrition Reviews* 75(11), 871–882.

Schuch, J., Jahn, S. and Boztuğ, Y. (2015). Abstract goal and external reference-point interaction in food decision making, *Advances in Consumer Research* 43, 572–573.

Stanovich, K. E. and West, R. F. (2000). Individual differences in reasoning: implications for the rationality debate?, *Behavioral and Brain Sciences* 23(5), 645–65; discussion 665.

Strack, F. and Deutsch, R. (2004). Reflective and impulsive determinants of social behavior, *Personality and Social Psychology Review* 8(3), 220–247.

Strack, F. and Deutsch, R. (2015). The duality of everyday life: dual-process and dual system models in social psychology. In: Mikulincer, M., Shaver, P. R., Borgida, E. and Bargh, J. A. (Eds). *APA Handbook of Personality and Social Psychology* (vol. 1). *Attitudes and Social Cognition*, Washington, DC: American Psychological Association, pp. 891–927.

Talati, Z., Pettigrew, S., Neal, B., Dixon, H., Hughes, C., Kelly, B. and Miller, C. (2017). 'Consumers' responses to health claims in the context of other on-pack nutrition information: A systematic review, *Nutrition Reviews* 75(4), 260–273.

Temple, N. J. (2020). Front-of-package food labels: A narrative review, *Appetite* 144, 104485.

Van de Veer, E., Van Herpen, E. and Van Trijp, H. C. M. (2016). Body and mind: mindfulness helps consumers to compensate for prior food intake by enhancing the responsiveness to physiological cues, *Journal of Consumer Research* 42(5), 783–803.

Van Herpen, E. and Van Trijp, H. C. (2011). Front-of-pack nutrition labels. Their effect on attention and choices when consumers have varying goals and time constraints, *Appetite* 57(1), 148–160.

VanEpps, E. M., Down, J. S. and Loewenstein, G. (2016). Calorie label formats: using numeric and traffic light calorie labels to reduce lunch calories, *Journal of Public Policy and Marketing* 35(1), 26–36.

Verplanken, B., Aarts, H. and Van Knippenberg, A. (1997). Habit, information acquisition, and the process of making travel mode choices, *European Journal of Social Psychology* 27(5), 539–560.

Werle, C. O. C., Pruski Yamim, A., Trendel, O., Roche, K. and Nadaud, P. (2022). When detailed information works better: comparison of three- and five-color/letter front-of-package nutrition labels, *Journal of Public Policy and Marketing* 41(2), 177–195.

Werner, A., Spiller, A. and Meyerding, S. G. H. (2020). The yoga of sustainable diets: exploring consumers mind and spirit, *Journal of Cleaner Production* 243, 118473.

Williams, P. (2005). Consumer understanding and use of health claims for foods, *Nutrition Reviews* 63(7), 256–264.

Wilson, A. L., Buckley, E., Buckley, J. D. and Bogomolova, S. (2016). Nudging healthier food and beverage choices through salience and priming. Evidence from a systematic review, *Food Quality and Preference* 51, 47–64.

Wood, W. and Neal, D. T. (2009). The habitual consumer, *Journal of Consumer Psychology* 19(4), 579–592.

© Burleigh Dodds Science Publishing Limited, 2024. All rights reserved.

Chapter 2

Neuropsychology of consumer food choice

Gordon R. Foxall, Cardiff University, UK and Reykjavik University, Iceland; Oscar Robayo-Pinzon, Universidad del Rosario, Colombia; and Sandra Rojas-Berrio, Universidad Nacional de Colombia, Colombia

1 Introduction

Food is essential to our survival, biological continuity and well-being. But it is also a source of pleasure and a context for social activity, both of which influence our patterns of consumption, all the way from underconsumption to overconsumption and even addiction. The former identifies food as an investment good; the latter, as very much a consumption good. Both involve consumer choice, which we may understand in broad terms as the selection of an action that competes with at least one other in terms of the degree of utility it promises to supply. Utility consists in the quality and quantity of the goods on offer (which comprise the product component of the marketing mix), the place and timing of the rewards involved (distribution), the personal and social meanings of the items in question (branding and promotion) and the exchange of values in the marketplace (price). A comprehensive understanding of the nature of consumer choice requires, therefore, an interdisciplinary approach that takes not only the economic and technical components of the mix into consideration but also the psycho-social aspects of the products and services on offer. An additional dimension that has recently come to the fore is the neurophysiological basis of consumer decision-making. This chapter is concerned principally with this facet of consumer choice, not in isolation but in concert with the economic, psychological and social aspects of consumption.

http://dx.doi.org/10.19103/AS.2023.0129.04
© Burleigh Dodds Science Publishing Limited, 2024. All rights reserved.

More specifically, the chapter aims to bring psychology and neurophysiology together to account for the relationships among consumers' thinking, emotions and behaviour, with the intention of understanding the nature of consumer choice. The use of the term 'neuropsychology' to designate this field of study requires further explication in view of the wide range of studies to which this term may be applied. Neuropsychology clearly links the brain and, especially, neuronal activity, with behaviour. It has been understood as the analysis of changes in behaviour in response to the lesioning of brain regions, especially as a result of accidents. This has in the past been a methodological requirement of research in this area as there are clearly limits to the ethical pursuit of investigations on intact humans. The advent of new techniques made available through electroencephalogram and brain imaging (functional magnetic resonance imaging) has significantly extended the range of investigatory routes to knowledge in this area. The neuropsychology of consumer choice is concerned with the foundations of consumer behaviour that are explicable in terms of neurophysiology. Perhaps the broadest understanding of neuropsychology is the study of brain–behaviour interactions. This is almost identical to biological psychology and is too all-embracing to guide the current discussion. We are certainly concerned with neural influences on sentience and behaviour (Bayle-Tourtoulou and Badoc, 2020).

However, the neuropsychology of food consumption is a vast subject (see, for instance, Avena, 2015; Cottone et al., 2019; Watts et al., 2022; Woods and Begg, 2016). It is necessary, therefore, to focus, and this chapter concentrates on how value is established at the neurophysiological level and how it is related to behaviour. This is especially pertinent to the analysis of consumer choice, which refers here to being faced with two or more options, each of which has its own set of short- and long-term consequences that are in conflict with one another. Situations of choice arise principally when the expected outcomes of purchasing one commodity are relatively immediate, while those of a competing purchase are delayed, e.g. the taste reward of consuming a hyper-palatable food now as opposed to a healthier life enjoyed in the longer term.

2 The continuum of consumer choice

2.1 Consumer choice and consumer behaviour

Consumer choice, as we employ the term here, differs from the usual notion of consumer behaviour. The latter term refers to all of the activities undertaken by a buyer and user of economic and social commodities; the latter, to the selection of such items that represent a conflict between the consequences of consumption that occur sooner and those that are delayed. More specifically, by consumer choice is meant the selection of a commodity or service that

© Burleigh Dodds Science Publishing Limited, 2024. All rights reserved.

represents either a smaller but more immediately available set of satisfactions or, alternatively, one for which the consumer must wait longer but which offers greater utility. The small but sooner reward (SSR) reflects impatience or impulsivity, an unwillingness to hold out for the superior option, while the larger but later reward shows a willingness to be patient and to act with self-control. Many consumers discount the future in ways that make the SSR the object of their consumption. A consumer may have a decided preference at the start of the day to eat only what is prescribed by her healthy-eating diet but, with the appearance of lunchtime, give in to more palatable but less wholesome fare. Or, resolving steadfastly to save up for a state-of-the-art sound system for her home, she may nevertheless purchase a less satisfactory alternative just as she has accumulated enough money to afford it. In both cases, the deciding factor is the subjective value of the less valuable item given the timing of its availability. While at the outset, the larger later reward was valued far more highly, just prior to the appearance of the SSR, its value increases dramatically, far outweighing that of the superior choice. The consumer may reverse her preference at this point almost without conscious involvement in the light of the instant gratification that the inferior item now affords (Foxall, 2023, 2024).

Not all consumer behaviour is of this kind, of course. Everyday brand choices of food products are made without such preference reversal on the basis of the consumer's experience with the product category and aspects of the marketing mix such as the chosen item's price compared with alternative brands that would be acceptable alternatives. Such a decision may be made quickly but it is hardly mindless or impulsive since it reflects an extensive consumption history in which competing brands have been evaluated in use. By contrast, finishing off a box of marshmallows on the excuse that so few are left to make a noticeable difference is certainly acting on impulse and, possibly, behaving compulsively.

Consumer behaviours differ in the extent of the temporal discounting that underlies them, that is the degree to which the consumer exhibits time preference in favour of more immediate consumption. Such revealed preferences range from routine consumption to compulsive and even addictive intake. Figure 1 illustrates this continuum of consumer choice. Everyday purchasing and consumption are minimally influenced by consideration of time discounting. Food buyers of a particular product category typically select among the brands they have bought and used previously. They do so on the basis of evaluative experience and it is comparatively rare for them to buy one brand exclusively over a series of shopping occasions. Although all brands attract a small percentage of the market who are sole purchasers, most consumers practise multi-brand purchasing, selecting one or other of a tried and trusted 'repertoire' of more or less functionally identical brands (Ehrenberg, 1988). Within this consideration set, the selection criterion on any one shopping trip may simply be the lowest price (Foxall and James, 2002, 2003). Moreover, the

© Burleigh Dodds Science Publishing Limited, 2024. All rights reserved.

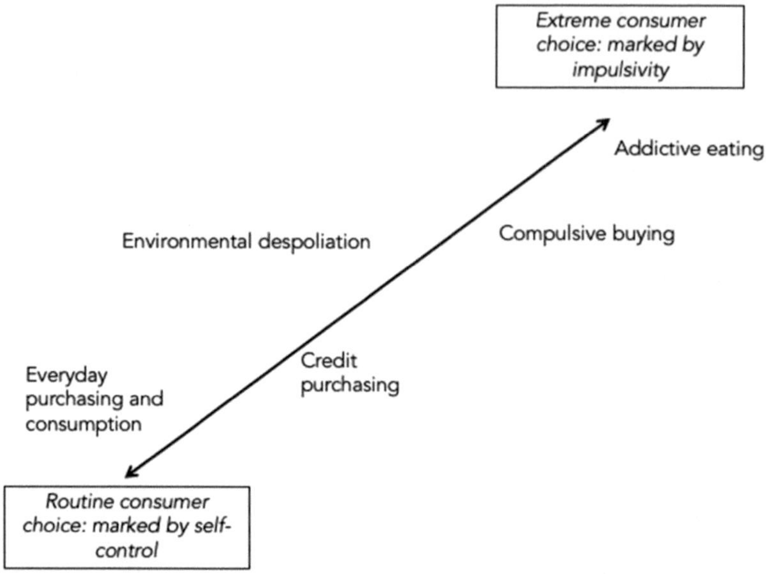

Figure 1 The continuum of consumer choice. Source: Foxall, G. R. (2016). *Addiction as Consumer Choice: Exploring the Cognitive Dimension.* London and New York: Routledge. Reproduced with permission.

rewards of making this selection are not confined to the economist's conception of utility (what we know as utilitarian or functional reinforcement) but include the social and reputational effects of buying and, ultimately, the consumer's self-esteem (i.e. informational or social reinforcement) (Foxall, 2020; Oliveira-Castro and Foxall, 2017; Oliveira-Castro et al., 2016a,b). The levels of utilitarian and informational reinforcement of the brands comprising the consumer's repertoire are generally equivalent, marking an apparently random selection at the point of purchase and a seamless pattern of consumption as one brand is substituted for another.

Less routine purchasing, say in preparation for a large dinner party or function may occasion buying that departs from this everyday model, perhaps by necessitating the acquisition of more expensive brands and even new product categories on credit. There is here some tendency to discount the future by incurring interest payments which, in the case of credit card purchases, might be substantial. Similar kinds of purchasing may simply result from a consumer's expanding the range of functional and social rewards they obtain, albeit at the cost of incurring higher charges not only as a result of increasing their utility and thereby paying higher prices but also as a result of making credit payments. The temporal discounting that is inherent in such consumption may well be shallow and occasional though it may be increasingly steep if the habits it is based on become stronger.

© Burleigh Dodds Science Publishing Limited, 2024. All rights reserved.

Increased consumption also results in the necessity of greater waste disposal, perhaps through the discarding of unconsumed food, perhaps as a result of clearing out the packaging that accompanies the primary product. This is a form of temporal discounting insofar as immediate consumption benefits mean abandoning waste, the disposal of which is ultimately costly. This becomes a matter of increasingly steep temporal discounting (see Foxall, 2016a). As one moves along the continuum of consumer choice in the direction of compulsive purchasing and addictive consumption, one is dealing less and less with the utilitarian reinforcement/functional benefits provided by foodstuffs and increasing towards the social and psychological benefits provided in the form of social and personal acceptance, social status and esteem and a kind of self-esteem, even though this may be fragile. This progression can be understood in terms of the neuropsychology of consumer choice.

2.2 Rewards of consumption

The outcomes of consumer behaviour, considered as operant response, are primarily reinforcement and punishment and, secondarily, behaviourally rewarding and sanctioning emotions (Foxall, 2017a). Reinforcers increase the rate of performance of the behaviours on which they are contingent; punishers reduce it. An extensional model of consumer choice can be constructed on the basis of objectively available reinforcers and punishers; within this depiction, the consumer situation comprises these outcomes of past behaviour and the stimuli that indicate the likely outcomes of further responding; this is the extensional consumer situation. An intentional model relies on concepts of subjective response to the contingencies of reinforcement and punishment, the intentional consumer situation which comprises the desires, beliefs, emotions and perceptions that represent these contingencies for the individual consumer. Reinforcement may be bifurcated into that which is utilitarian, denoting the functional outcomes of consumer behaviour, the benefits directly obtained from a product or service, and that which is informational, denoting the social outcomes such as prestige and esteem. While a purely extensional depiction of consumer choice incorporates the latter in the form of observed 'thank yous' and smiles, an intentional account employs consumers' feelings of self-esteem as outcomes of their consumer choices (Foxall, 2017b, 2020). Consumers purchase a combination of utilitarian and informational reinforcement with each transaction they make and most products and services yield social and personal values as well as functional benefits. Consumer choice is, therefore, a selection among differing quantities of functional and social values provided by goods.

Utilitarian reinforcement in the case of food products consists in the functional properties of these items, which accommodate both the body's need

© Burleigh Dodds Science Publishing Limited, 2024. All rights reserved.

for homeostasis in its operations and the hedonic outcomes of consuming palatable foods. Informational reinforcement provided by food products includes the social cachet of acquiring, owning and consuming food of high quality and prestige, being among others in high-end restaurants and sometimes deriving from the sheer quantity of what one ingests. We can identify and measure these reinforcers extensionally, i.e. in terms of their observable physical components and effects, or intentionally, i.e. in terms of their reported effects on the consumer's desires, beliefs, emotions and perceptions and – ultimately – their self-esteem. We must bear in mind that what reinforces today may punish tomorrow: the pleasures of immediate gastronomic gratification may yield in time to a wider waistline, lack of physical fitness, obesity and even serious illness. It is the trade-off between these earlier and later consequences of consumption that is the heart of the temporal discounting that defines consumer choice.

2.3 Valuation

The conception of consumer choice as a means of selecting among different courses of action each of which promises a specific payoff means that the individual consumer requires a system of valuation. It must be one that permits rapid decision-making when circumstances demand a behavioural response and more deliberated action when the long-term consequences of choice are important. 'Value' has a number of closely related meanings. It refers to the worth established intersubjectively through market exchange, the agreed price of a commodity. It also has a fully subjective import which relates to a personal evaluation of an action or commodity. We assume that market exchanges occur only because the price is equal to the subjective worth attributed to whatever is exchanged; often this personal and private estimation exceeds in some sense the pecuniary outlay and receipt achieved in the course of the transaction. The attainment of value, embodied in a product or service, is the mainspring of the actions of both consumer and marketer. The opportunity to gain value is, therefore, a prime motivator of economic and social behaviour and a key source of this motivation is the neural valuation accorded potential behaviours.

The determination of value may be a matter that the consumer deliberates over cognitively, making fine judgements of the incentives and costs that are likely to follow each of a number of competing actions. Sometimes, however, it is necessary to act more rapidly in order to capture the benefits of owning and consuming economic and social goods. On other occasions, the choice appears to be spontaneous, an automatic and instant response to an opportunity. In all of these cases, there is an underlying neural activity based on the value system, which reflects experience in the form of the consumer's learning history and its neurophysiological substrates. Such a system of valuation is provided by the consumer's neurophysiology. Specifically, it inheres in the rate of firing of

© Burleigh Dodds Science Publishing Limited, 2024. All rights reserved.

dopaminergic neurons. Brain cells, or neurons, are centrally concerned with communication among the functional parts of the brain. Although neurons have a great variety of forms, they typically comprise a cell body with a dendritic tree (that provides inputs to cellular activity) and an axon (which conducts the outputs). The functions of the neuron are concerned with the communication of information; each neuron inputs data from presynaptic brain cells, integrates it into a valuation of potential behaviours and outputs this information to postsynaptic cells.[1]

Glimcher (2009, 2010) proposes a concept of value which is measured as action potentials per second, i.e. the mean firing rate of specific populations of neurons, a measure assumed to be consistent with choice. This neuronal value is encoded by the mean activity of the medial frontal cortex (in the case of a good) and the ventral striatum (in the case of an action). The neurotransmitter, dopamine, originates in the midbrain regions, the substantia nigra pars compacta (SNpc) and the ventral tegmental area (VTA). This nigrostriatal pathway delivers dopamine (DA) to the dorsal striatum. This dopamine pathway is involved particularly in motor control and, crucially from our perspective, the establishment of the stimulus-response links that are central to habit formation (Horstmann, 2016, p. 43). Two additional dopamine pathways (Fig. 2) are implicated in behavioural reward and motivation. The first of these, the mesolimbic pathway, takes dopamine from the VTA to the nucleus accumbens (NAc) in the ventral striatum and from there to the limbic system [comprising the amygdala, the hippocampus and the prefrontal cortex (PFC)]. The second is the mesocortical pathway which conveys DA from the VTA to the frontal cortex.

Valuation is the outcome of an assessment of the likely benefits and costs of an action or commodity; motivation results from such calculations and leads to decisions that maximise the expected value of behaviours. When there is a choice of behaviours, that which promises the highest value will be selected.

2.4 Sources of value in food consumption

The value of an encountered foodstuff may be computed in terms of its taste and/or its nutritional contribution (Gottfried et al., 2003).

2.4.1 Motivational value

The two primary sense modalities through which one encounters foods are (a) sight and (b) smell, through which the food is identified and its potential

[1] Numerous basic neurophysiology texts described the structure and functions of neurons; see, for instance, Rudi (2020). A succinct overview of the brain can be found in O'Shea (2005). In the specific context of consumer choice, I describe these processes in *Addiction as Consumer Choice: Exploring the Cognitive Function* (Foxall, 2016a) and *Neuropsychology of Consumer Choice: Implications for a Unified Theory of Consumer Choice* (Foxall, 2023).

© Burleigh Dodds Science Publishing Limited, 2024. All rights reserved.

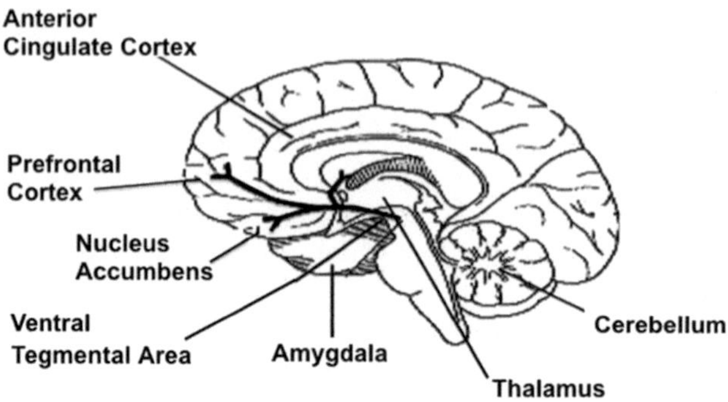

Figure 2 The mesolimbocortical pathway. Learning relationships between environmental events and rewards is facilitated by the dopamine system. The rate of action potentials of DAergic cells is sensitive to both the potential appearance and the actual appearance of a reward. Environmental events such as discriminative stimuli and motivating operations (which might take the form, for instance, of smelling a desirable food). The figure shows brain structures relevant to the dopamine reward system. The dark lines indicate dopamine pathways (i) from the ventral tegmental area to the nucleus accumbens (the mesolimbic pathway) and (ii) from the ventral tegmental area to the prefrontal cortex (the mesocortical pathway). The mesolimbic dopaminergic system comprises the midbrain ventral tegmental area (VTA) and its projections to the nucleus accumbens (NAc) and to the limbic system of the forebrain, primarily the amygdala and the hippocampus. The mesocorticolimbic system is implicated in the transmission of dopamine from the VTA to the frontal cortex. Together these systems are known as the dopaminergic mesocorticolimbic system/pathway. Source: Foxall, G. R. (2016). *Addiction as Consumer Choice: Exploring the Cognitive Dimension.* London and New York: Routledge. Reproduced with permission.

reinforcement value is calculated. Food stimuli excite visual receptors in the retina the outcome of which is the transmission of information to primary and secondary visual areas in the occipital and temporal lobes via the thalamus. Information is also relayed to the amygdala which registers the behavioural significance of the stimuli. Information reaches the orbitofrontal cortex (OFC) through both of these routes. Ortho-nasal odour information (information entering the nose via nostrils) reaches primary olfactory neurons and then the olfactory bulb. This information is transduced to central piriform and entorhinal cortices; it is conducted from the former to the posterior OFC. The function of the OFC is to use this information to arrive, via comparison with the already-learned properties of foods, at a motivational value that influences the probability of approach behaviour in the form of food seeking/acquisition and ingestion. The strength of the approach is determined, therefore, by the organism's learning history, the potential benefits of obtaining and eating the detected food, and the probable costs of doing so. The anterior insula is responsible for the generation of homeostatic signals that encourage or discourage food seeking, acquisition and ingestion (Horstmann, 2016; Hortmann et al., 2011).

© Burleigh Dodds Science Publishing Limited, 2024. All rights reserved.

2.4.2 Nutritional value

The ingestion of food rapidly provides information with respect to taste, temperature, texture, viscosity and nutritional value. Flavour, in particular, leads to affective reactions and this link is stronger in the obese than the non-obese. The anterior insula acts now to integrate information allowing a homeostatic computation of the initial need to ingest and the need to continue ingesting. Homeostatic information is thus updated in the insular cortex and hypothalamus. An important element in motivational value is incentive salience which refers to the tendency towards approach based on the probable reinforcement that is contingent upon it and the environmental stimuli that predict such reinforcement (Watts et al., 2022, p. 698). Salience refers to whatever brings an item to the fore, standing out from others. The brain 'prefers' to process items that are highly salient in terms of, e.g. colour, orientation, size, pitch and velocity, i.e. arousal in terms of Mehrabian and Russell's (1974) classification of emotions as pleasure, arousal and dominance. In the context of incentive salience, the relevant property of a stimulus inheres in its reward-predicting value, i.e. its stimulation of wanting. Stimuli that demonstrate incentive salience command attention, induce approach behaviour and are relatively easily detected in the environment. Being sensitive to incentive salience implies more than having learned an association between the performance of an action and its consequences. Even when the learning of such operant contingencies has been accomplished, the probability of the actions being repeated is likely to be increased only if the consequence is valued at a level sufficient to command compliance with what has been learned (Watts et al., 2022, p. 699). This value, incentive salience, is the outcome of dopaminergic action that results in wanting the good on offer (Berridge, 2004).

Horstmann speaks of the appetitive brain network which is involved in homeostatic regulation by generating arousal, directing attention to stimuli that predict food and implementing learning and memory of nutritional value and food availability. In regard to valuation, it weighs the costs of effort involved in attaining the reward against the value of the reward, '*thus providing absolute decision values*, and it integrates information about short- and long-term energy stores with the available energy choices' (Horstmann, 2016, p. 42, emphasis added). Dopamine is the key neurotransmitter in the appetitive brain network. The connection of the network with the hippocampus, 'the brain's primary homeostatic center', is significant. Moreover, 'The amygdala and posterior OFC provide the current incentive value of food items' (ibid.). The amygdala 'attributes values to sensory stimuli and relays this information to the OFC and ventromedial prefrontal cortex (PFC) which in turn computes an absolute decision-value from the costs associated with obtaining the food and the value signal assigned by the amygdala' (ibid.). Representations of foods are made in

© Burleigh Dodds Science Publishing Limited, 2024. All rights reserved.

the anterior insular cortex and central region of the frontal operculum and these are integrated with the homeostatic state. The anterior insula is concerned with nutritional value and these representations. 'Therefore, it possibly mediates the conditioning to food cues. The striatum provides the link between motivational state and action, since it plays a crucial role in incentive learning and motivated behavior' (ibid.).

3 The continuum of food consumption

Historically, more effort has been put into studying the physiological factors associated with food choice, while far fewer resources have been devoted to exploring the environmental and contextual factors that lead people to make decisions about the type and quantity of food eaten on a daily basis (Just et al., 2009). It is this joint approach that can best provide us with an interpretation of food choice that accounts for the current complex environment we face as consumers of food products. If we add this complexity to the incredibly vast variety of habits and preferences not only at the individual but also at the social and even cultural level, we find ourselves with an interdisciplinary explanation that combines knowledge from different areas of study, both basic and applied. The organising model for our discussion, the continuum of food consumption, is shown in Fig. 3.

3.1 Normal eating

This wide variety of preferences dates to the origins of the human species, when, by populating the most dissimilar territories across the planet, communities began a long sequence of experiences and knowledge related to food sources

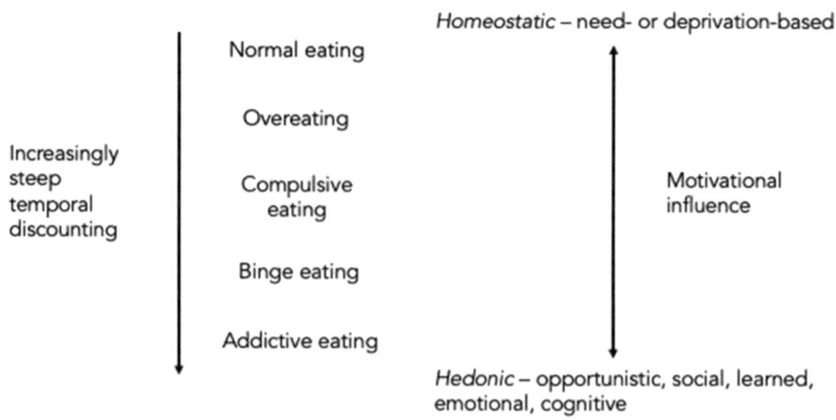

Figure 3 The continuum of food consumption.

© Burleigh Dodds Science Publishing Limited, 2024. All rights reserved.

and ways of consumption, preservation and preparation. This experience has led to the development of a very important feature of the sensory-perceptual process of food choice, learning-dependent flexibility. It has not been possible, therefore, for humans to have genetically pre-established neural circuits that allow us to recognise and evaluate food sources without previous experience with them (Myers, 2015). This flexibility in the preference for a variety of foods reflects our characterisation as generalist omnivores. The wide range of possible food sources, coupled with the need for experience to achieve habit formation, has made it difficult to establish estimates of the energy density and caloric content of foods. For example, in one series of studies, participants were asked to make consumption choices between a normal (control) and a 'low-fat' alternative. A 'healthy' halo effect was observed, as people who preferred the low-fat option, despite being informed that the calorie content was only 11 percent lower than the control option, ended up consuming 34 percent more calories when compared to those who preferred the control option (Wansink and Chandon, 2006).

At this point, it is worth considering what are eating behaviours and how do they occur? First, the timing of eating behaviour must be taken into account, whereby the body's physiological signals, the circadian cycle and external signals, such as looking at the time on a clock and determining that it is time to have a meal, are important factors. Eating behaviours can be defined by the primary unit used for multi-species studies of eating, the meal. This unit comprises the composition of the foods, the time of day they are eaten, the number of portions and the frequency with which they occur in a unit of time determined by the researcher. For humans, several factors in addition to physiological processes and signals influence the amount and timing of meals (Leech et al., 2015).

The temporal organisation of meals is of great interest in understanding the phenomenon of eating. Three types of meals have been proposed: deficit-induced eating, which is reactive and occurs when there is a negative energy balance, resulting from a condition of voluntary deprivation or lack of access to food sources. Therefore, it is also known as homeostatic eating. On the other hand, habitual eating consists of the distribution of meals throughout the day, socially and culturally constructed on the basis of regular moments in which certain combinations of foods are ingested that are available in different contexts (home, work, etc.). Finally, opportunistic eating occurs when there is an activation of a memory associated with a particularly preferred food or through direct contact with a cue or stimulus in the context that indicates the availability of food, which facilitates a consummatory response. This is the type of behaviour most closely associated with the concept of hedonic eating (Watts et al., 2022).

© Burleigh Dodds Science Publishing Limited, 2024. All rights reserved.

Hedonic eating has been the subject of much research that has sought to establish the effect of the obesogenic context, particularly in the USA since the early 1980s. This context has been characterised by the proliferation of ultra-processed foods 'engineered' to be tastier in order to optimise the consumer's pleasurable experience. In addition, the scale of industrial production has made these high-calorie foods progressively cheaper, with the result that a large proportion of the population, paradoxically those with low incomes, have these foods rich in sugars, fats and carbohydrates as their main eating option (Myers, 2015). This pattern of associations between stimuli that signal the availability of food with sensory characteristics that predict high caloric effects can, at least for certain individuals, generate negative consequences that have been compared to the addictive effects of psychoactive substances or 'drugs', and which manifest themselves in actions of constant food seeking, excessive attention to stimuli that signal the presence of tasty food and the loss of control over the amount and frequency with which these foods are consumed. The nutritional and sensory characteristics of these foods, coupled with the ease of access and comparatively lower costs compared to healthier foods, are contributing to neurobehavioural, affective and motivational responses that are leading more and more people to consume more food to a level that exceeds normal physiological energy requirements, which will be discussed in the following sections.

3.2 Overeating

Overeating could be considered an eating disorder, as well as anorexia and bulimia; any of this set of distortions has been shown to cause impairments to the neural processes associated with reinforcement, as well as changes in the neurochemistry of the brain (Avena et al., 2013). Overeating will affect the metabolic functioning of the endocrinological system in a snowball effect causing the organism to function negatively, which may further hinder the management of one's own food intake behaviour; the underlying cause is an imbalance in the brain's response to reward stimuli. Although environmental factors including culture, media, composition, taste and flavour send signals to the brain about the need for food and so motivate its intake, the fact of being hungry and, in this context, overeating depends also on the mood and psychological conditions of the consumer such as depression, anxiety and stress. Overeating is not something that has shown correspondence with consumer self-perception or initial body mass index (Murakami et al., 2022); however, persistence in this behaviour leads to its further increase in the long term both in developing and developed countries (Lewis, 2013). Misinteractions between corticolimbic structures and the hypothalamus may lead to overeating and as a consequence obesity (Berthoud, 2007; Lewis, 2013).

© Burleigh Dodds Science Publishing Limited, 2024. All rights reserved.

Since the hypothalamus regulates appetite, neural mechanisms can stimulate or reduce food consumption; it is also important to note that the hypothalamus regulates appetite as well as supporting stress reactivity (Leitch and Geliebter, 2015; Vainik et al., 2019). However, in the short term, although it starts before tasting a portion of food with a reaction according to the consumer's previous knowledge and cognition about food that allows identifying when food may be excessively caloric, it is the stomach that can send a satiety signal to the brain, given its interaction with the processes conducted by the small intestine and the hormonal system – especially insulin – which is also responsible for the sensation of satiety. Another relevant hormone system for overeating is the dopamine mesolimbic circuit associated with reward sensitivity since the pleasurable eating of foods that are delicious to the palate elicits an increase of this hormone which stimulates reward by activating the mesolimbic system.

One of the future challenges for the area of neurosciences and consumer behaviour in the field of food consumption is to differentiate and recognise in depth the neural circuits and their functions in relation to these processes, since according to the available evidence (Rumiati and di Pellegrino, 2016), they are closely interrelated and underlie the homeostatic and hedonic mechanisms that regulate food intake. On the other hand, although people struggle with obesity and habits such as overeating, there are external factors that are influential in maintaining weight loss including social support but there are more individual factors in this battle which involve dealing with stress, internal motivation, self-efficacy and most importantly coping with negative emotions appropriately (Elfhag and Rössner, 2005).

3.3 Compulsive eating

As mentioned earlier, the contemporary environment to which the consumer is exposed is saturated with signals indicating the availability of highly palatable food, which has been designed to generate a hedonic sensory experience unlike anything seen before in the history of the human species. This environmental design of eating has been associated with the high rates of obesity globally over the last 30 years (Gordon et al., 2018). The concern of health authorities has led to an exponential growth in studies on the factors influencing obesity. A highly influential finding is that an increasing number of neuroimaging studies indicate that obesity and binge eating are associated with changes in dopaminergic circuitry and that certain foods elicit a greater than normal response in some areas of the brain related to reward systems, which has facilitated comparison with psychoactive substance addiction processes (Tang et al., 2012).

Within this approach to food addiction, an interaction has been proposed between homeostatic mechanisms (given that food, unlike other addictive

© Burleigh Dodds Science Publishing Limited, 2024. All rights reserved.

substances, is essential for survival) and hedonic mechanisms, given the predisposition we have as a species towards a preference for sweet and palatable foods. Added to this interaction is the role of inhibitory systems, which seek to control the amount of food ingested at each meal (Leitch and Geliebter, 2015). However, this balance has been disrupted in individuals with eating disorders because the brain responds in an exacerbated way to signals that indicate pleasurable consequences. In addition, as more and more food is consumed, there is a decreased sensitivity to the brain's reward from eating (Nora D Volkow et al., 2013). In summary, food addiction involves a bias towards cues that indicate the presence of tasty food while there is an alteration in the reward mechanism that makes it increasingly necessary to consume larger portions of food to obtain a pleasurable sensation.

This is manifested in increased food intake, which, as it becomes more frequent, eventually increases the reinforcing value of eating by generating conditioned associations to the environmental stimuli surrounding the eating behaviour, while facilitating a rewarding experience through the neural mechanism of dopamine. There is then a circular process that increases the incentive value of a set of foods, such that each time they are eaten, dopamine release is stimulated, and as this behaviour is repeated, the association between the pleasant consequence and the environmental stimuli present is strengthened, such that these stimuli alone can generate dopamine release.

This conditioned overeating relates to the difference between 'liking' and 'wanting'. While liking is the positive emotional response to a pleasant stimulus that provides a pleasurable consequence, wanting is a process that departs from liking and is characterised by a deep motivation to obtain a reward, such as occurs in drug addiction (Berridge and Kringelbach, 2015; Berridge and Robinson, 1995). There is evidence that for motivational processes related to food and drug addiction, the prefrontal cortex plays a relevant role in both animals and humans. Similarly, norepinephrine (NE) transmission in the medial PFC is related to the behavioural aspects of drug abuse, as well as being a crucial mechanism for motivated eating behaviours (Ventura et al., 2007, 2008). Continuous stimulation of the dopaminergic circuit can lead to an unrestrained search for pleasurable consequences and to a sensitisation of the wanting mechanism. This leads to compulsive eating involving an abnormally high incentive value that leads to eating for pleasure rather than for need.

3.4 Binge eating

The main feature of binge eating disorder (BED) is a series of recurrent binge eating occasions in the absence of inappropriate compensatory behaviours, such as vomiting or excessive exercise. It is therefore very common for BED to be part of other eating disorders, such as bulimia nervosa, which do include

© Burleigh Dodds Science Publishing Limited, 2024. All rights reserved.

such behaviours. These occasions of BED are characterised by eating an amount of food in a given period of time that exceeds what would normally be considered appropriate under these conditions, coupled with a sense of loss of control over the situation (APA, 2013). Common underlying factors have been suggested between binge eating and substance addiction, such as impulsivity and compulsivity (Tomasi and Volkow, 2013), as well as alterations in some neurotransmitter systems, including dopamine, serotonin, GABA and glutamate receptors (Leitch and Geliebter, 2015).

Research into the aetiology of this condition indicates that one of the most studied hypotheses involves dysfunction in behavioural control and processing of reward signals. It has been identified that alterations in the frontostriatal neural circuits responsible for response processes to reward signals and self-regulation may play an important role in the development and maintenance of this condition (Berner and Marsh, 2014; O'Hara et al., 2015). Binge eating is thus considered to be a pattern of choice with a strong tendency to prefer immediate craving satisfaction, coupled with a strong sensitivity to cues that indicate reward through food (Brooks et al., 2011). Given the earlier, the presence of high levels of temporal discounting should be considered as a risk factor for the occurrence of a binge eating pattern (Schag et al., 2013).

With respect to neurobiological findings, it has been found that obese binge eaters show significant activation of frontal and prefrontal cortical areas in the presence of food cues, compared to non-binge eaters. Relationships have also been found between alterations in striatal dopamine mechanisms and the presence of compulsive behaviours in the proximity of palatable food (Schienle et al., 2009). Such alterations in reward-related responses to food stimuli affect homeostatic processes, increasing the probability of weight gain (Lutter and Nestler, 2009). These and other findings stem from a line of research that has privileged the exploration of reward circuits in brain regions associated with hedonic food intake. Particularly promising has been the manipulation of dopaminergic neurons in mice, leading to an increase in hyperphagia and a greater preference for sugary and high-fat foods (Burger et al., 2015). Furthermore, according to Volkow et al. (2001), other results indicate that reduced D2 receptor action in some cortical areas does not cause binge eating but may increase its frequency once such a pattern has been established. That is, an anomaly in dopaminergic circuitry caused by initial episodes of binge eating makes it very difficult to reverse this dynamic once these pattern has been initiated (Corwin et al., 2011).

3.5 Addictive eating

Addictive eating is the extreme end of the continuum raised in this chapter, a form of addictive behaviour that should be a major concern of public policy.

© Burleigh Dodds Science Publishing Limited, 2024. All rights reserved.

In fact, in some ways, obesity and addiction and 'junk food' ingestion have links in the literature that need to be studied from the neurosciences since hedonic and reward systems in the brain are configured in the same way for fast food and conventional drugs (Finlayson, 2017; Fraser, 2013). Nevertheless, the great distinction between addiction to both drugs and gambling is that these are not necessary to life itself, while food is an essential requirement for survival; in fact, even without being hungry or needing food, the act of imagining and mental re-creation the flavours of one's favourite dishes may lead to the desire to eat them. Hence, food, even without being present, sends signals to the brain that are mediated by tastes and the cultural construction of flavour. Likewise, the sensation of hunger increases the taste for food possibly due to the neurobiological reaction promoted by the orexin processed in the hypothalamus (hypocretin) (Peciña and Berridge, 2015), where the hyperliking or hyperwanting for food makes the difference and marks the addiction for food, given the insufficient feedback mechanisms for recognising, controlling and stopping calorie intake.

Hyperpalatable foods are those that have high levels of sugar, fat and/or sodium, and their excessive consumption elicits a sense of reward that goes beyond the need to eat (Gearhardt and Schulte, 2021; Pursey et al., 2021), which from a neurobiological perspective apparently underlies the need to survive a possible famine (Sullivan et al., 2015); nevertheless, in gender comparison females tend to have higher values than males on emotional dysregulation and addictive behaviours related to eating (Forsén Mantilla et al., 2022). Additionally, addictive eating presents neuroadaptations in reward and decision-making circuits and previous evidence has identified the physical and social contexts that drive reward mechanisms and emotions to dependence behaviour (Ho and Verdejo-Garcia, 2021), as well as triggering factors such as stress leading to alterations in the hypothalamic–pituitary–adrenal (HPA) axis, glucose, insulin and appetite-related hormones (Yau and Potenza, 2013).

4 Treatment of extreme food consumption

We conclude this chapter with a short discussion of the treatment of extreme consumption of foods in the form of pharmacological and non-pharmacological interventions.

4.1 Pharmacological

Since 2003, when obesity was established as an epidemic by the World Health Organization (WHO, 2003), the amount of research aimed at establishing an aetiology of overeating has increased dramatically, as has the use of this knowledge for the development of pharmacological and behavioural

© Burleigh Dodds Science Publishing Limited, 2024. All rights reserved.

interventions (Potenza, 2013). A common approach in this research is the parallel that has been drawn between gambling, as the first behavioural addiction, and overeating and obesity, in that it is characterised by patterns of impulsive behaviour and the activation of neural reward pathways similar to those produced by substances of abuse. In addition, some treatments for substance addiction have shown some positive results when applied to samples of individuals with obesity and overeating (Adams et al., 2019).

In the last decade, several treatments for obesity have been tested that target the neural areas associated with reward sensitivity and inhibitory control, as well as the neural circuits and neurotransmitters involved in these processes. In particular, raclopride and some antipsychotics have been studied for the dopaminergic system, nalmefene and naloxone for the opioid system and baclofen for the GABA-ergic system (Sullivan et al., 2015). A recent review states that lisdexamfetamine is the only approved pharmacological treatment for BED, while stating that for most medications, including weight loss drugs, there is no robust evidence of efficacy in BED. It also identifies drugs that act via noradrenaline, GABA and opioid receptors as having a minimally positive effect. Finally, the evidence indicates that treatment of BED should focus on reducing impulsivity by increasing the cognitive ability to restrict consumption of palatable food. (Heal and Gosden, 2022).

4.2 Non-pharmacological

One of the most widely accepted perspectives on the aetiology of overeating and obesity states that some individuals develop this pattern of behaviour as a way of coping with negative emotional states ranging from boredom and loneliness to anxiety and depression. Therefore, a joint approach between pharmacological and behavioural treatments may be the most effective way to achieve long-term change (Sullivan et al., 2015). However, in terms of the most commonly applied intervention approach, the standard method for weight loss is to combine a healthy diet with regular physical activity, which has been shown to have poor adherence rates and may even induce weight gain in individuals (Pietiläinen et al., 2012). This leads us to consider an alternative interpretation, which is that this approach aims to address the consequence of overeating rather than the underlying cause. Therefore, interventions aimed at reducing impulsivity and lack of inhibitory control might have better results (Adams et al., 2019).

Among the most innovative treatments are cognitive interventions, which seek to train individuals with overeating through multiple trials in which stimuli that signal palatable food are associated with another type of experience, in order to modify this strongly rooted attentional bias. However, the evidence obtained indicates very little modification of these biases (Smith et al., 2020). Other studies

© Burleigh Dodds Science Publishing Limited, 2024. All rights reserved.

have taken stimulus conditioning to the affective level, for instance, Hollands et al. (2011) found that participants who were repeatedly exposed to snack images together with negative body images were more likely to select fruit over an unhealthy food. However, the results of these treatments remain inconclusive in terms of an actual and sustained reduction in unhealthy food choices.

Another alternative that is gaining ground is the application of non-surgical brain stimulation in the prefrontal area. It is assumed that this neuromodulation-based intervention can enhance learning effects, reducing unhealthy food consumption and craving (Alonso-Alonso, 2013). The most commonly used methods are transcranial magnetic stimulation (TMS) and transcranial direct current stimulation (tDCS), which modify neural activity by increasing dopamine levels. TMS involves the application of electromagnetic pulses that reach the cerebral cortex, inducing small electrical currents that are assumed to generate short-term changes in cortical excitability (Maizey et al., 2013). To enhance the effects beyond the period of application, trains of pulses are used, known as repetitive TMS (rTMS). This variation of the technique has shown positive results in reducing cravings for substances of abuse such as cigarettes and alcohol (Bolloni et al., 2016). However, the use of tDCS as a treatment for food cravings has shown a reduction in reported efficacy, according to a meta-analysis that included eight experiments (Lowe et al., 2017). From the previous discussion, it is noted that there is no robust evidence for the effectiveness of these alternative treatments and that further research on the combination of pharmacological and behavioural approaches is therefore required.

5 Conclusion

The necessity and pleasures of eating form the focus of extensive investigation in the social, behavioural and cultural sciences. Our aim has been to draw attention to the neurophysiological dimensions of this form of consumption, not in order to supplant these perspectives but in order to provide a launch pad for further research that incorporates cross-disciplinary insights.

6 Where to look for further information

- Altheimer, G. and Urry, H. L. (2019). Do Emotions Cause Eating? The Role of Previous Experiences and Social Context in Emotional Eating. *Current Directions in Psychological Science* 28, 234–240.
- Avena, N. M. (2015). *Hedonic Eating: How the Pleasure of Food Affects Our Brains and Behavior*. Oxford University Press.
- Cottone, P., Sabino, V., Moore, K. F. and Koob, G. F. (Eds.). (2019). *Compulsive Eating Behavior and Food Addiction: Emerging Pathological Constructs*. Academic Press.

© Burleigh Dodds Science Publishing Limited, 2024. All rights reserved.

- Foxall, G. R. (2024). *The Continuum of Consumer Choice*. Routledge.
- Miljkovic, D. (2021). Nonconvex consumer preferences, compulsive eating and food addiction. *British Food Journal* 123, 355–371.
- Watts, A. G., Kanoski, S. E., Sanchez-Watts, G. and Langhans, W. (2022). The physiological control of eating: signals, neurons, and networks. *Physiological Reviews* 102(2), 689–813. https://doi.org/10.1152/physrev.00028.2020.

7 References

Adams, R. C., Sedgmond, J., Maizey, L., Chambers, C. D. and Lawrence, N. S. (2019). Food addiction: implications for the diagnosis and treatment of overeating. *Nutrients* 11(9). https://doi.org/10.3390/nu11092086.

Alonso-Alonso, M. (2013). Translating tDCS into the field of obesity: Mechanism-driven approaches. *Frontiers in Human Neuroscience* 7, 512. https://doi.org/10.3389/fnhum.2013.00512.

APA (2013). *Diagnostic and Statistical Manual of Mental Disorders: DSM-5* (vol. 5). American Psychiatric Association, Washington, DC.

Avena, N. M., Murray, S. and Gold, M. S. (2013). Comparing the effects of food restriction and overeating on brain reward systems. *Experimental Gerontology* 48(10), 1062–1067. https://doi.org/10.1016/j.exger.2013.03.006.

Bayle-Tourtoulou, A.-S. and Badoc, M. (2020). *The Neuro-Consumer: Adapting Marketing and Communication Strategies for the Subconscious, Instinctive and Irrational Consumer's Brain*. Routledge.

Berner, L. A. and Marsh, R. (2014). Frontostriatal circuits and the development of bulimia nervosa. *Frontiers in Behavioral Neuroscience* 8, 395. https://doi.org/10.3389/fnbeh.2014.00395.

Berridge, K. C. and Kringelbach, M. L. (2015). Pleasure systems in the brain. *Neuron* 86(3), 646–664. https://doi.org/10.1016/j.neuron.2015.02.018.

Berridge, K. C. and Robinson, T. E. (1995). The mind of an addicted brain: neural sensitization of wanting versus liking. *Current Directions in Psychological Science* 4(3), 71–76. https://doi.org/10.1111/1467-8721.ep10772316.

Berthoud, H. R. (2007). Interactions between the "cognitive" and "metabolic" brain in the control of food intake. *Physiology and Behavior* 91(5), 486–498. https://doi.org/10.1016/j.physbeh.2006.12.016.

Bolloni, C., Panella, R., Pedetti, M., Frascella, A. G., Gambelunghe, C., Piccoli, T., Maniaci, G., Brancato, A., Cannizzaro, C. and Diana, M. (2016). Bilateral transcranial magnetic stimulation of the prefrontal cortex reduces cocaine intake: A pilot study. In: *Frontiers in Psychiatry* 7. Available at: https://www.frontiersin.org/article/10.3389/fpsyt.2016.00133, 133.

Brooks, S. J., O'Daly, O. G., Uher, R., Friederich, H. C., Giampietro, V., Brammer, M., Williams, S. C. R., Schiöth, H. B., Treasure, J. and Campbell, I. C. (2011). Differential neural responses to food images in women with bulimia versus anorexia nervosa. *PLoS ONE* 6(7), e22259. https://doi.org/10.1371/journal.pone.0022259.

Burger, K. S., Kantak, P. A. and Sanders, A. J. (2015). Brain-based drivers and consequences of hedonic overeating. In: *Hedonic Eating: How the Pleasure of Food Affects Our Brains and Behavior*. (pp. 107–123). Oxford University Press.

© Burleigh Dodds Science Publishing Limited, 2024. All rights reserved.

Corwin, R. L., Avena, N. M. and Boggiano, M. M. (2011). Feeding and reward: Perspectives from three rat models of binge eating. *Physiology and Behavior* 104(1), 87–97. https://doi.org/10.1016/j.physbeh.2011.04.041.

Cottone, P., Sabino, V., Moore, K. F. and Koob, G. F. (Eds.). (2019). *Compulsive Eating Behavior and Food Addiction: Emerging Pathological Constructs*. Academic Press.

Ehrenberg, A. S. C. (1988). *Repeat Buying*. North-Holland.

Elfhag, K. and Rössner, S. (2005). Who succeeds in maintaining weight loss? A conceptual review of factors associated with weight loss maintenance and weight regain. *Obesity Reviews: an Official Journal of the International Association for the Study of Obesity* 6(1), 67–85. https://doi.org/10.1111/j.1467-789X.2005.00170.x.

Finlayson, G. (2017). Food addiction and obesity: Unnecessary medicalization of hedonic overeating. *Nature Reviews. Endocrinology* 13(8), 493–498. https://doi.org/10.1038/nrendo.2017.61.

Forsén Mantilla, E., Clinton, D., Monell, E., Levallius, J. and Birgegård, A. (2022). Impulsivity and compulsivity as parallel mediators of emotion dysregulation in eating-related addictive-like behaviors, alcohol use, and compulsive exercise. *Brain and Behavior* 12(1), e2458. https://doi.org/10.1002/brb3.2458.

Foxall, G. R. (2016a). *Addiction as Consumer Choice: Exploring the Cognitive Dimension*. Routledge.

Foxall, G. R. (2017a). *Context and Cognition in Consumer Psychology: How Perception and Emotion Guide Action*. Routledge.

Foxall, G. R. (2017b). *Advanced Introduction to Consumer Behavior Analysis*. Edward Elgar Publishing.

Foxall, G. R. and James, V. K. (2002). Behavior analysis of consumer brand choice: a preliminary analysis. *European Journal of Behavior Analysis* 2, 209–220.

Foxall, G. R. and James, V. K. (2003). The behavioral ecology of brand choice: how and what do consumers maximize? *Psychology and Marketing* 20(9), 811–836.

Foxall, G. R. (2020). *Intentional Behaviorism: Philosophical Foundations of Economic Psychology*. Academic Press.

Foxall, G. R. (2023). The neurophysiological Behavioral Perspective Model and its contribution to the intentional behaviorist research program. *Frontiers in Human Neuroscience* 17, 1190108.

Foxall, G. R. (2024). The Continuum of Consumer Choice. London and New York: Routledge.

Fraser, S. (2013). Junk: overeating and obesity and the neuroscience of addiction. *Addiction Research and Theory* 21(6), 496–506. https://doi.org/10.3109/16066359.2012.749868.

Gearhardt, A. N. and Schulte, E. M. (2021). Is food addictive? A review of the science. *Annual Review of Nutrition*, 41. https://doi.org/10.1146/annurev-nutr-110420-111710.

Glimcher, P. (2009). Choice: toward as a standard back-pocket model. In: Glimcher, P., Camerer, C., Fehr, E. and Poldrack, A. (Eds). *Neuroeconomics: Decision-Making and the Brain* (pp. 503–521). Elsevier.

Glimcher, P. W. (2010). *Foundations of Neuroeconomic Analysis*. Oxford University Press.

Gordon, E. L., Ariel-Donges, A. H., Bauman, V. and Merlo, L. J. (2018). What is the evidence for "food addiction?" A systematic review. *Nutrients* 10(4), 477. https://doi.org/10.3390/nu10040477.

© Burleigh Dodds Science Publishing Limited, 2024. All rights reserved.

Heal, D. J. and Gosden, J. (2022). What pharmacological interventions are effective in binge-eating disorder? Insights from a critical evaluation of the evidence from clinical trials. *International Journal of Obesity* 46(4), 677-695. https://doi.org/10.1038/s41366-021-01032-9.

Ho, D. and Verdejo-Garcia, A. (2021). Interactive influences of food, contexts and neurocognitive systems on addictive eating. *Progress in Neuro-Psychopharmacology and Biological Psychiatry* 110, 110295. https://doi.org/10.1016/j.pnpbp.2021.110295.

Hollands, G. J., Prestwich, A. and Marteau, T. M. (2011). Using aversive images to enhance healthy food choices and implicit attitudes: an experimental test of evaluative conditioning. *Health Psychology: Official Journal of the Division of Health Psychology, American Psychological Association* 30(2), 195–203. https://doi.org/10.1037/a0022261.

Horstmann, A. (2016). The brain's got a taste for good food. In: Avena, N. M. (Ed.). *Hedonic Eating: How the Pleasure of Food Affects Our Brains and Behavior* (pp. 39–55). Oxford University Press.

Leech, R. M., Worsley, A., Timperio, A. and McNaughton, S. A. (2015). Understanding meal patterns: definitions, methodology and impact on nutrient intake and diet quality. *Nutrition Research Reviews* 28(1), 1–21. https://doi.org/10.1017/S0954422414000262.

Leitch, M. and Geliebter, A. (2015). Overeating and binge eating. In: Avena, N. M. (Ed.). *Hedonic Eating: How the Pleasure of Food Affects Our Brains and Behavior* (pp. 85–106). Oxford University Press.

Lewis, G. F. (2013). Devastating metabolic consequences of a life of plenty: Focus on the dyslipidemia of overnutrition. *Clinical and Investigative Medicine. Medecine Clinique et Experimentale* 36(5), E242–E247. https://doi.org/10.25011/cim.v36i5.20123.

Lowe, C. J., Vincent, C. and Hall, P. A. (2017). Effects of noninvasive brain stimulation on food cravings and consumption: A meta-analytic review. *Psychosomatic Medicine* 79(1). Available at: https://journals.lww.com/psychosomaticmedicine/Fulltext/2017/01000/Effects_of_Noninvasive_Brain_Stimulation_on_Food.2.aspx, 2–13.

Lutter, M. and Nestler, E. J. (2009). Homeostatic and hedonic signals interact in the regulation of food intake. *The Journal of Nutrition* 139(3), 629–632. https://doi.org/10.3945/jn.108.097618.

Maizey, L., Allen, C. P. G., Dervinis, M., Verbruggen, F., Varnava, A., Kozlov, M., Adams, R. C., Stokes, M., Klemen, J., Bungert, A., Hounsell, C. A. and Chambers, C. D. (2013). Comparative incidence rates of mild adverse effects to transcranial magnetic stimulation. *Clinical Neurophysiology : Official Journal of the International Federation of Clinical Neurophysiology* 124(3), 536–544. https://doi.org/10.1016/j.clinph.2012.07.024.

Murakami, K., Shinozaki, N., Yuan, X., Tajima, R., Matsumoto, M., Masayasu, S. and Sasaki, S. (2022). Food choice values and food literacy in a nationwide sample of Japanese adults: associations with sex, age, and body mass index. *Nutrients* 14(9). https://doi.org/10.3390/nu14091899.

Myers, K. (2015). Why do we eat what we eat? In: Eating, H. (Ed.). (pp. 9–38). Oxford University Press. https://doi.org/10.1093/med/9780199330454.003.0002.

O'Hara, C. B., Campbell, I. C. and Schmidt, U. (2015). A reward-centred model of anorexia nervosa: A focussed narrative review of the neurological and psychophysiological

© Burleigh Dodds Science Publishing Limited, 2024. All rights reserved.

literature. *Neuroscience and Biobehavioral Reviews* 52, 131–152. https://doi.org/10.1016/j.neubiorev.2015.02.012.

Oliveira-Castro, J. M. and Foxall, G. R. (2017). Consumer maximization of utilitarian and informational reinforcement: comparing two utility measures with reference to social class. *The Behavior Analyst* 40(2), 457–474.

Oliveira-Castro, J. M., Cavalcanti, P. R. and Foxall, G. R. (2016a). What consumers maximize: brand choice as a function of utilitarian and informational reinforcement. *Managerial and Decision Economics* 37(4–5), 360–371.

Oliveira-Castro, J. M., Cavalcanti, P. and Foxall, G. R. (2016b). What do consumers maximize? The analysis of utility functions in light of the Behavioral Perspective Model. In: Foxall, G. R. (Ed.). *The Routledge Companion to Consumer Behavior Analysis* (pp. 202–212). Routledge.

Peciña, S. and Berridge, K. (2015). *Food "Liking" and "Wanting": A Neurobiological Perspective* 125–145. Oxford University Press.

Pietiläinen, K. H., Saarni, S. E., Kaprio, J. and Rissanen, A. (2012). Does dieting make you fat? A twin study. *International Journal of Obesity* 36(3), 456–464. https://doi.org/10.1038/ijo.2011.160.

Potenza, M. N. (2013). Biological contributions to addictions in adolescents and adults: Prevention, treatment, and policy implications. *The Journal of Adolescent Health: Official Publication of the Society for Adolescent Medicine* 52(2) (Suppl. 2), S22–S32. https://doi.org/10.1016/j.jadohealth.2012.05.007.

Pursey, K. M., Skinner, J., Leary, M. and Burrows, T. (2021). The relationship between addictive eating and dietary intake: A systematic review. *Nutrients* 14(1). https://doi.org/10.3390/nu14010164.

Rumiati, R. I. and di Pellegrino, G. (2016). Thoughts for food in cognitive neuroscience: an introduction to the special issue. *Brain and Cognition* 110, 1–3. https://doi.org/10.1016/j.bandc.2016.09.007.

Schag, K., Schönleber, J., Teufel, M., Zipfel, S. and Giel, K. E. (2013). Food-related impulsivity in obesity and binge eating disorder--A systematic review. *Obesity Reviews: an Official Journal of the International Association for the Study of Obesity* 14(6), 477–495. https://doi.org/10.1111/obr.12017.

Schienle, A., Schäfer, A., Hermann, A. and Vaitl, D. (2009). Binge-eating disorder: Reward sensitivity and brain activation to images of food. *Biological Psychiatry* 65(8), 654–661. https://doi.org/10.1016/j.biopsych.2008.09.028.

Smith, E., Treffiletti, A., Bailey, P. E. and Moustafa, A. A. (2020). The effect of attentional bias modification training on food intake in overweight and obese women. *Journal of Health Psychology* 25(10–11), 1511–1521. https://doi.org/10.1177/1359105318758856.

Sullivan, E., Tulloch, A., Gearhardt, A. and Avena, N. M. (2015). Food addiction. In: *Hedonic Eating: How the Pleasure of Food Affects Our Brains and Behavior* (pp. 185–207). Oxford University Press.

Tang, D. W., Fellows, L. K., Small, D. M. and Dagher, A. (2012). Food and drug cues activate similar brain regions: A meta-analysis of functional MRI studies. *Physiology and Behavior* 106(3), 317–324. https://doi.org/10.1016/j.physbeh.2012.03.009.

Tomasi, D. and Volkow, N. D. (2013). Striatocortical pathway dysfunction in addiction and obesity: Differences and similarities. *Critical Reviews in Biochemistry and Molecular Biology* 48(1), 1–19. https://doi.org/10.3109/10409238.2012.735642.

© Burleigh Dodds Science Publishing Limited, 2024. All rights reserved.

Vainik, U., García-García, I. and Dagher, A. (2019). Uncontrolled eating: A unifying heritable trait linked with obesity, overeating, personality and the brain. *The European Journal of Neuroscience* 50(3), 2430–2445. https://doi.org/10.1111/ejn.14352.

Ventura, R., Latagliata, E. C., Morrone, C., La Mela, I. and Puglisi-Allegra, S. (2008). Prefrontal norepinephrine determines attribution of "high" motivational salience. *PLoS ONE* 3(8), e3044. https://doi.org/10.1371/journal.pone.0003044.

Ventura, R., Morrone, C. and Puglisi-Allegra, S. (2007). Prefrontal/accumbal catecholamine system determines motivational salience attribution to both reward- and aversion-related stimuli. *Proceedings of the National Academy of Sciences* 104(12), 5181–5186. https://doi.org/10.1073/pnas.0610178104.

Volkow, N. D., Chang, L., Wang, G. J., Fowler, J. S., Ding, Y. S., Sedler, M., Logan, J., Franceschi, D., Gatley, J., Hitzemann, R., Gifford, A., Wong, C. and Pappas, N. (2001). Low level of brain dopamine D2 receptors in methamphetamine abusers: Association with metabolism in the orbitofrontal cortex. *The American Journal of Psychiatry* 158(12), 2015–2021. https://doi.org/10.1176/appi.ajp.158.12.2015.

Volkow, N. D., Wang, G. J., Tomasi, D. and Baler, R. D. (2013). The addictive dimensionality of obesity. *Biological Psychiatry* 73(9), 811–818. https://doi.org/10.1016/j.biopsych.2012.12.020.

Wansink, B. and Chandon, P. (2006). Can "low-fat" nutrition labels lead to obesity? *Journal of Marketing Research (JMR)* 43(4), 605–617. Available at: http://10.0.5.229/jmkr.43.4.605.

Wansink, B., Just, D. R. and Payne, C. R. (2009). Mindless eating and healthy heuristics for the irrational. *American Economic Review* 99(2), 165–169. https://doi.org/10.1257/aer.99.2.165.

Watts, A. G., Kanoski, S. E., Sanchez-Watts, G. and Langhans, W. (2022). The physiological control of eating: Signals, neurons, and networks. *Physiological Reviews* 102(2), 689–813. https://doi.org/10.1152/physrev.00028.2020.

WHO (2003). *Obesity and Overweight.* https://www.who.int/news-room/fact-sheets/detail/obesity-and-overweight.

Woods, S. C. and Begg, D. P. (2016). Regulation of the motivation to eat. *Current Topics in Behavioral Neurosciences* 27, 15–34.

© Burleigh Dodds Science Publishing Limited, 2024. All rights reserved.

Chapter 3

Geographical patterns of food-purchasing behaviour: the example of sub-Saharan Africa

John Kuada, African Business Education & Research, Denmark

1 Introduction

Food security remains a major challenge in most sub-Saharan African (SSA) countries and there are no indications that the situation will improve in the immediate future. Available evidence indicates that households in the sub-continent averagely spend 40% of their income on food - varying from 18% in South Africa to about 50% in Rwanda (Cleland and Machiyama, 2017). Food demand is also expected to increase with the consistent increase in population, which is projected to be more than 2 billion by 2050 (United Nations Population Division, 2021; World Bank Data, 2022). Furthermore, local food production in most of the countries has not matched increases in demand for nearly half a century (OECD/FAO, 2020). These conditions suggest that increases in global food prices are likely to have serious repercussions on food security of individual households, especially in the rural areas where over 50% of the populations still live (Groth and May, 2017).

The problem of food security is directly linked to poverty alleviation since food insecurity makes people resign to their living conditions and sets negative spiral in motion, thereby holding their creativity in check (Letelier et al., 2003). This means any effort made to improve food security is itself growth-propelling since it unleashes hitherto untapped psychological and physical human

http://dx.doi.org/10.19103/AS.2023.0129.11
© Burleigh Dodds Science Publishing Limited, 2024. All rights reserved.

resources within a community and thereby helps transform a negative spiral into a positive one. Based on this understanding, it can be argued that food marketing process improvements in SSA constitute not only a key variable in the global food and nutrition security equation but have also important poverty alleviation implications.

Despite the food security challenge, there have been some significant shifts in food consumption and marketing processes in SSA, fuelled primarily by population and demographic changes combined with the modernization of the food distribution system and the influence of foreign consumption habits and health considerations. The changes also offer reasons to reassess previous perspectives on food marketing and its role in African economies. It is against this background that the relevance of the present chapter must be assessed. It seeks to provide an overview of the multiplicity of factors that are shaping the food marketing and purchasing behaviours of consumers in SSA and the implications they carry for policy, strategy, and future research in the field.

In terms of structure the discussions continue after this brief introduction with a review of some theoretical perspectives that have guided empirical investigations into food marketing processes in SSA. This is followed by discussions of some major shifts in consumer preferences and purchase behaviours. We then examine changes in the food distribution infrastructure, highlighting the roles that modern food retail firms are playing in the process. Finally, we explore the implications of these changes for policy, strategy, and future research.

2 Overview of prior theoretical arguments and perspectives

A key assumption in marketing theory is that consumers behave rationally and given the chance they will choose the best alternative goods and services that match their level of income and preference. This holds true in food-purchasing decisions as well, with consumers evaluating product features, quality, and available brands before making their purchase decisions (Bett et al., 2013). In this decision-making process consumers usually make a trade-off between the price they are expected to pay and the sacrifices they must make in terms of money, time, and effort. Thus, the challenges of achieving the twin goals of low cost and high quality of food products and delivery services have encouraged studies into food market infrastructure – i.e. food system's physical and institutional infrastructure that links food producers (including farmers) to consumers. (IFAMA, 2017; Rana and Paul, 2017). Physical infrastructure studies have examined the changing roles of supply chain members (i.e. retailers, wholesalers, and distributors) in the marketing systems in Africa (van Berkum et al., 2017). It has been shown that with seasonal harvests, it is

© Burleigh Dodds Science Publishing Limited, 2024. All rights reserved.

necessary for the distribution system to have adequate storage facilities to ensure regular supply of locally produced food items (Wamalwa et al., 2019). The concept of institutional market infrastructure has been applied to describe quality standards, price information systems, contracts, purchasing processes, competition rules, and national and international regulations (Osakwe and Anaza, 2018; van Berkum et al., 2017). Seen from a rational economic perspective, factors including supply rigidities in key food producing regions combine with low stocks and exchange rate fluctuations to influence food prices globally (Reardon et al., 2007), and these invariably affect the prices of food items in Africa's retail markets (Reardon et al., 2003).

The rational behavioural argument suggests that demand patterns change with changing income levels. That is, economic growth and rising levels of income in a country influence consumers' lifestyle and broaden the market opportunity frontier within which retailers can operate, as previously lower-income consumers adopt the tastes of high-income consumers (van Berkum et al., 2017). Building on the earlier understanding, some African studies have suggested that the growing middle class in some countries and the general improvements in living standards, particularly among the urban populations, have led to changes in their consumption patterns that create opportunities for modern retail outlets, including supermarkets (Hattingh et al., 2012; Lysonski and Durvasula, 2013; Wamalwa et al., 2019).

Some segments of the food marketing literature have relaxed the assumption that consumers are entirely rational in their food-purchasing decisions. The central argument in this strand of research is that consumers' beliefs and attitudes tend to mediate the rational choice matrix. Leaning on the theory of reasoned action and its extension in the form of the theory of planned behaviour (Ajzen and Fishbein, 1980; Ajzen, 1991), these scholars argue that for some groups of consumers, beliefs about the nutritional quality and health effects of a particular food item may be a stronger determinant of choice than small price differences. In the same vein, social, cultural, and religious factors may impact the beliefs held by a person about a particular food item and influence their choices. Furthermore, people's habits evolve with age and time, resulting in changing needs for products and services. For these reasons, marketers are enjoined to leverage their understanding of consumers' attitudes to predict their behaviour (Rana and Paul, 2017; Rahman and Yu, 2019).

Studies focusing on the performance of food retailers have identified consumer satisfaction as a critical determinant of performance, just as in the marketing of other product categories (Huddleston et al., 2009). Customer satisfaction is generally described in the marketing literature as customers' post-consumption experience (in relation to expectations) and their fulfilment response following the consumption experience. It is frequently argued that when perceived performance exceeds a consumer's pre-purchase expectation,

© Burleigh Dodds Science Publishing Limited, 2024. All rights reserved.

a positive disconfirmation results and the consumer could be highly satisfied or even delighted. However, when post-purchase performance falls below pre-purchase expectation, a negative disconfirmation occurs, and the customer is dissatisfied. In a situation where the product or service performs as expected, a simple confirmation occurs (Kuada, 2016). Customer satisfaction has been found to impact customer loyalty, and this can substantially increase customer retention and reduce marketing costs (Rahman, and Yu, 2019). By understanding the infrastructural weaknesses inherent in a food marketing system, food retailers may more readily spot any evidence of dissatisfaction among their customers and make appropriate strategic adjustments to their services to minimize the dissatisfactions and maintain the overall value of the services they offer.

The centrality of customer satisfaction to the performance of modern food retailers that are entering SSA food marketing systems has encouraged some investigations into the extent to which consumers on the continent respond to Western-inspired marketing strategies (Kuada and Buatsi, 2005). Some studies have found the middle-income African consumers to be as sophisticated as consumers in the developed countries and tend to demand superior quality products and services offered in clean environments. They are also described as being more open-minded than other segments of populations, and more socially connected, more spontaneous, and more willing to try new products (Onyemah and Akpa, 2016). Nevertheless, the general conclusion from these studies is that marketing strategies aimed at reducing consumers' dissatisfaction must be contextualized since African consumers demonstrate weak responses to Western marketing strategies (Nandonde and Kuada, 2016a). For example, Deon (2011) found that the use of sales promotions to encourage impulse buying in South Africa have had limited impact, partly because the shoppers plan their shopping reasonably well and make use of an informal or formal shopping list for shopping purposes. Similarly, Lee et al. (2011) found that Mozambican youth exhibit tendencies of developing consumer-related skills, knowledge, and attitudes through socialization processes (e.g. mimicking peers) rather than falling prey to Western-inspired marketing stimuli. The wisdom offered by these studies have implications for strategies that foreign food retailers can adopt on their entry into SSA.

Some strands of the food marketing research have suggested that households' food-shopping motives may go beyond sociocultural considerations and include hedonic motives such as consumers' desire to satisfy sensory needs that the shopping environment may offer – with many middle-income shoppers expecting to enjoy the thrill of shopping. Consequently, modern retail outlets such as supermarkets in both developed and developing countries tend to adopt various strategy combinations to position themselves in the minds of consumers (Nandonde and Kuada, 2016b). The overriding

© Burleigh Dodds Science Publishing Limited, 2024. All rights reserved.

objective of these strategies is to increase positive shopping experience, convenience, and perceived 'value for money' among consumers (Motoki and Sugiura, 2018). This segment of consumers tends to be attracted to retail outlets with appealing physical features like lighting, architecture, space, and decorations. Space is particularly important in Muslim-dominant countries where women may worry about physical contacts with men and therefore prefer retail outlets where they can move more freely (Ashraf et al., 2022).

As argued subsequently, supermarkets have become an integral part of SSA food marketing infrastructure, mainly serving the middle-class market segment. The marketing literature refers to the increasing adoption supermarkets in developing countries as *supermarketization* (Anand, 2009). In Africa, this process of supermarketization has been spearheaded by foreign retail companies (Nandonde and Kuada, 2016b). For the middle-class urban household, supermarkets appear to be the natural place to buy health- and nutrition-sensitive imported food items since such outlets carry a greater variety of processed foods, the prices of some of which may be lower than smaller local products (Reardon et al., 2007).

The introduction of increasing number of foreign food items to African consumers via the supermarkets has encouraged scholars to examine Africans' general attitudes towards these food items and the manner in which their introduction tends to influence the consumption and production of local food items and the overall food security in SSA. These studies lean on the theoretical construct of 'foreign country effect' – i.e. consumers' dispositions towards foreign product. The general understanding presented in this strand of research is that preconceptions and attitudes towards people of a given country tend to influence consumers' evaluation of products coming from the country. For some products, 'country factor' may have a positive impact on consumer perception when sold abroad; for others, the impact may be negative. Studies of the country factor are generally anchored in two other constructs: (1) the psychosocial construct of ethnocentrism, which describes the belief that consumers in a given country hold about goods and services produced in that country as being superior to those produced in other countries and (2) the country-of-origin effect describing the positive or negative image of products originating from a given country among the host country consumers. Ethnocentric orientations of consumers are influenced by such demographic variables as age, gender, education, and income, as well as such socio-psychological factors as openness to foreign cultures and patriotism. For example, it has been shown that females, in general, tend to be more conservative, more conformist, more patriotic, less individualistic, and more concerned about preserving social harmony and promoting positive feelings among group members or domestic products (Saffu and Walker, 2006; Dubihlela, and Ngxukumeshe, 2016). Females therefore tend to exhibit greater

© Burleigh Dodds Science Publishing Limited, 2024. All rights reserved.

ethnocentric tendencies than males. Since food purchase decisions are most often made mainly by female household heads in Africa, it is generally believed that preferences will be tilted in favour of local food items.

The rest of the chapter draws on the theoretical perspectives outlined in the preceding section to examine the main divers of food consumption and purchase decisions of African households as well as the changing patterns of food distribution. We argue subsequently that changes in food consumption and purchasing behaviours induced by demographic and urban transitions have introduced changes into the food distribution landscape. Together, they have further widened the already yawning gap between the poorer and richer segments of African populations. These changes therefore have implications for food security policies and strategies.

3 Key drivers of food consumption decisions and habits

There have been noticeable shifts in food consumption and purchasing behaviours in Africa during the past three decades, particularly in the urban areas. Available evidence points at the following four factors as driving this shift:

1. Population and demographic changes;
2. Economic growth;
3. Increased variety of imported food items; and
4. Growing health and environmental consciousness.

The population of sub-Saharan Africa (SSA) has grown from 186 million to 856 million people between 1950 and 2010 and to 1.1 billion people in 2021. Annual population growth rate is at 2.7% and the forecast is that there will be as many as 2.7 billion people on the continent by 2060 (The Economist, 2020; Statistica, 2020). In addition to the overall population growth, there has been substantial movement of people from rural to urban areas. The urbanization rate is expected to increase up to 46% by 2030 and 57% by 2050, with West Africa showing a faster urban growth than East Africa (Groth and May, 2017). This population growth has resulted in increased demand for food, even though consumption per head has not changed in any significant way.

In terms of economic growth, SSA countries have experienced some modest growth gains during the past three decades. Out of 54 countries, 24 of them more than doubled their per capita income over the 1990–2010 period (African Development Bank Report, 2014). Together, the continent's growth reached 4.8% in 2013 (from 4.2% in 2012) and 4.5% in 2021 – a decline attributable to high volatility and uncertainties within the world economy (IMF, 2022). The modest growth has resulted in an equally modest rise in household income, especially for those living in the urban areas. As a result, urban

© Burleigh Dodds Science Publishing Limited, 2024. All rights reserved.

household food consumption has shifted away from stables such as tubers (e.g. yams, cocoyam, cassava, and plantain) and towards meat, fish, oils, fresh produce, and processed foods (van Berkum et al., 2017). However, the fact that about 95% of African agriculture is rainfed and thus extremely vulnerable to climate change implies that food deficits are covered by imports. Thus, Africa has been a net importer of food during the past half a century, with net imports in real terms growing at 3.4% per year (Cleland and Machiyama, 2017).

Economic growth has also produced a noticeable middle-class market segment consisting of households who have their own food storage facilities (including fridges and deep freezers), as well as personal transportation. Furthermore, an increasing number of women now work outside the home and therefore spend more time in transit to and from work. This tends to affect their food-purchasing decisions with respect to the types of food items they choose to buy, their shopping frequency, and the places from where they tend to make their food purchases (Beckline and Kato, 2014). It has also been noted that despite the modest economic gains in some SSA countries, most households remain poor. Their low level of income implies that they are generally price-sensitive, and the level of food prices still affects food affordability and therefore food consumption as well as the basket of food items that most households choose to buy.

Demographic characteristics of shoppers have also been found to influence the types of food items that people buy, from where they buy them, as well as their frequency of purchase. A study in Ghana by Bannor et al. (2022) revealed that while male shoppers are more likely to purchase fresh fruits and vegetables in supermarkets, female shoppers buy such items from traditional markets, apparently because they emphasize cost and freshness to a larger extent than male shoppers. The male supermarket shoppers were also relatively younger on average (33 years) than their female counterparts (48 years). In addition to this, supermarket shoppers have relatively higher level of formal education than those buying from traditional markets.

It is also important to remember that women tend to play roles both inside and outside of the home, and this 'double burden' makes more claims on their time than for men. Added to this, traditional gender norms can make mobility a challenge for women who may be expected to remain within or close to the home, and in some cultures their movement outside is limited and often dependent upon their husband's approval. This can have implications for their ability to shop at distant locations.

Some studies have noted changes in dietary habits of the middle-class urban households. As noted earlier, items such as dairy products, edible oils and fats, and meat and meat products now feature prominently in the shopping baskets of middle-class consumers. However, these changes have not been entirely positive. For example, a study by Pradeilles et al. (2021)

© Burleigh Dodds Science Publishing Limited, 2024. All rights reserved.

revealed a significant rise in overweight and obesity among households, with its prevalence being twice as high among urban-dwelling women compared to men. Although some consumers have acquired taste for unhealthy food items, others have demonstrated noteworthy health consciousness with regard to their choices of food items and are willing to pay premium prices for them (Kuada and Bujac, 2018). For example, Kenyan urban consumers were willing to pay a lot more for new gravity-driven membrane drinking-water filters introduced to the market because they were perceived to have extra health advantages (Amugsi et al., 2017). Similarly, Yahaya et al. (2015) assessed consumer motivation and willingness to pay for 'safer' vegetables from the use of treated (as compared to non-treated) wastewater in urban/peri-urban vegetable production in Ghana. They found that consumers were willing to pay an average amount of $2.40 per month for a technology change that would result in the production of the 'safer' vegetables. Furthermore, income and gender turned out to be key demographic factors that influenced consumers' willingness to pay the relatively higher amounts for the 'safer' vegetables – females and consumers in the higher income bracket were more willing to pay the higher prices. Similarly, Bett et al. (2013) also studied the Kenyan consumers' responsiveness to an increase in prices of the indigenous chicken products and how much they are willing to pay for them in the market. The results showed that consumers were willing to pay about 25% more for local chicken and eggs than the imported ones, due to their perceived superior quality. Along the same lines, Dubihlela and Ngxukumeshe (2016) found a significant association between eco-friendly retail product attributes and South African consumers' purchasing intentions. Furthermore, Thanika et al. (2012) found that most consumers in Mauritius showed great interest in the protection of the environment, and this appeared to impact their purchase decisions with regard to green products.

Regarding ethnocentrism, John and Brady (2011) found Mozambican consumers to be moderately ethnocentric and to have relatively unfavourable attitudes towards South African food items. Although Saffu and Walker (2006) found Ghanaian consumers to be moderately ethnocentric, they were willing to buy Ghanaian-made products only when these products were of comparable quality and price to foreign-made products. Furthermore, a study by Pentz et al. (2013) in South Africa revealed a positive relationship between age and consumer ethnocentrism and a negative relationship between income and ethnocentrism with regard to food items.

In sum, the food marketing literature has registered noticeable changes in the consumption patterns and dietary habits in most African countries, although the degree to which major factors are shaping these changes in each country has not been fully investigated.

© Burleigh Dodds Science Publishing Limited, 2024. All rights reserved.

4 Changes in food distribution infrastructure

Food retail accounts for more than half of all retail sales in major African countries. For this reason, the changing food-purchasing intentions and habits outlined previously have significant impact on the retail landscape in Africa. African food market infrastructure is now made up of a mix of traditional retail markets (i.e. the so-called wet shops), supermarkets, convenience stores, and micro-retailers (i.e. street vendors and peddlers), all of them coexisting and playing complementary roles as they serve different market segments.

4.1 Traditional food retail outlets

The traditional food retailers serve low-income shoppers who cannot afford to purchase more than their daily food requirements and have none or limited storage facilities (Deon, 2011). Their frequency of shopping is therefore high, and the variety of food items purchased is limited (Lysonski and Durvasula, 2013; Onyemah and Akpa, 2016). These households also have low shopping outreach, buying their food items from outlets close by. The general assessment is that with the exception of few countries such as South Africa, these traditional retail outlets will remain at the core of African retail food landscape for the foreseeable future (Wamalwa et. al., 2019).

To most Western visitors, the traditional food retail markets remain unhygienic and are often characterized by utter confusion. The retailers usually operate from stalls that they may construct themselves or are rented from local institutions that own them. Other traders sell their items in the open sun. The passages between the stalls are very often thronged with sellers who heap their offerings on trays placed on boxes. It is not unusual to find markets extending beyond their original boundaries with sellers displaying their products on pavements and other similar public spaces. Each market may have specific days on which to operate.

Despite these apparent weaknesses, the traditional outlets remain resilient because they serve their target market segment better than modern retail outlets. Not only do they offer the proximity, flexibility, and convenient operating hours needed to serve their customers, they develop stronger relationships with them, offering some households credit when household income dries out and survival may depend on such a credit (Deon, 2011).

The continued dominance of traditional retail outlets in the overall food retail landscape has partly been explained by the generally skewed income distribution on the continent. For example, in South Africa, the top 10% of the population account for 47% of income, with a typical household in the highest income group earning 71 times more than a household in the lowest income group (Cleland and Machiyama, 2017; Groth et al., 2017). Hence, it

© Burleigh Dodds Science Publishing Limited, 2024. All rights reserved.

is not surprising that prices in the modern retail outlets are normally higher than those for similar products sold in traditional stores (Battersby, 2014).

4.2 Modern retail outlets

As intimated earlier, an increasing number of modern retail outlets now serve the growing middle-class populations, particularly in the urban centres of Africa (Nandonde and Kuada, 2016b). The growing importance of the modern retail outlets suggests that their development also has a fundamental impact on the shape and functioning of the food value chain in most African countries. It is therefore purposeful to examine their development process more closely.

Existing evidence suggests that modern retail outlets (spearheaded by supermarkets) have entered Africa's food retail infrastructure in waves, starting in South Africa and then moving to Kenya, Zambia, and Zimbabwe and spreading subsequently to West African countries such as Nigeria and Ghana (Reardon et al., 2007; Nandonde and Kuada, 2016b). They have become the preferred outlet for middle-class households due to their ability to provide wide assortments of relatively good quality products in convenient environments to their clients. However, they are highly concentrated in terms of operations. For example, in South Africa, five main outlets (referred to as the Big 5 – i.e. Shoprite, Pick n Pay, Spar, Woolworths, and Walmart's Cambridge Foods) account for over 60% of total sales from the modern outlets (Ntloedibe, 2018).

Despite the growing importance of modern foreign outlets, their entry and operations in African markets have not been smooth sailing and even the larger retailers have been surprised at the challenges that the new markets have posed. Many of them were not adequately prepared to tackle a diverse range of unfamiliar consumer behaviours, as well as the surprisingly stiff local competition. In countries such as South Africa, local retailers have been successful in using their local knowledge and networks to outcompete the foreign-owned supermarkets, thereby growing to become strong regional players in the food market. Factors such as 'emotional proximity' between customers and local retailers as well as the empathy that their frontline employees show to customers have made them preferred outlets for many customers (Klemz et al., 2006).

In addition to the aforementioned factors, past research has also shown that the shopping habits of the high-income and middle-class households make the supermarkets appropriate outlets for their purchases. For example, it has been noted that this category of shoppers tends to shop less frequently, buy larger quantities and varieties during each shopping trip, and have higher outreach (Nandonde and Kuada, 2016b). Furthermore, unlike the time-pressed consumers in the developed world who are generally in a hurry and are constantly looking for ways to save time, the middle class in Africa may delegate

© Burleigh Dodds Science Publishing Limited, 2024. All rights reserved.

the purchasing tasks to their house assistants or choose to shop at less busy times if the female head of the household decides to join the shopping activity (Promozione, 2013). They are also found to use shopping lists to simplify their food-shopping experience and ensure that their shopping requirements are met within a relatively shorter time frame. House assistants typically keep to the shopping list given to them. However, when their female head of the household join them for shopping, they may engage in unplanned spending and may take advantage of special offers that may be in the shops (Nandonde and Kuada, 2016a).

Generally, the supermarket chains have maintained their own centralized distribution and modern warehouse systems which supply products to their outlets located in different African countries. Regional supermarkets such as Shoprite from South Africa tend to import produce from their home countries, but they also source locally from a few well-organized out-growers. In this way, the regional retailers provide pathways for food producers in their countries to enter regional food value chains and thereby improve their food marketing competencies in the process. For example, previous studies have shown that South African retailers operating in other African countries tend to source about 80% of the merchandise from South Africa (Biénabe et al., 2011). Furthermore, a study conducted in Zambia by Muradzikwa (2002), estimated that 60% of the food items sold by Shoprite in Zambia were imported from South Africa.

Hitherto, supermarkets regard it advantageous to enter into non-committal relationships with larger farmers – i.e. they receive supplies from them without formal contracts but requiring them to deliver their harvest directly to stores and be paid after sales of their produce. They also require them to clean the produce before delivery; leafy vegetables also have to be sorted and bundled ready for supermarket shelves (Battersby et al., 2015). In this way, the supermarkets exhibit coercive channel power over the farmers.

The relative difficulties supermarkets are facing in some African food markets may be attributed to the dissatisfactions that some local partners have with the non-committal relationship offered to them. It also speaks to the fact that foreign retailers may face difficulties operating in Africa if they assume that they can succeed using strategies that are not as good as or better than those used in the developed countries. Some wonder if local adaptations of their strategies are worth the incremental costs (Weatherspoon and Reardon, 2003).

4.3 Other retail outlets

The lack of zoning and enforcement of laws in most African countries has allowed the establishment of micro-retailers, petty traders, and hawkers to sell food items in public spaces or along roadsides. Some position themselves

© Burleigh Dodds Science Publishing Limited, 2024. All rights reserved.

strategically close to modern retail outlets, especially those selling locally produced fresh fruits and vegetables that are not carried by the modern outlets (Beckline and Kato, 2014).

In addition to the supermarkets and the micro-retailers, online food marketing has also made its presence on the food retail landscape. Taking South Africa as example, there is evidence indicating that it has well-advanced online shopping culture with well-established infrastructure and effective laws and regulations (Neven et al., 2006). The growing popularity of online shopping with certain market segment is attributable to the splendid development of mobile money payment systems in several countries, better information sharing, faster time to market, and more efficient supply chains, while customers have benefited from the convenience of purchasing at any time of day. The expectations conveyed in the practitioner publications is that online food shops will increase in number in different African countries but will take different shapes across the continent in the coming years – all depending on each nation's level of digital maturity and economic development.

5 Implications

The discussions in the previous section are important for several reasons. First, we have listed factors including urbanization, population increases, growing middle class, climate change, and the emergence of modern food retail outlets as some of the key factors influencing food marketing and purchasing behaviour in Africa. Arguably, the consequences of each of these factors in isolation is fairly manageable since different parts of the world have experienced their incidence and SSA could learn from them. The real challenge lies in the fact that their influences are concurrent and experience from other countries appears not to provide any viable pathway for positive and enduring change. Taking climate change as an example, expert assessment is that Africa will experience an increased average temperature of 3–6°C by the end of the century (Ofori et al., 2021). Consequently, the length of growing season of agricultural products is projected to decrease, with a resultant decrease in crop yield. While agricultural economists have recommended the adoption of improved storage facilities at both farm level and within the distribution channels and/or the application of appropriate post-harvest food preservation and processing technologies to reduce food losses (Kumar and Kalita, 2017), marketing scholars have suggested the adoption of improved contractual relationships between modern retail outlets and local farmers as part of the solution to the food deficit problem (Weatherspoon and Reardon, 2003). It has also been noted that as the urban areas expand, they tend to encroach

© Burleigh Dodds Science Publishing Limited, 2024. All rights reserved.

on productive agricultural areas, thereby further reducing the already limited local food production and supplies to the traditional markets. Thus, the multidimensionality of the relationships between the factors calls for a nexus approach that addresses the interconnectedness, synergies, and trade-offs existing between them. Herein lies the real challenge to policymakers.

Second, intra-channel relationships have presented significant challenges to the efficiency and effectiveness of the evolving food distribution system. As noted earlier, one aspect of the challenge is the terms of agreement between local farmers and the modern retail outlets, including supermarkets. The most critical contract attributes are related to delivery systems and payment mode. As argued earlier, the most prevalent arrangement is for supermarkets to place orders with farmers through phone calls a few days prior to required delivery instead of requiring them to supply fixed quantities on fixed delivery dates. Consequently, farmers face problems of planning and higher operational costs with these arrangements. In addition to these, they are also paid for quantities that the supermarkets are able to sell, with farmers taking back unsold deliveries. This means nearly all of the supermarket's marketing risk are transferred to the farmers. There is therefore a need to negotiate other risk-sharing modalities that can give local farmers greater degree of reliability in their earnings.

Third, past research has identified weak contract enforcement mechanisms in Africa as a major challenge facing local farmers' value chain integration process (Kuada, 2015). In the absence of credible contracts, the economic literature shows that business partners require minimal levels of trust and confidence that commitments will be honoured, for them to engage in an enduring relationship. However, the general view is that African business culture is trust deficient (Fafchamps, 1996; Kuada, 2020). Policy interventions that can strengthen institutional trust-building mechanisms (e.g. legal systems) as well as other non-personal routines and traditions are needed for purposes of reducing intra-channel transactional costs.

Fourth, the discussions also reveal significant gaps in the contemporary food marketing knowledge in Africa. One gap that requires filling is how gender impacts food purchasing in Africa. Researchers need to remember that African women continue to perform dual roles as caregivers and as workers or producers in their households. As noted earlier, the duality of roles impacts their shopping habits in a manner that has not received adequate attention. We also know very little about what consumers want and why they consider specific food items as important to them as individuals and as households. Finally, past research has suggested that urban and rural consumers in Africa may have differing motivations to buy different food items, but the mechanisms underlying these differences are less known. Addressing all these issues will

© Burleigh Dodds Science Publishing Limited, 2024. All rights reserved.

require the adoption of a transformative approach to food marketing research that gives greater emphasis to the voices of the consumers.

6 Conclusion

The discussions in this chapter suggest that divergent constellations of factors affect African consumers' food consumption and purchase decisions. As such, it is erroneous to assume that there is only one possible or even one preferred road to improving the food retail infrastructure in Africa. We have also noted that supermarkets have not been able to transform the food retail market rapidly and radically and traditional outlets will continue to play a strong role in the food marketing systems. The juxtaposition of modern and traditional outlets has been possible because consumers hold the perception that the two major categories of outlets serve different market segments. Finally, the food insecurity problem in Africa is likely to be further exacerbated in the coming decades, partly due to climate change. All these present enormous challenges to African policymakers.

7 Where to look for further information

- Abbott, J. D. (1987). *Agricultural Marketing Enterprises for the Developing World*. Cambridge: Cambridge University Press.
- Dolan, C. and Humphrey, J. (2004). Changing governance patterns in the trade in fresh vegetables between Africa and the United Kingdom. *Environment and Planning A: Economy and Space* 36(3), 491–509.
- Kabasa, J. D., Kirsten, J. and Minde, I. (2015). Implications of changing agri-food system structure for agricultural education and training in Sub-Saharan Africa. *Journal of Agribusiness in Developing and Emerging Economies* 5(2), 190–199.
- Kuada, J. and Hinson, R. (2014). *Service Marketing in Ghana: A Customer Relationship Management Approach*. London: Adonis & Abbey Publishers Ltd.
- Nandonde, F. A. and Stanton, J. L. (2022). *Supermarket Retailing in Africa*. London: Routledge Contemporary Africa.
- Nyikahadzoi, K. A., Adekunle, A., Fatunbi, O. and Zamasiya, B. (2017). Promoting production and marketing of root crops in Southern Africa in a changing climate using integrated agricultural research for development. *African Journal of Food, Agriculture, Nutrition, and Development: AJFAND* 17(1), 11787.
- Smale, M., Simpungwe, E., Birol, E., Kassie, G. T., de Groote, H. and Mutale, R. (2015). The changing structure of the maize seed industry in Zambia: prospects for orange maize. *Agribusiness (New York, N.Y.)* 31(1), 132–146.

© Burleigh Dodds Science Publishing Limited, 2024. All rights reserved.

8 References

African Development Bank (2014). *The Bank's Human Capital Strategy for Africa (2014–2018)*. OSHD Department, May 2014.

Ajzen, I. and Fishbein, M. (1980). *Understanding Attitudes and Predicting Social Behaviour*, Prentice-Hall, Englewood Cliffs, NJ.

Ajzen, I. (1991). The theory of planned behavior, *Organizational Behavior and Human Decision Processes* 50(2), 179–211.

Amugsi, D. A., Dimbuene, Z. T., Mberu, B., Muthuri, S. and Ezeh, A. C. (2017). Prevalence and time trends in overweight and obesity among urban women: an analysis of demographic and health surveys data from 24 African countries, 1991–2014, *BMJ Open* 7(10), e017344 doi: 10.1136/bmjopen-2017-017344.

Anand, J. (2009). Supermarketization, consumer choices, and the changing food retail market structure: the case of Citlalicalli, Mexico. In: Dr Wood, D. (Ed.). *Economic Development, Integration, and Morality in Asia and the Americas* (Research in Economic Anthropology, Volume 29), Emerald Group Publishing Limited, pp. 63–88.

Ashraf, S., Williams, A. M. and Bray, J. (2022). Female Muslim identity and modest clothing consumption in the UK, *Journal of Islamic Marketing*. doi: 10.1108/JIMA-05-2021-0167.

Bannor, R. K., Amfo, B., Oppong-Kyeremeh, H. and Kyire, S. K. C. (2022). Choice of supermarkets as a marketing outlet for purchasing fresh agricultural products in urban Ghana, *Nankai Business Review International*, 2040. doi: 10.1108/NBRI-08-2021-0059.

Battersby, J. (2014). The need for diverse and responsive food system, *Urban Agriculture Magazine* 27, 12–20.

Battersby, J., Moseley, W. And Peyton, S. (2015). Implications of supermarket expansion on urban food insecurity in Cape Town, South Africa, *African Geographical Review* 34(1), 36–54.

Beckline, M. and Kato, M. S. (2014). Assessing the impact of consumer behaviour on food security in South-West Cameroon, *Journal of Food Security* 2, 87–91.

Bett, H. K., Peters, K. J., Nwankwo, U. M. and Bokelmann, W. (2013). Estimating consumer preferences and willingness to pay for the underutilised indigenous chicken products, *Food Policy* 41, 218–225.

Biénabe, E., Vermeulen, H. and Bramley, C. (2011). The food 'quality turn' in South Africa: an initial exploration of its implications for small-scale farmers' market access, *Agrekon* 50(1), 36–52.

Cleland, J. and Machiyama, K. (2017). The challenges posed by demographic change in sub-Saharan Africa: A concise overview: challenges posed by demographic change in sub-Saharan Africa, *Population and Development Review* 43, 264–286.

Deon, T. (2011). The prevalence of impulsive, compulsive and innovative shopping behaviour in the economic retail hub of South Africa: A marketing segmentation approach, *African Journal of Business Management* 5(14), 5424.

Devi Juwaheer, T., Pudaruth, S. and Monique Emmanuelle Noyaux, M. (2012). Analysing the impact of green marketing strategies on consumer purchasing patterns in Mauritius, *World Journal of Entrepreneurship, Management and Sustainable Development* 8(1), 36–59.

Dubihlela, J. and Ngxukumeshe, T. (2016). Eco-friendly retail product attributes, customer attributes and the repurchase intentions of South African consumers, *The International Business & Economics Research Journal (Online)* 15(4), 163-n/a.

© Burleigh Dodds Science Publishing Limited, 2024. All rights reserved.

Fafchamps, M. (1996). The enforcement of commercial contracts in Ghana, *World Development* 24(3), 427–448.

Groth, H. and May, J. F. (2017) *Africa's Population: In Search of a Demographic Dividend*, Springer International Publishing, Cham, Switzerland.

Hattingh, D., Russo, B., Sun-Basorun, A. and Van Wamelen, A. (2012). *The Rise of the African Consumer*, McKinsey Company, London.

Huddleston, P., Whipple, J., Nye Mattick, R. and Jung Lee, S. (2009). Customer satisfaction in food retailing: comparing specialty and conventional grocery stores, *International Journal of Retail and Distribution Management* 37(1), 63–80.

IFAMA (2017). Bucking the trend of Africa's Food Trade Deficit, Panel Discussion, International Food and Agribusiness Management Association (IFAMA) World Conference, Miami, FL. Available at: http://www.bfap.co.za/documents/ Articles%20and%20Conference%20Papers/IFAMA%20Africa%20Panel%202017%20final.pdf.

IMF regional economic outlook (2022). Available at: https://www.imf.org/en/Publications/REO/SSA/Issues/2022/04/28/regional-economic-outlook-for-sub-saharan-africa-april-2022.

John, A. V. and Brady, M. P. (2011). Consumer ethnocentrism and attitudes toward South African consumables in Mozambique, *African Journal of Economic and Management Studies* 2(1), 72–93.

Klemz, B. R., Boshoff, C. and Mazibuko, N. (2006). Emerging markets in black South African townships: small local independently owned versus large national retailers, *European Journal of Marketing* 40(5/6), 590–610.

Kuada, J. and Buatsi, S. N. (2005). Market orientation and management practices in Ghanaian firms: revisiting the Jarworski and Kohli framework, *Journal of International Marketing* 13(1), 58–88.

Kuada, J. (2015). *Private Enterprise-Led Economic Development in Sub-Saharan Africa: the Human Side of Growth*, Palgrave Macmillan, London.

Kuada, J. (2016). *Marketing Decisions and Strategies: An International Perspective*, Adonis&Abbey Publishers Ltd, London.

Kuada, J. (2020). Culture and economic development in Africa – opportunities and challenges, *African Journal of Religion, Philosophy and Culture* 1(1), 83–99.

Kuada, J. and Bujac, A. I. (2018). Understanding consumer buying behavior in Africa. In: Tesar, G., Anderson, S., Traore, H. and Graff, J. (Eds). *Marketing Management in Africa*, Routledge, New York, pp. 207–223.

Kumar, D. and Kalita, P. (2017). Reducing postharvest losses during storage of grain crops to strengthen food security in developing countries, *Foods* 6(1), 8. doi: 10.3390/foods6010008.

Lee, P. K. C., Cheng, T. C. E., Yeung, A. C. L. and Lai, K. (2011). An empirical study of transformational leadership, team performance and service quality in retail banks, *Omega* 39(6), 690–701. doi: 10.1016/j.omega.2011.02.001.

Letelier, M. F., Flores, F. and Spinosa, C. (2003). Developing productive customers in emerging markets, *California Management Review* 45(4), 77–103.

Lysonski, S. and Durvasula, S. (2013). Nigeria in transition: acculturation to global consumer culture, *Journal of Consumer Marketing* 30(6), 493–508.

Motoki, K. and Sugiura, M. (2018). Disgust, sadness, and appraisal: disgusted consumers dislike food more than sad ones, *Frontiers in Psychology* 9, 76. doi: 10.3389/fpsyg.2018.00076.

© Burleigh Dodds Science Publishing Limited, 2024. All rights reserved.

Nandonde, F. A. and Kuada, J. (2016a). Modern food retailing buying behaviour in Africa: the case of Tanzania, *British Food Journal* 118(5), 1163–1178.

Nandonde, F. A. and Kuada, J. (2016b). International firms in Africa's food retail business-emerging issues and research agenda, *International Journal of Retail and Distribution Management* 44(4), 448–464.

Neven, D., Reardon, T., Chege, J. and Wang, H. (2006). Supermarkets and consumers in Africa: the case of Nairobi, Kenya, *Journal of International Food and Agribusiness Marketing* 18(1–2), 103–123.

Ntloedibe, M. (2018). *The South African Retail Foods Industry* (USDA Foreign Agricultural Services Report). Available at: https://agriexchange.apeda.gov.in › Reports.

OECD/FAO (2020). *OECD-FAO Agricultural Outlook 2020–2029*, Food and Agriculture Organization, Rome/OECD Publishing, Paris.

Ofori, S. A., Cobbina, S. J. and Obiri, S. (2021). Climate change, land, water, and food security: perspectives from sub-Saharan Africa, *Frontiers in Sustainable Food Systems* 5. doi: 10.3389/fsufs.2021.680924.

Onyemah, V. and Akpa, S. O. (2016). Open air markets: uniquenesses about African marketing channels, *International Marketing Review* 33(1), 112–136.

Osakwe, C. N. and Anaza, N. A. (2018). Understanding marketing resources and size in agro-based enterprises, *Marketing Intelligence and Planning* 36(2), 230–244.

Pentz, C., Terblanche, N. S. and Boshoff, C. (2013). Measuring consumer ethnocentrism in a developing context: an assessment of the reliability, validity and dimensionality of the CETSCALE, *Journal of Transnational Management* 18(3), 204–218.

Pradeilles, R., Irache, A., Wanjohi, M. N., Holdsworth, M., Laar, A., Zotor, F., Tandoh, A., Klomegah, S., Graham, F., Muthuri, S. K., Kimani-Murage, E. W., Coleman, N., Green, M. A., Osei-Kwasi, H. A., Bohr, M., Rousham, E. K., Asiki, G., Akparibo, R., Mensah, K., Aryeetey, R., Bricas, N. and Griffiths, P. (2021). Urban physical food environments drive dietary behaviours in Ghana and Kenya: A photovoice study, *Health and Place* 71, 102647. doi: 10.1016/j.healthplace.2021.102647.

Promozione, V. (2013). Overview of the South African retail market. Available at: https://www.businesswire.com/news/home/20211111005817/en/South-Africa-Food-Wholesale-and-Retail-Market-Report-2021.

Rahman, O. and Yu, H. (2019). Key antecedents to the shopping behaviours and preferences of aging consumers. A qualitative study, *Journal of Fashion Marketing and Management* 23(2), 193–208.

Rana, J. and Paul, J. (2017). Consumer behavior and purchase intention for organic food: a review and research agenda, *Journal of Retailing and Consumer Services* 38, 157–165.

Reardon, T., Henson, S. and Berdegue, J. (2007). Proactive fast-tracking diffusion of supermarkets in developing countries: Implications for market institutions and trade, *Journal of Economic Geography* 7(4), 399–431.

Reardon, T., Timmer, C. P., Barrett, C. B. and Berdegue, J. (2003). The rise of supermarkets in Africa, Asia and Latin America, *American Journal of Agricultural Economics* 85(5), 1140–1146.

Saffu, K. and Walker, J. (2006). The country-of-Origin effects and consumer attitudes to "buy local campaign: The Ghanaian case, *Journal of African Business* 7(1–2), 183–199.

Statistica (2020) Available at: https://www.statista.com/statistics/805619/population-growth-in-sub-saharan-africa/.

The Economist (2020). Available at: https://www.economist.com/special-report/2020/03/26/africas-population-will-double-by-2050.

© Burleigh Dodds Science Publishing Limited, 2024. All rights reserved.

van Berkum, S., Achterbosch, T. and Linderhof, V. (2017). Dynamics of food systems in Sub-Saharan Africa. Implications for consumption patterns and farmers' positions in food supply chains. (Wageningen University Research Technical Report). Available at: https://www.researchgate.net/publication/318125647.

Wamalwa, H., Upadhyaya, R., Kamau, P. and McCormick, D. (2019). Strategies of Kenyan firms: A case study of food processing firms in Nairobi, *African Journal of Economic and Management Studies* 10(4), 507–520.

Weatherspoon, D. D. and Reardon, T. (2003). The rise of supermarkets in Africa: implications for agrifood systems and the rural poor, *Development Policy Review* 21(3), 333–355.

World Bank (2022). Population – sub-Saharan Africa. Available at: https://data.worldbank.org.

Yahaya, I., Yamoah, F. A. and Adams, F. (2015). Consumer motivation and willingness to pay for "safer" vegetables in Ghana, *British Food Journal* 117(3), 1043–1065.

© Burleigh Dodds Science Publishing Limited, 2024. All rights reserved.

Chapter 4

Sustainable food consumption attitudes and behavior: generational cohort differences

Irene (Eirini) Kamenidou, International Hellenic University, Greece; and George Menexes and Stergios Gkitsas, Aristotle University of Thessaloniki, Greece

1 Introduction

Consumer food consumption choices clearly influence production in the agricultural sector (Mehrabi et al., 2022; Popp et al., 2010). By influencing food production, these choices also affect wider domains such as the economy, public health, and the environment (FAO, 2018). In particular agriculture and food processing have been seen to have a major negative impact on the environment, linked to the global expansion of more intensive types of agricultural production after the Second World War (Reisch et al., 2013; Bacon and Krpan, 2018).

Problems arising from current agricultural production methods include their contribution to climate change (Chabbi et al., 2017; Kristiansen et al., 2021), water pollution and scarcity (Rosegrant et al., 2009; Xiao et al., 2021; Xu et al., 2019), habitat and biodiversity loss (Dudley and Alexander, 2017; Tscharntke et al., 2021), and soil degradation (Alvarado et al., 2021; Chabbi et al., 2017). In terms of climate change, food production, distribution, and preparation account for about one-third of total anthropogenic greenhouse gas

http://dx.doi.org/10.19103/AS.2023.0129.15
© Burleigh Dodds Science Publishing Limited, 2024. All rights reserved.

(GHG) emissions (Crippa et al., 2021). At the same time, the global population is projected to reach 9.7 billion by 2050 (UN, 2019), which means there will be approximately 2 billion more people to feed. With the pressure to increase agricultural production, there is an urgent need for more sustainable methods of food production, distribution, and consumption.

Consumer demand influences production in the agricultural sector, in terms of both what is produced and the methods of production required to meet consumer expectations of quality, choice, and price (Mehrabi et al., 2022; Popp et al., 2010). As has been noted, consumers' food consumption choices also have a broader impact on the economy, environment, and individual and public health (FAO, 2018). In the latter case, current dietary choices in developed countries (Kowalsky et al., 2022) have been linked to chronic diseases such as type II diabetes, stroke, heart disease, and more than one-third of cancers (Giosuè et al., 2022). In addition, emerging economies are changing their diets to follow the animal-based, high-calorie diets of developed countries (Allievi et al., 2015), with repercussions both for public health and the environment.

Given current trends in food consumption, especially in developing countries, these problems are expected to worsen. Therefore, adopting sustainable food consumption (SFC) is essential to preserve the environment for future generations as well as contribute to other goals such as improved public health (FAO, 2018). The importance of adopting a SFC is underscored by the United Nations 2030 Agenda (United Nations, 2015), in which sustainable consumption and production are directly and indirectly included in several of the 17 Sustainable Development Goals (e.g. SDG 2 'end hunger, achieve food security, and improve nutrition, and promote sustainable agriculture'; SDG 12 'ensure sustainable consumption and production patterns'; SDG 13 'Take urgent action to combat climate change and its impacts').

This chapter explores generational cohorts' attitudes and behavior toward different dimensions of SFC. It also examines differences in attitudes to SFC between generational cohorts. It then develops a typology characterizing participants based on their attitudes toward SFC, based on generational, demographic, and other socioeconomic characteristics, which can be used to better understand consumption behavior.

Greece provides an attractive case study for two reasons. First, Greece has a long tradition of adopting the Mediterranean diet which is associated with both improved health and more sustainable production (Pairotti et al., 2015). This makes it particularly interesting in tracking shifting attitudes to SFC, including shifts away from the Mediterranean diet (da Silva et al., 2009). Second, since Greece has been in economic recession for more than 10 years, it has allowed researchers to explore the way low income in particular affects attitudes to SFC (Held and Haubach, 2017).

© Burleigh Dodds Science Publishing Limited, 2024. All rights reserved.

2 Defining sustainable food consumption

Trudel (2019, p. 1) considers sustainable (consumer) behavior as 'behavior that attempts to satisfy present needs while simultaneously benefiting or limiting environmental impact'. However, there is currently no single fully accepted definition of SFC (Aguirre Sánchez et al., 2021). This means that SFC is defined in many different ways (Hoek et al., 2021). Reisch (2010, p. 1) argues that 'For food consumption to be sustainable it has to be safe and healthy in amount and quality; and it has to be realized through means that are economically, socially, culturally and environmentally sustainable – minimizing waste and pollution and not jeopardizing the needs of others'. Verain et al. (2021, p. 2) define sustainable food as 'food that is produced, processed, packaged, transported and traded with respect for people, animals and the environment, without compromising future generations'.

Sustainable food consumption can be seen as an outcome of sustainable food choices and diets (UK Parliament, 2011). Food and Agriculture Organization (FAO; 2012, p. 7) defines sustainable diets as those with 'low environmental impacts which contribute to food and nutrition security and to healthy life for present and future generations'. According to Willett et al. (2019), healthy diets from sustainable food systems can be achieved not just through changes in agricultural production but also through pervasive changes in dietary habits and other aspects of consumer behavior such as waste reduction.

Sustainable food consumption has been associated with high consumption of plant-based foods (European Public Health Association, 2017; Van Loo et al., 2017) or alternative protein sources such as insect-based foods which are seen as having less environmental impact than conventional agriculture (Dobermann et al., 2017). It has also been linked to low consumption of meat (European Public Health Association, 2017; Green et al., 2018) and highly processed foods (Green et al., 2015). It has also been connected to choosing locally sourced, seasonal products (Garnett, 2006; Kowalsky et al., 2022) and consuming products with a low ecological, carbon and water footprint (characteristics associated, e.g. with organic methods of production) (Galli et al., 2020; Vanham et al., 2018).

High consumption of plant-based foods and low or no meat consumption are becoming increasingly popular as consumer approaches to achieve SFC, especially in Western societies, with movements such as vegetarianism and veganism becoming more mainstream (Niederle and Schubert, 2020; Twine, 2018). Vegetarian and vegan dietary choices emphasize both on improving health and on more sustainable food production methods with lower environmental footprints, particularly in contrast to livestock production which is associated with high GHG emissions contributing to climate change (Aleksandrowicz et al., 2016; Twine, 2018).

© Burleigh Dodds Science Publishing Limited, 2024. All rights reserved.

Other factors which determine consumer concepts of SFC include whether the food is organic (Seconda et al., 2017). There is an ample literature on organic food consumption as a dimension of SFC (e.g. Azzurra et al., 2019; Thøgersen, 2010). Organic agriculture is an environmentally friendly farming system that excludes or largely limits the use of synthetic fertilizers, insecticides, and other chemicals (Nosi et al., 2020). Organic products are perceived both to be safer and healthier and to have been produced in a more environmentally friendly way (Thøgersen, 2010), even if the actual sustainability of organic food production is still a matter of debate (Annunziata et al., 2019).

Another aspect of SFC highlighted by researchers is whether food is locally produced, based on the perception that locally produced food will have a lower environmental footprint than imported food which relies on complex, resource-intensive global supply chains for its distribution (Paloviita, 2010; Pradhan et al., 2020). Bianchi and Mortimer (2015, p. 2284) define local food as 'food produced, retailed and consumed in a specific geographical area'. Local food is also associated with short supply chains (Scalvedi and Saba, 2018). Short supply chains have been described as systems where 'the number of intermediaries is minimized, the ideal being a direct contact between the producer and the consumer' e.g. farmers' markets where farmers sell their produce directly to consumers (Kneafsey et al., 2013, p. 13). However, as with organic production, the reality can be more complex: depending on conditions in the producing country, locally produced food may, e.g. have a larger water footprint than imported crops (Mekonnen and Hoekstra, 2011). These complications in definitions, perception, and the realities of what constitutes SFC create a challenging environment for consumer food choices.

3 Generational cohorts

Demographic and socioeconomic characteristics are known to influence consumption behavior (e.g. Nguyen et al., 2013). Among these characteristics, age seems to play an important role in food consumption choices (Popkin et al., 2005; Nardi et al., 2019), including attitudes to sustainability (Diamantopoulos et al., 2003). The concept of generational cohorts can be used to investigate attitudes to SFC between different age groups since such cohorts are seen to share similar experiences and attitudes (e.g. Kol et al., 2023; Zhang et al., 2022).

A generational cohort is a group of individuals that share similar historical experiences (Strauss and Howe, 1991; Schewe et al., 2000; Wey Smola and Sutton, 2002). Cohorts are seen to be significantly influenced by external events that took place when they 'came of age', generally between the ages of 17 and 23 (Jurkiewicz and Brown, 1998; Meredith and Schewe, 1994). Generational cohorts are distinct from generations (approximately 20–25 years) since they are defined by a particular set of external events (which may occur over differing

© Burleigh Dodds Science Publishing Limited, 2024. All rights reserved.

time spans) and the effects these events have on their subsequent attitudes and behaviors (Hauck and Stanforth, 2007; Rogler, 2002; Wey Smola and Sutton, 2002). On this basis, generational cohorts have been used as a target group for marketing and communication campaigns (Schewe et al., 2000; Schewe and Meredith, 2004).

There is no consensus in the literature on naming and defining generational cohorts (e.g. in terms of dates). The oldest living (after 1900) generational cohort has been defined as the GI Generation (also known as the First World War Generation, Great Generation, Pre-Depression Generation, and Veterans or Federation Generation). This generational cohort is defined as comprising people born from 1901 to 1924 (McCrindle and Wolfinger, 2009). The next generational cohort identified is the Silent Generation (also known as Matures, Veterans, Builders, Traditionalists, or the Swing Generation), comprising people born (depending on different authors) between

- 1920 and 1945 (Jackson et al., 2011);
- 1925 and 1942 (Collins and Tilson, 2001; Jurkiewicz and Brown, 1998; Parry and Urwin, 2011; Strauss and Howe, 1991);
- 1925 and 1945 (McCrindle and Wolfinger, 2009); and
- 1930 and 1945 (Littrell et al., 2005; Williams and Page, 2011).

Members of this generational cohort experienced the consequences of global events such as the Great Depression of the 1930s, the Second World War, and its Cold-War aftermath. These experiences are associated with more conservative social attitudes focused on stability as well as reliance on traditional media such as newspapers and television for information (Littrell et al., 2005; Williams and Page, 2011).

The next generational cohort identified is the Baby Boomer Generation (also known as Boomers, Me Generation, Baboo, Love Generation, Woodstock Generation, and Sandwich Generation), comprising of people born between

- 1943 and 1960 (Collins and Tilson, 2001; Parry and Urwin, 2011; Strauss and Howe, 1991); and
- 1946 and 1964 (Jackson et al., 2011; Littrell et al., 2005; McCrindle and Wolfinger, 2009; Norum, 2003; Schewe et al., 2000; Wey Smola and Sutton, 2002; Williams and Page, 2011).

This cohort was the product of the spike in births in the immediate aftermath of the Second World War. They were influenced by global events such as the post-war economic boom in many developed countries, profound changes in social attitudes (e.g. in attitudes to marriage and divorce), the development of a distinctive youth culture (e.g. with the rise of pop music and festivals such as

© Burleigh Dodds Science Publishing Limited, 2024. All rights reserved.

Woodstock) as well as a greater willingness to question established authority (e.g. the Civil Rights movement and opposition to the Vietnam War in the USA) (Wey Smola and Sutton, 2002). Baby Boomers are identified as generally more affluent, independent, and individualistic, with a strong interest in self-actualization and personal growth, as well as a greater willingness to embrace new technologies (Littrell et al., 2005; Jackson et al., 2011; Pantano et al., 2020). Attitudes to SFC vary between a greater awareness of social issues and a focus on conspicuous consumption and freedom of choice in food purchasing (Severo et al., 2018; Wright et al., 2003).

A third generational cohort is Generation X (also known as the Thirteenth Generation, Baby Busters, or the Lost Generation), consisting of people born between

- 1961 and 1975 (Adams, 2000);
- 1961 and 1981 (Collins and Tilson, 2001; Parry and Urwin, 2011);
- 1965 and 1975 (Littrell et al., 2005);
- 1965 and 1977 (Morton, 2003; Norum, 2003; Wey Smola and Sutton, 2002);
- 1965 and 1980 (Jackson et al., 2011; McCrindle and Wolfinger, 2009); and
- 1961 and 1981 (Strauss and Howe, 1991).

Members of this cohort experienced events such as the global economic recessions of the early 1980s and 1990s, the rise of neo-liberalism and social changes such as increasing rates of divorce, single-parent families, and increasing female participation in the workplace. This led to the image of 'latchkey' children lacking traditional mechanisms of parental support and supervision (Lyons et al., 2007; Schewe et al., 2000; Morton, 2003). Generation X has been characterized as well-educated, technologically and media savvy, but more cynical and skeptical in outlook than Baby Boomers (Jackson et al., 2011; Littrell et al., 2005). In terms of shopping habits, Generation X consumers are seen to balance convenience with a strong awareness of what others think, including environmental issues (Reisenwitz and Iyer, 2009; Severo et al., 2018).

Members of Generation Y (or Millennials born just before the new millennium; Nexters; Echo Boomers; or Generation Me) consist of those born between

- 1977 and 1987 (Norum, 2003);
- 1978 and 1994 (Morton, 2002; Williams and Page, 2011);
- 1980 and 1994 (McCrindle and Wolfinger, 2009);
- 1981 and 1996 (Dimock, 2019; Jackson et al., 2011);
- 1982 and 2000 (Collins and Tilson, 2001); and
- 1982– (Strauss and Howe, 1991).

© Burleigh Dodds Science Publishing Limited, 2024. All rights reserved.

Global events experienced by this group include the financial crash of 2007 and subsequent recession in many developed countries (significantly affecting the economic prospects of many younger adults), the destruction of the Twin Towers in New York and the subsequent 'war on terror' (Strauss and Howe, 2000; Morton, 2002). Generation Y is regarded as the first generation to grow up entirely in the age of the Internet and social media (Norum, 2003). It is perceived to be consumer-oriented and sophisticated when shopping (Jackson et al., 2011; Wolburg and Pokrywczynski, 2001). Like Baby Boomers, Generation Y is socially and environmentally conscious and potentially more sensitive to issues such as multiculturalism (Hewlett et al., 2009). They have been found to be aware of SFC issues (Bollani et al., 2019) and are seen as more likely to purchase green products (Průša and Sadílek, 2019).

The youngest adult generational cohort is Generation Z, which is also known as post-Millennials, the iGeneration, or Homeland Generation (Dimock, 2019), consisting of individuals variously defined as born

- between 1995 and 2009 (Strauss and Howe, 2020; McCrindle and Wolfinger, 2009);
- after 1995 (Dimock, 2019; Wood, 2013 and Adecco, 2015 in Iorgulescu, 2016); and
- after 2000 (e.g. Bennett et al., 2012).

With the launch of the iPhone in 2007, Generation Z was able to connect more or less continuously to the Internet through mobile devices, WiFi, and high-bandwidth cellular services (Dimock, 2019). This means Generation Z is fully immersed in social media services such as Facebook and Twitter (Singh and Dangmei, 2016). Combined with continuing economic insecurity in many countries, Generation Z has been characterized as acutely image conscious as well as more anxious and pessimistic than their parents (Bennett et al., 2012). As a generation growing up in the face of climate change, they tend to be more aware of SFC issues although, as digitally sophisticated consumers, they also value issues such as convenience (Bogueva et al., 2022; Kymäläinen et al., 2021).

4 Research on sustainable food consumption attitudes and behaviors among generational cohorts

There is a wide range of literature dealing with issues relating to SFC which has been reviewed by several authors (e.g. Aguirre Sánchez et al., 2021; Lorenz and Langen, 2018; Molin et al., 2021; Rödiger and Hamm, 2015). There have been a number of studies, e.g. dealing with changing preferences for plant-based (e.g. Knaapila et al., 2022) and insect-based foods (Grasso et al., 2019),

© Burleigh Dodds Science Publishing Limited, 2024. All rights reserved.

reducing meat consumption (Szczebyło et al., 2022), and choosing organic and/or seasonal products (Amvrosiou et al., 2017; Bonaccio et al., 2016). Other studies have explored the range of motivations and barriers to making more sustainable food choices (Smith and Paladino, 2010; Wang and Scrimgeour, 2021). Researchers have also sought to create profiles and/or develop typologies of sustainable food consumers (e.g. Baudry et al., 2016; Perea, 2014).

Some studies have examined the influence of age on SFC (e.g. Eldesouky et al., 2020; Matacena et al., 2021; Molinario et al., 2020; Niva et al., 2014; Thøgersen, 2017), though not in the form of generational cohorts. However, there have been a number of multigenerational and intergenerational cohort studies in areas such as behavior in the workplace (e.g. Stevanin et al., 2020), travel and tourism preferences (e.g. McKercher et al., 2020), and education (e.g. Ostermann et al., 2019). There are also studies on generational cohort differences in purchasing non-food products (e.g. Ivanova et al., 2019; Calvo-Porral and Pesqueira-Sanchez, 2019).

Some generational cohort studies have also reviewed differences between cohorts in broader food purchase patterns across the food and beverage sector (e.g. Arenas-Gaitán et al., 2022; Balenović et al., 2021). There are fewer studies that focus on generational cohorts and SFC (e.g. Arenas-Gaitán et al., 2022; Dabija et al., 2020; Vilceanu et al., 2019). Most of these studies focus on a single cohort, with a particular emphasis on Generation Y/Millennials (Kaczorowska et al., 2017; Bollani et al., 2019; Knaapila et al., 2022; Muposhi and Dhurup, 2016; Ntanos et al., 2014; Pomarici and Vecchio, 2014; Szczebyło et al., 2022). Generation Z has also attracted increasing attention as a cohort (e.g. Bogueva and Marinova, 2020; Julianty et al., 2021; Kymäläinen et al., 2021; Su et al., 2019).

A number of generational cohort studies have compared differences between cohorts in attitudes to SFC. Bogueva et al. (2022), e.g. compared preferences within Generation Y and Generation Z consumers for alternative plant-based protein foods. A few studies have examined cohort or multigenerational differences (≥2 generational cohorts) in attitudes to SFC (e.g. Di Vita et al., 2021a; Kamenidou et al., 2021; Kita et al., 2021; Kol et al., 2023; Matacena et al., 2021). Consumer segmentation based on issues of SFC has also been widely studied (e.g. Gazdecki et al., 2021; Sarti et al., 2018; Verain et al., 2015), though few studies focus on generational cohorts.

Studies developing consumer typologies based on issues of SFC are relatively rare (e.g. Balderjahn et al., 2018; Funk et al., 2021), with one study focusing on generational cohorts (Su et al., 2019). Table 1 presents the studies that focus on generational cohorts and SFC that are relevant to this chapter (generational cohort differences, segmentation, and typology based on SFC). As Table 1 suggests, SFC related to generational cohort analysis is relatively understudied.

© Burleigh Dodds Science Publishing Limited, 2024. All rights reserved.

Table 1 Sustainable food consumption (SFC) cohort differences and cohort segmentation

Authors	Cohorts studied	SFC issue studied	Country -data	Findings
SFC and cohort differences				
Satinover Nichols and Wehr Holt (2023)	BB, Gen X, Y, and Z	Attitudes toward SFC and sustainability	USA (n = 1250)	Older cohorts (BB) attitudes of SFC and sustainability are less favorable as compared to the younger ones (Gen Z and Millennials).
Kol et al. (2023)	BB, Gen X and Y	Attitudes to local food purchases	Israel n = 672	Gen Y has a higher relationship between pro-environmental behavior and intention to buy local food products. This did not translate into actual purchasing behavior.
Kamenidou et al. (2021)	SG, BB, Gen X, Y, and Z	Adopted or willing to adopt SFC behavior	Greece n = 1561	Older cohorts are adopting SFC behavior compared to the younger ones.
Matacena et al. (2021)	SG, BB, Gen X, Y, and Z	Impact of the COVID-imposed lockdown on health, social, and environmental sustainability of diets	Italy n= 2288	BB and Gen Z improved dietary health and sustainability due to lockdown.
Kamenidou et al. (2020a)	SG, BB, Gen X, Gen Y, and Gen Z	Organic food purchasing behavior	Greece n = 1562	Attitudes were positive for all cohorts. Gen X had the most positive, and Gen Z had the least positive attitude. Older generations were most likely to purchase organic foods.
SFC and cohort segmentation (based on:)				
Mamalis et al. (2023)	Gen Z	Organic food decision-making style (DMS) in the context of SFC	Greece (n = 1040)	Three groups: 'high-quality consumers', 'ecological and health-conscious organic food consumers', and 'conventional food consumers'.
Kamenidou et al. (2022)	Gen Z, Gen Y	Environmental benefits from specific consumption patterns	Greece (n = 513)	Three groups: 'neutrals', the 'mindful and relatively optimistic', and the 'pessimists'.

(*Continued*)

© Burleigh Dodds Science Publishing Limited, 2024. All rights reserved.

Table 1 (*Continued*)

Authors	Cohorts studied	SFC issue studied	Country -data	Findings
Di Vita et al. (2021a)	Gen X and Gen Y (millennials) 58+	If clementines from integrated farming are perceived as desirable alternatives to organic and conventional fruits	Italy $n = 345$	4 groups: 'price sensitive', 'geographical origin and mild sustainability concerned', 'pro-organic', and 'skeptics'.
Di Vita et al. (2021b)	Gen X and Gen Y/ Millenials, BB, SG	The role that local, organic, and traditional attributes have on consumers' preferences	Italy $n = 709$	4 groups: 'low-cost and undifferentiated olive oil lovers'. 'organic olive oil lovers'. 'high-end metropolitan consumers'. 'local olive oil and sustainability-oriented consumers'.
Kita et al. (2021)	SG, BB, Gen X, Gen Y, and Gen Z	Organic food consumption	Slovakia $n = 1373$	5 clusters (not named). Gen Z and Gen Y consume more organic food than older individuals.
Kamenidou et al. (2019)	Gen Z	SFC behavior	Greece $n =$ 252 university students	Two groups: 'the under-consideration students' and 'the negatively positioned students'.
Su et al. (2019)	Gen Z	Environmental consciousness	USA $n = 812$	3 groups: 'sustainable moderates', 'sustainable activists', and, 'sustainable believers'.
Kumar & Smith (2018)	BB, Gen X, and Gen Y	Attitudes toward local food and segmentation based on their food-related lifestyle	Online panel $n = 502$	4 groups: 'impromptu novelty', 'explorer', 'uninvolved connoisseur', 'involved information seeker', and 'apathetic local food consumer'.
Ntanos et al. (2014)	Y/Millennials	Perceptions on the organic products	Greece $n = 163$	Two groups: Not named.

© Burleigh Dodds Science Publishing Limited, 2024. All rights reserved.

SFC and cohort typology (based on:)				
Su et al. (2019)	Gen Z	16 environmental consciousness items (of which perception of sustainable food attributes, by measuring environmental protection and product characteristics). (items/variables not associated with the ones in the study)	USA $n = 812$	Group 1: 'sustainable moderates (environmentally practical: moderate; environmentally supportive: moderate; environmentally social: low)'. Group 2: 'sustainable activists (environmentally practical: high; environmentally supportive: high; environmentally social: moderate)'. Group 3: 'sustainable believers (environmentally practical: high; environmentally supportive: high; environmentally social: low)'.

BB = Baby Boomers; Gen = generation; SG = Silent Generation.

© Burleigh Dodds Science Publishing Limited, 2024. All rights reserved.

Some general conclusions can be drawn from the earlier analysis. Interest in SFC and generational cohort research is growing but is still in its infancy. Studies that address SFC and generational cohorts focus primarily on one cohort, and only a handful of studies address cohort differences and segmentation, while one study has addressed consumer cohort typologies and SFC.

It is important to note that prior studies reach to contradictory conclusions. Studies claim that younger cohorts (Generation Y and Generation Z) are environmentally friendly and have positive attitudes toward sustainable consumption. However, when the SFC settings of these cohorts were examined, conflicting results were found. For example, Julianty et al. (2021) found that Indonesian Generation Z have positive attitudes toward organic food consumption. Kamenidou et al. (2019) found that this cohort (as well as Generation Y) limits their actual behavior to eating seasonal fruits and vegetables and buying regional products. On the other hand, Pólya and Máté (2021) found that environmental awareness is important to Generation Z, but they tend to be more price-conscious in terms of actual behavior. These conflicting results and the limited overall number of studies suggest the need for further research regarding SFC and cohort differences.

5 Case study: sustainable food consumption and generational cohorts in Greece

The generational cohort case study conducted by the authors is focused on the five main generational cohorts in Greece. As calculated by the Greek Statistical Authority (GSA, 2020), the population of Greece on 1 January 2020 was 10 718 565. The GSA divides the population into 5-year groups, from 0 to 5 years of age up to 100+. Based on these categories, the population of the 5 cohorts examined in 2020 (year of the research) was 8 835 055 (adding all categories from 20 to 99 years old and including those born in 2002 and 2001, reflecting the 18- and 19-year-old, i.e. adults of the Generation Z cohort).

5.1 Sample definition

The sampling unit was any adult member of the five main generational cohorts in Greece. Generational cohort theory and applications are in their infancy in Greece as compared to other countries (e.g. USA, Australia, New Zealand), and there is no research that focuses on defining each cohort in the Greek context. Therefore, the cohort year range used in this study is based on previous literature on generational cohorts (e.g. Collins and Tilson, 2001; Dimock, 2019; Kamenidou et al., 2020a; McCrindle and Wolfinger, 2009; Morton, 2003; Norum, 2003; Strauss and Howe, 2020; Wey Smola and Sutton, 2002; Williams

© Burleigh Dodds Science Publishing Limited, 2024. All rights reserved.

and Page, 2011), taking into account events that influenced cohorts in their formative years (17–23 years old).

The five Greek generational cohorts studied are as follows:

1 Silent Generation, born between 1925 and 1945 (McCrindle and Wolfinger, 2009);
2 Baby Boomers, born between 1946 and 1964 (Littrell et al., 2005; McCrindle and Wolfinger, 2009; Norum, 2003; Schewe et al., 2000; Wey Smola and Sutton, 2002; Williams and Page, 2011);
3 Generation X, born between 1965 and 1977 (Morton, 2003; Norum, 2003; Wey Smola and Sutton, 2002);
4 Generation Y (or Millennials), born between 1978 and 1994 (Norum, 2003; Morton, 2002; Williams and Page, 2011); and
5 Generation Z, born 1995 and 2009 (Strauss and Howe, 2020; McCrindle and Wolfinger, 2009).

In the case of Generation Z, only the adult members were approached, i.e. those born during 1995–2002 (being 18 years old in 2020 when the study was undertaken). The GI Generation born before 1925 was not included given the very small number of those still living and difficulty tracing them. The generational cohort periods were selected based on key events in Greek history as well as the broader global events seen as affecting these cohorts.

Some of the key events that the Greek Silent Generation witnessed were the occupation of Greece during the Second World War and the subsequent civil war between communists and nationalists which lasted until 1949, leaving a legacy of bitter political division. They also witnessed Greece and Turkey becoming members of NATO (1952) and the assassination of Nikos Belogiannis (a leading member of the Greek Communist Party) the same year. Members of this cohort were also affected by a struggling economy during the 1950s, including the devaluation of the drachma (1954). They witnessed the establishment of the Center Union Party (1961) and the student movement in favor of a more democratic government (1962).

Some of the major events that shaped the Baby Boomers were the assassination of the socialist politician Grigoris Lambrakis (1963), the April 21 coup establishing a military junta in Greece (1967–1974), and the crushing of the insurrection against the junta carried out by students from the Polytechnic University of Athens (1973), which contributed to the collapse of the junta in 1974. They also experienced the invasion of Cyprus by Turkey (1974), witnessed the subsequent referendum to establish a new republican constitution in 1975, and the election of Greece's first socialist Prime Minister (1981). This generation also witnessed the country becoming a member of the European Community (subsequently the European Union) in 1981.

© Burleigh Dodds Science Publishing Limited, 2024. All rights reserved.

Significant events witnessed by the Generation X include the Sismik Turkey-Greece incident, bringing the two countries to the brink of war (1987), the austerity program (1985–1989) which resulted in significant social unrest, and the assassination of the Greek liberal politician Pavlos Bakoyannis by the terrorist group N17 (1989). Additionally, they witnessed the European Union recognizing the Former Yugoslav Republic of Macedonia (1992) and the death of the world-famous Greek actress, singer, political activist, and Minister of Culture: Melina Mercouri (1994).

Events significant experienced by Generation Y included the Imia crisis with Turkey (1996), the death of the first socialist Prime Minister of Greece (1996) and the 2004 Summer Olympics in Athens. Generation Y cohort was particularly affected by the global financial crisis of 2007 which crippled the Greek economy. This forced the government to negotiate a loan from the International Monetary Fund, requiring a program of austerity measures which triggered widespread unrest and an exodus of younger people looking for better opportunities (Kamenidou et al., 2017; Aravossitas and Sugiman, 2019).

As to the youngest adult cohort, the years shaping the characteristics and values of the Generation Z members are still ongoing. They have perhaps been most affected by the ongoing economic crisis in Greece, which led to significant political instability, and the entry of the extreme right party (Chrisi Augi-Golden Dawn) in 2012 in the Greek parliament (2012–2019). They also experienced the election of the left-wing anti-austerity SYRIZA party in 2015 and a series of renegotiations by successive governments with the country's creditors (2012, 2015, and 2017). Most recently, this generation has experienced the onset of the coronavirus disease 2019 (COVID-19) pandemic with its resulting deaths and illness as well as a national lockdown in 2020 and 2021.

5.2 Methodology

In the first stage of the study, an extensive literature review provided insights into the key dimensions of SFC and dietary patterns and provided the basis for the survey. In the second phase, qualitative research was conducted with 15 participants to confirm the questions used. This involved a non-probability sampling method (Berdie et al., 1986); 1 focus group with 5 participants (1 per cohort) and 10 in-depth interviews (2 participants per cohort).

Interviews focused on their attitudes toward different aspects of SFC. The conversations were recorded (with the respondent's consent), transcribed, and cross-referenced with questions in the survey. This resulted in adding new questions. Participants were sent a copy of their responses and the way researchers interpreted these responses within the survey design so that they could validate their views had been correctly understood (Creswell and Miller, 2000).

© Burleigh Dodds Science Publishing Limited, 2024. All rights reserved.

A third phase involved pilot testing the questionnaire. The pilot test was conducted in two ways: electronically via the Internet (85 participants) and aided self-administered or face-to-face interviews for older people or people who did not have access to the Internet (33 participants). This resulted in minor changes to make the questionnaire more comprehensible to all participants (grammar and syntax). The questionnaire included issues relating to consent and confidentiality (Priporas et al., 2015).

In the fourth stage, the questionnaire was finalized and sent out: data collection was done (mainly) online and in person (Lee et al., 2017; Perry et al., 2014). The research took place from the beginning of March 2020 to December 2020 and resulted in a valid sample of $n = 2853$ participants (including a period of complete lockdown in response to the COVID-19 pandemic). It is pointed out that from 13 March the country had a total lockdown (Kamenidou et al., 2020b). On 4 May, some restrictions were altered, while new restrictions were imposed again on 7 November, which were removed in 2021.

The validity and the reliability of the two scales used (SFC and actual behavior/intension to comply) were examined using item analysis and factor analysis based on Principal Components Analysis (PCA). Groups of participants were compared either with the analysis of variance (ANOVA) method, followed by Tukey's post hoc multiple comparisons of means (in case of quantitative dependent variables) or with the chi-squared test (in case of qualitative dependent variables). In order to develop a typology of the participants relative to their attitudes toward SFC, Hierarchical Cluster Analysis (HCA) was utilized on the *z*-scores of the participants on the dimensions derived from PCA concerning the SFC scale. The squared Euclidian distance was used as a measure of dissimilarity among participants. The clusters' joining method was based on the Ward's criterion (Hair et al., 2018). The visual inspection of the cluster's dendrogram (not shown) revealed a number of major groupings (types) of participants.

Table 2 shows the key elements covered in the SFC questionnaire, the studies on which these elements were based, and the scale used.

The 7-point Likert scale had the following options '1 = Completely disagree, up to, 7 = Completely agree, with 4 = the neutral point, i.e. neither disagree nor agree'. The Likert-type scale had the following options:

- 1 = I am not doing this, and I am not willing to do it ever;
- 2 = I am not doing this, and I am not willing to do so in the far future (long-run);
- 3 = I am not doing this, and I am not willing to do this in the near future;
- 4 = I am not doing this and will think about it in the future if I will do it (neutral);
- 5 = I am doing this occasionally (whenever I can);

© Burleigh Dodds Science Publishing Limited, 2024. All rights reserved.

Table 2 Key elements of the questionnaire of the study

Key elements of the study	Adopted from	Scale used
Low-processed products/natural produce		
Agricultural products and foods that are produced or prepared in a natural way taste better than other foods	Questions developed based on qualitative research and the papers of: Cousin and Siegrist in Tobler et al., 2011 Román et al., 2017	7-point Likert scale
Agricultural products and foods that are produced or prepared in a natural way are better for my health		
Agricultural products and foods that are produced or prepared in a natural way do not contain chemicals or other residues that are harmful to the environment		
Plant-based food		
It is good for the environment to eat more plant-based foods	Reipurth et al., 2019	7-point Likert scale
It is healthy to eat a plant-based diet		
In-season fruits and vegetables		
In-season fruits and vegetables are best for the environment	Siegrist et al., 2015	7-point Likert scale
Seasonal fruits and vegetables taste better		
By buying seasonal fruits and vegetables, I can save money		
Low meat consumption		
Reducing meat consumption is better for health	Siegrist et al., 2015	7-point Likert scale
Reducing meat consumption is better for the environment		
Organic food consumption		
Organic food is healthier than conventionally grown food because it has no pesticide residues	Lea and Worsley, 2005	7-point Likert scale
Organic foods are better for the environment than conventionally grown foods		
Local products		
It is important for me to support local farmers when buying agricultural products and food	Tanner and Wölfing Kast, 2003	7-point Likert scale
It is good to support domestic agriculture by buying local - domestic products		
Consumers must show solidarity with the farmers of their place		

(*Continued*)

© Burleigh Dodds Science Publishing Limited, 2024. All rights reserved.

Table 2 (*Continued*)

Key elements of the study	Adopted from	Scale used
Short supply chain		
It is better to buy directly from producers than from popular stores or supermarkets	Qualitative research	7-point Likert scale
Purchasing food products directly from producers reduces the environmental impact because it reduces the number of intermediaries and means of transport involved		
Actual SFC behavior and intention to comply		
Purchase agricultural products and food produced in my area where I live or in my country	Qualitative research	7-point Likert-type scale
Eat only seasonal fruits and vegetables	Kamenidou et al., 2019	
Consumed organic food		
Eat less meat, maximum once or twice per week		
Avoid consumption of any imported agricultural products and foodstuffs		
Avoid buying or consuming ready-made prepacked foods		
Purchase fruit and vegetables in bulk form		
Consume sustainably farmed fish		
Eat meat types with lower environmental impact		
Consume organic meat		
Eat plant-based meat substitutes		

- 6 = I am doing this already very often; and
- 7 = I am already doing this, and I consider that I will continue doing it forever'. The scale was considered as a summated scale in order for reliability analysis to be executed (Chon, 1990).

Questionnaire content validity was ensured by referring to previous research and using pilot testing (Kent, 1993). Convergent and discriminant validity were determined using exploratory factor analysis based on PCA with Varimax Rotation (retaining items with loading on factor >0.55) provided with seven dimensions of the SFC attitude scale:

- Local products;
- In-season fruits and vegetables;

© Burleigh Dodds Science Publishing Limited, 2024. All rights reserved.

- Low-processed products/natural produce;
- Reduced meat consumption;
- Short supply chain;
- Organic food consumption (OFC); and
- Plant-based diet (PBD).

In the factor analysis the Kaiser–Meyer–Olkin measure of sampling adequacy (KMO) and the Bartlett's test of sphericity (BTS) indicators are presented, as well as significance (*P*) (KMO = 0.912; BTS = 35194.964; df = 153; $P < 0.001$; 82.78% of the total variance; Cronbach's α of each dimension ranging from 0.618 to 0.923).

The same procedure delivered two dimensions of the SFC behavior/intention to comply scale:

- Adoption of a sustainable purchasing and eating behavior; and
- Adoption of an alternative meat diet.

(KMO = 0.900; BTS = 14858.885; df = 55; $P < 0.001$; 59.17% of the total variance; Cronbach's α of each dimension ranging from 0.767 to 0.879). These dimensions were consistent with those established by the authors of the adopted scales and were therefore named accordingly. For all these dimensions, convergent validity was ensured based on the equation of Fornell and Larcker (1981). Discriminant validity was also calculated (Fornell and Larcker, 1981) and established for both questions and all nine dimensions. The overall reliability of each scale yielded a satisfactory result (Spector, 1992), with SFC attitudes overall α = 0.895 and SFC behavior/intention to comply scale α = 0.902.

This study was approved by the Coordinating Committee of the Dept. of Management Science and Technology, School of Business and Economics, International Hellenic University, under reference number 3/13-2-2020 and conducted with the direct consent of each respondent.

5.2 Results

The mean age of the participants was 43.34 years (SD = 19.62) (minimum 18–93 maximum) and was distributed among five cohorts. Generational cohorts were spread as follows:

- Generation Z: 27.3% (n = 779);
- Generation Y: 26.1% (n = 744);
- Generation X: 17.4% (n = 497);

© Burleigh Dodds Science Publishing Limited, 2024. All rights reserved.

- Baby Boomers: 21.5% (n = 612); and
- Silent Generation: 7.7% (n = 221).

Females (53.3%) were overrepresented compared to male subjects (46.7%), with about one-half of the sample married (49.8%), and only 28.4% with children (<18 years old) in their household. Moreover, the majority have secondary education (44.0%), are employees (29.7%), reside in the city (65.6%), and have a net family monthly income ranging from €1000.01 to €2000.00.

Table 3 presents the central tendency (mean and median values) of the attitude scores of each generational cohort as well as for the total sample toward the different dimensions of SFC examined in the questionnaire. In columns 2–6, the numbers represent the mean value (MV) per cohort of responses on the Likert scale, while the seventh and eighth column presents the MV of the total sample and the median of the total sample.

The three statements with the highest mean score across the total sample (thus expressing the most positive attitude) are as follows:

- 'Seasonal fruits and vegetables taste better' (MV = 5.62);
- 'Agricultural products and foods produced or prepared in a natural way are better for my health' (MV = 5.57); and
- 'Seasonal fruits and vegetables are best for the environment' (MV = 5.42).

In contrast, the three statements with the most negative attitudes are the ones referring to PBDs:

- 'It is healthy to eat a plant-based diet' (MV =3.25);
- 'It is good for the environment to eat more plant-based food' (MV = 3.76); and
- 'Organic food is healthier than conventionally grown food because it does not contain pesticide residues' (MV=4.00)

In general, regarding participants' answers on the 7-point Likert scale, 8 out of 17 statements have a median of 6.00 which is the highest observed, with no median of 7.00, implying that participants generally have a positive attitude on only half of the statements.

Table 4 shows participants' compliance in adopting a SFC behavior, per cohort as well as for the total sample. Eleven statements were presented on a 7-point Likert-type scale. Table 4 reveals that the statement with the most complying behavior is eating seasonal fruits and vegetables, with participants' behavior being 'I am doing this already very often' (median = 6.00). Six statements that refer to SFC behavior but are not associated with

© Burleigh Dodds Science Publishing Limited, 2024. All rights reserved.

Table 3 Participants attitudes (mean values) toward SFC dimensions

	Gen Z	Gen Y	Gen X	BB	SG	MV TS	Med. TS
Local products							
Consumers must show solidarity with the farmers of their place	4.84	4.92	5.48	5.30	5.52	5.29	6.00
It is good to support domestic agriculture by buying local - domestic products	5.01	5.14	5.64	5.42	5.39	5.27	6.00
It is important for me to support local farmers when buying agricultural products and food	5.12	5.15	5.61	5.35	5.44	5.12	5.00
In-season fruits and vegetables							
Seasonal fruits and vegetables taste better	5.15	5.31	5.74	5.61	5.52	5.62	6.00
Seasonal fruits and vegetables are best for the environment	5.37	5.53	5.95	5.75	5.67	5.42	6.00
By buying seasonal fruits and vegetables. I can save money	4.69	5.09	5.65	5.46	5.42	5.18	6.00
Low process - natural							
Agricultural products and foods that are produced or prepared in a natural way taste better than other foods	4.85	5.31	5.62	5.41	5.55	5.28	6.00
Agricultural products and foods that are produced or prepared in a natural way are better for my health	5.39	5.55	5.86	5.56	5.63	5.57	6.00
Agricultural products and foods that are produced or prepared in a natural way do not contain chemicals or other residues that are harmful to the environment	4.87	5.02	5.53	5.36	5.53	5.18	6.00
Reduced meat consumption							
Reducing meat consumption is better for health	4.36	4.69	5.06	5.08	5.07	4.77	5.00

(Continued)

© Burleigh Dodds Science Publishing Limited, 2024. All rights reserved.

Table 3 (*Continued*)

	Gen Z	Gen Y	Gen X	BB	SG	MV TS	Med. TS
Reducing meat consumption is better for the environment	4.32	4.57	4.93	5.06	4.89	4.69	5.00
Short supply chain							
It is better to buy directly from producers than from popular stores or supermarkets	4.60	4.64	5.09	4.96	4.96	4.80	5.00
Purchasing food products directly from producers reduces the environmental impact, because it reduces the number of intermediaries and means of transport involved	4.68	4.78	5.39	5.23	4.94	4.97	5.00
Organic food consumption							
Organic food is healthier than conventionally grown food because it has no pesticide residues	4.06	3.75	4.05	4.11	4.20	4.00	4.00
Organic foods are better for the environment than conventionally grown foods	4.34	4.44	4.59	4.75	4.69	4.53	4.00
Plant-based diet							
It is good for the environment to eat more plant-based foods	3.76	3.78	3.55	3.74	4.19	3.76	4.00
It is healthy to eat a plant-based diet	3.22	3.38	3.14	3.10	3.60	3.25	3.00

BB = Baby Boomers; Gen X = Generation X; Gen Y = Generation Y; Gen Z = Generation Z; Med. TS = median value of the total sample; MV = mean value; MV TS = mean value of the total sample; SFC = sustainable food consumption; SG = Silent Generation. Source: The authors.

meat substitutes or sustainable meat have a median of 5.00, meaning that they occasionally adhere to the statements' behaviors, 'I do this occasionally (whenever I can)'. Lastly, four statements related to sustainable meat and meat substitutes for protein intake have a median of 4.00, i.e. 'I am not doing this and will think about it in the future if I will do it'.

Assessing generational cohort differences in SFC attitude and behavior was based on the seven dimensions relating to SFC attitudes and the two dimensions relating to SFC behavior. To do this, a series of one-way ANOVA were conducted, in which generational cohorts constitute the independent variable and each dimension of SFC attitudes and actual behavior constitute the dependent variables (based on the mean factor scores of dimensions

© Burleigh Dodds Science Publishing Limited, 2024. All rights reserved.

Table 4 Actual behavior and intention to comply toward a SFC diet (mean values)

Statements I…….	Gen Z	Gen Y	Gen X	BB	SG	MV TS	Med. TS
Adoption of a sustainable purchasing and eating behavior							
Purchase agricultural products and food produced in my area where I live or in my country	4.82	5.00	5.33	5.36	5.21	5.10	5.00
Eat only seasonal fruits and vegetables	5.19	5.16	5.47	5.43	5.33	5.29	6.00
Consume organic food	4.26	4.32	4.79	4.71	5.03	4.52	5.00
Eat less meat, maximum once or twice per week	4.13	4.37	4.54	4.70	4.86	4.44	5.00
Avoid consumption of any imported agricultural products and foodstuffs	4.02	4.35	4.57	4.57	4.84	4.38	5.00
Avoid buying or consuming ready-made prepacked foods	4.25	4.28	4.59	4.77	5.08	4.49	5.00
Purchase fruit and vegetables in bulk form	4.75	4.70	5.08	5.05	5.26	4.90	5.00
Adoption of an alternative meat diet							
Consume sustainably farmed fish	3.97	4.25	4.45	4.56	4.49	4.29	4.00
Eat meat types with lower environmental impact	4.00	4.18	4.55	4.52	4.58	4.30	4.00
Consume organic meat	3.94	3.89	4.51	4.47	4.43	4.18	4.00
Eat plant-based meat substitutes	3.45	3.63	3.78	3.48	3.37	3.55	4.00

BB = Baby Boomers; Gen X = Generation X; Gen Y = Generation Y; Gen Z = Generation Z; Med. TS = median value of the total sample; MV TS = mean value of the total sample; SFC = sustainable food consumption; SG = Silent Generation. Source: The authors.

derived from factor analysis). The results show that for all dimensions referring to attitudes and actual behavior, there are statistically significant differences ($P < 0.001$ in all cases). Therefore, the null hypothesis is rejected in all cases and differences in attitudes and behavior exist between at least two cohorts (Table 5).

To further investigate where generational differences originate, multiple comparisons of means employing the post hoc Tukey's comparison test were used (Table 5). The first column displays the nine dimensions that were tested for cohort differences. Each cohort column shows the MV per dimension. Within each row different letter(s) next to the MV identifies statistically significant differences, starting with 'a' for the highest MV.

© Burleigh Dodds Science Publishing Limited, 2024. All rights reserved.

Table 5 Multiple comparisons of means with Tukey's test

Dimensions measured	Generation Z	Generation Y	Generation X	Baby Boomers	Silent Generation	ANOVA *P*-value
SFC attitudes						
Local products	4.99b	5.07b	5.58a	5.36a	5.45a	<0.001
In-season fruits and vegetables	5.07d	5.31c	5.78a	5.61ab	5.53bc	<0.001
Low-processed products/natural produce	5.04c	5.29b	5.67a	5.45ab	5.57a	<0.001
Reduced meat consumption	4.34c	4.63b	4.99a	5.07a	4.98a	<0.001
Short supply chain	4.64c	4.71c	5.24a	5.09ab	4.95b	<0.001
Organic food consumption	4.20b	4.10b	4.32ab	4.43a	4.45a	<0.001
Plant-based diet	3.49b	3.58b	3.34b	3.42b	3.89a	<0.001
SFC behavior/intention to comply						
Adoption of a sustainable purchasing and eating behavior	4.49b	4.60b	4.91a	4.94a	5.09a	<0.001
Adoption of an alternative meat diet	3.84b	3.99b	4.32a	4.26a	4.22a	<0.001

In the same row, mean values followed by different letter or letters are statistically significantly different, at significance level *a* = 0.05, according to the results of the Tukey's test. Source: The authors.

© Burleigh Dodds Science Publishing Limited, 2024. All rights reserved.

Table 6 Clusters' centroids (mean values) and comparison relative to the seven dimensions of SFC

Dimensions of SFC	Cluster 1 n = 1224 (42.9%)	Cluster 2 n = 1167 (40.9%)	Cluster 3 n = 462 (16.2%)	Effect size (partial eta squared)	ANOVA P-value
Local products	6.22	5.00	3.17	0.529	<0.001
In-season fruits and vegetables	6.51	5.15	3.13	0.626	<0.001
Low-processed products/ natural produce	6.47	5.09	3.01	0.599	<0.001
Reduced meat consumption	5.82	4.27	3.02	0.469	<0.001
Short supply chain	5.81	4.64	3.05	0.420	<0.001
Organic food consumption	4.67	4.17	3.42	0.099	<0.001
Plant-based diet	3.05	4.20	2.96	0.128	<0.001

Table 7 Clusters' profiles (mean values) and comparison relative to the two dimensions of actual behavior/intention to comply

Dimensions of actual behavior/intention to comply	Cluster 1 n = 1224 (42.9%)	Cluster 2 n = 1167 (40.9%)	Cluster 3 n = 462 (16.2%)	Effect size (partial eta squared)	ANOVA P-value
Adoption of a sustainable purchasing and eating behavior	5.51	4.55	3.15	0.450	<0.001
Adoption of an alternative meat diet	4.62	3.97	2.94	0.233	<0.001

From Table 5, it is evident that, overall, the three older cohorts have a more positive attitude and behavior for all dimensions of SFC examined, while the two younger cohorts have the most negative attitudes and behavior.

The objective of the study was to produce a consumer typology based on SFC attitudes and behavior. The basic idea behind the typology developed by means of HCA was to reveal latent structures among participants relative to their attributes toward SFC without making any *a priori* assumption concerning the mechanism (generational cohorts and other demographic and socioeconomic characteristics) by which data were generated. In addition, this typology could offer important insights in order to propose specific and focused policy implications for targeting each consumer typo to maintain or produce positive behavioral change. The HCA revealed three major groupings (types) of the participants. Based on the data presented in Tables 6–8, the profiles of the three clusters are drawn.

© Burleigh Dodds Science Publishing Limited, 2024. All rights reserved.

Table 8 Cluster's profiles relative to demographic and socioeconomic characteristics of the participants (%)

	Percentages (%)		
Cluster profile	Cluster 1	Cluster 2	Cluster 3
Generational cohort (χ^2 = 120.364; df = 8; P < 0.001)			
Silent Generation	8.0	9.3	3.1
Baby Boomers	22.4	22.1	17.1
Generation X	24.3	11.4	14.3
Generation Y	24.2	26.0	31.2
Generation Z	21.0	31.2	34.3
Gender (χ^2 = 112.750; df = 2; P < 0.001)			
Male	35.2	55.4	55.0
Female	64.8	44.6	45.0
Marital status (χ^2 = 94.546; df = 6; P < 0.001)			
Single	29.2	44.7	48.5
Married	59.5	44.0	39.4
Divorced	4.6	5.1	3.9
Widowed	6.7	6.2	8.2
Children in the household (χ^2 = 15.324; df = 2; P < 0.001)			
Yes	27.2	26.7	35.9
No	72.8	73.3	64.1
Area of residence (χ^2 = 76.881; df = 4; P < 0.001)			
City	58.0	71.3	71.2
Town	16.3	15.8	13.4
Village	25.7	12.9	15.4
Education (χ^2 = 124.568; df = 8; P < 0.001)			
Primary	18.2	13.2	0.9
Secondary	42.1	45.7	44.8
Postsecondary	11.0	10.8	21.6
Graduate	23.8	27.0	30.1
Postgraduate	4.9	3.3	2.6
Profession (χ^2 = 74.058; df = 8; P < 0.001)			
Employee (public-private)	27.0	28.4	40.5
Businessman (including farmers)	22.1	15.4	16.9
Worker	3.3	2.2	5.0
Dependent (housekeeper, student, unemployed)	27.3	30.2	26.6
Pensioner	20.3	23.8	11.0

(Continued)

© Burleigh Dodds Science Publishing Limited, 2024. All rights reserved.

Table 8 (*Continued*)

	Percentages (%)		
Cluster profile	Cluster 1	Cluster 2	Cluster 3
Net monthly family income (in Euros) (χ^2 = 127.030; df = 8; $P < 0.001$)			
≤350.00	4.3	6.0	6.1
350.01–1000.00	44.8	28.5	27.5
1000.01–2000.00	37.8	50.8	45.7
2000.01–3000.00	11.3	9.1	17.7
≥3000.01	1.8	5.6	3.0

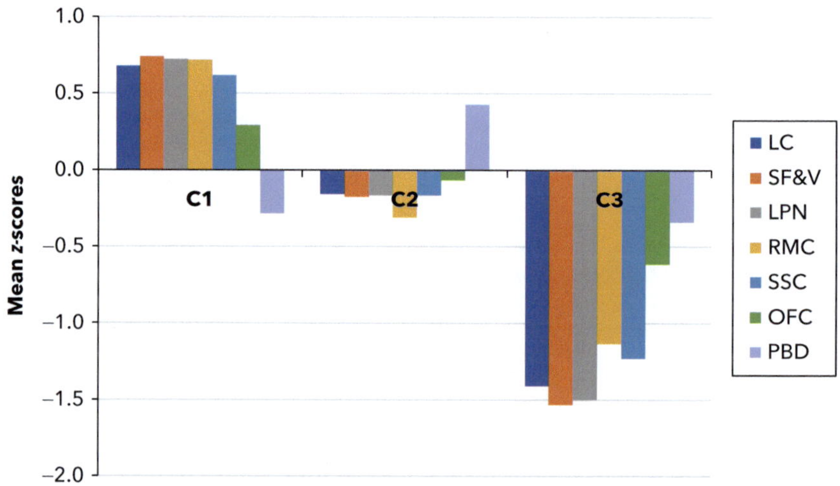

Figure 1 Cluster's centroids (mean values) relative to the *z*-scores of the seven dimensions of SFC. LC = local products; LPNP = low-processed products/natural produce; OFC = organic food consumption; RBD = plant-based diet; RMC= reduced meat consumption; SSC = short supply chain; SF&V = in-season fruits and vegetables.

To explore the specific characteristics of each cluster, the socioeconomic and demographic profiles of each cluster were examined (Table 8).

In summary, comparing the three clusters or types (Fig. 1) based on their *z* scores, cluster 1 shows more positive attitudes toward all dimensions of SFC except for the seventh dimension (PBD), cluster 2 shows a tendency toward neutral attitudes except for the seventh dimension (PBD), and cluster 3 presents, comparatively to the other two clusters, negative attitudes toward all the dimensions of SFC.

Summarizing the data presented in Tables 6–8 and in Fig. 1, the consumers typology is highlighted in Table 9. Based on the earlier analysis, cluster 1 could be named as 'Half-way sustainables', cluster 2 as 'Convenient sustainables', and cluster 3 as 'Nihilists'.

© Burleigh Dodds Science Publishing Limited, 2024. All rights reserved.

A general conclusion deriving from the earlier typology is that no cluster has an absolute SFC behavior. Cluster 1 has people across almost all generational cohorts, and they are the ones that have the most sustainable attitude and behavior as compared to others. Cluster 2 mainly incorporates three cohorts and expresses SFC toward dimensions that are convenient to them and do not require extra effort. Cluster 3 is the young cohort segment and is the negative positioned consumers. The second and third clusters are also characterized as the 'city people'.

5.3 Discussion

Generational cohort differences revealed that the three older cohorts have a more positive attitude and behavior toward SFC compared to the younger cohorts (which is also demonstrated when observing attitude and behavior per cohort per individual SFC dimension). Older cohorts' attitude and behavior are statistically different from younger cohorts. There are several potential reasons for these differences. Firstly, the diet of older cohorts, especially the older Baby Boomers as well as the Silent Generation, was based on the Mediterranean diet which included a high plant-food component, moderate fish and dairy (mainly yogurt) consumption, low red meat consumption as well as olive oil as a fat resource (Salas-Salvadó and Papandreou, 2020; Trichopoulou et al., 2003). On the other hand, younger cohorts have shifted away from the Mediterranean diet toward a more 'western' high-calorie, meat, and processed food type of diet. More broadly, these differences reflect a shift from a more traditional predominantly agricultural society to a more diverse and urban society (Halstead, 1996). In 1951, e.g. the agricultural sector employed 59.5% of the working population, while in 1981 this had declined to 37.6% (Kyrkilis and Simeon, 2015). Another potential factor was the survey took place during a period of restrictions due to the COVID-19 pandemic, including two lockdowns, which impacted behavior such as increased reliance on snacking and comfort food (Fountoulakis et al., 2021; Matacena et al., 2021).

As has been noted, there was a strong preference for locally sourced products. This reflects a number of factors, including the recurring economic problems faced by Greece, amplified by the COVID-19 pandemic, which have increased support for and reliance on local produce as safer and easier to access (Irshad, 2010). These findings are in line with Kyroglou and Henn (2022) who that young Greeks (18–29 years of age) expressed a desire to support their national economy in their purchasing behavior. This behavior links to a preference for seasonal produce among respondents and which is seen as a key aspect of SFC (Tobarra et al., 2018). This is a well-established behavior in Greece linked to the traditional Mediterranean diet and facilitated by a climate that allows the production of a range of fresh produce throughout the

© Burleigh Dodds Science Publishing Limited, 2024. All rights reserved.

Table 9 Typology of consumers and their SFC

Cluster 1 (1224; 42.9%)	Cluster 2 (1167; 40.9%)	Cluster 3 (462; 16.2%)
Half-way sustainables	**Convenient sustainables**	**Nihilists**
First-level profile	**First-level profile**	**First-level profile**
• Very strong positive attitudes: LP, SF&V, LPP-NP • Positive attitudes: RMC, SSC • Tending to somewhat strong attitudes: OFC • Negative attitudes: PBD	• Somewhat positive attitudes: LP, SF&V, LPP-NP • Indifferent attitudes: RMC, SSC, OFC, PBD	• Somewhat negative attitudes towards all dimensions of SFC: LP, SF&V, LPP-NP, RMC, SSC, OFC, PBD
Second-level profile	**Second-level profile**	**Second-level profile**
• 4 cohorts dominate almost equally: BB, Gen X, Gen Y, and Gen Z Mainly: • Females • Married • Without children • Secondary education • Employees, dependent on others, businessman and pensioners • Low class income ≥ €1000.00 • Reside in city	• 3 cohorts dominate: BB, Gen Y, and Gen Z Mainly: • Males • Single and married • Without children • Secondary education • Dependent on others, employees and pensioners • Medium class income (€000.01–€2000.00) • Reside in city	• Gen Z and Gen Y Mainly: • Males • Single • Without children • Secondary education • Large percentage of high educated • Employees • Medium income families • Reside in city

© Burleigh Dodds Science Publishing Limited, 2024. All rights reserved.

Compared to groups 2 and 3: Smallest % of: • City people • High educated Compared to groups 2 and 3: Largest % of: • Females • Gen X • Married • Primary education • Businessman (including farmers) • People in rural and semi-rural areas • Low-income families	Compared to groups 1 and 3: Smallest % of: • Gen X Compared to groups 2 and 3: Largest % of: • SG participants • Dependent on others and employees and pensioners • Medium-income families • Very high-income families (≥€3000.00)	Compared to groups 1 and 2: Smallest % of: • SG and BB • Low educated (primary) • Pensioners Compared to groups 2 and 3: Highest % of: • Gen Y and Gen Z • Single and widowed • Children in the household • Postsecondary education and high educated ones • Employees • High-income families (€2000.01–€3000.00)
Third-level profile	**Third-level profile**	**Third-level profile**
• Tendency to adopt a sustainable purchasing and eating behavior • Non-adoption of an alternative meat diet at the moment, maybe in the future	• Slight tendency to adopt a sustainable purchasing and eating behavior at the moment • Non-adoption of an alternative meat diet at the moment, maybe in the future	• Non-adoption of a sustainable purchasing and eating behavior, nor willing to change in the near future • Non-adoption of an alternative meat diet, nor willing to change in the near future

BB = Baby Boomers; Gen X = Generation X; Gen Y = Generation Y; Gen Z = Generation Z; LP = local products; LPP-NP= low-processed products/natural produce; Med. TS = median value of the total sample; MV TS = mean value of the total sample; OFC = organic food consumption, PBD = plant-based diet; RMC = reduced meat consumption; SF&V = in-season fruits and vegetables; SFC = sustainable food consumption; SG = Silent Generation; SSC = short supply chain.

© Burleigh Dodds Science Publishing Limited, 2024. All rights reserved.

year (Serra-Majem et al., 2004). This culinary tradition has also been passed down from one generation to the other in the extended families that are still characteristic of the Greek society (Siatitsa, 2021). Other research also supports an emphasis on purchasing seasonable produce as a key aspect of SFC (Greer et al., 2018; Tsakiridou et al., 2011). A preference for local, seasonal produce also links to a preference for less highly processed foods, reflected, e.g. in the research by Jorge et al. (2020) who found that millennials preferred natural producing methods and minimal processing of food.

Although it has been associated with SFC, reducing meat consumption is much less widely accepted by consumers and has been harder to encourage (Borusiak et al., 2022; Harguess et al., 2020; Bogueva and Marinova, 2020). In countries such as Greece, meat consumption has been associated with rising incomes and the adoption of the lifestyles of wealthier countries such as the United States. Meat consumption has been linked to higher social status (Chan and Zlatevska, 2019), masculinity (Chan and Zlatevska, 2019; Lax and Mertig, 2020), and high income (McMichael et al., 2007). For some social groups, a PBD is associated with poverty (Beverland, 2014). As Lazaridis (2003) points out, the proportion of Greek food budgets spent on meat increased from 14.5% in 1960 to 30% in 1983. This is mirrored by the noticeable shift away from the Mediterranean diet in many countries, as measured by the FAO Mediterranean Adequacy Index, with Greece showing the largest shift (da Silva et al., 2009).

Referring to organic food consumption, the result of this study is in line with other research in the UK and China which suggests that older consumers have a more positive attitude to organic food (Geen and Firth, 2006; Xie et al., 2020) and have been a major factor in transforming a niche sector into a well-established market (Feil et al., 2020; Oroian et al., 2017). This is linked to trust in organic produce and its certification (Kamenidou et al., 2017; Nuttavuthisit and Thøgersen, 2017; Wu et al., 2021) which is lower in younger consumers in this study. A preference for purchasing organic produce is also associated with wealthier consumers who tend to come from older cohorts (Fotopoulos and Krystallis, 2002; Kamenidou et al., 2019; Torjusen et al., 2001; Zagata, 2012). This reflects wider research that older cohorts demonstrate more SFC behaviors (Kamenidou et al., 2021). It is also consistent with the meta-analysis by Nardi et al. (2019) which suggests that younger cohorts had a greater relationship between intention and behavior, meaning that a less favorable attitude to SFC is more consistently translated into actual (less) purchasing behavior.

Funk et al. (2021) point out that identifying the various consumer types is very significant for developing and implementing customized intervention strategies, while Sütterlin et al. (2011) and Funk et al. (2021) also assert that attitudinal and behavioral variables are very useful for policy making. Three consumer types were developed based on attitudes of SFC in association with their generational cohorts and other demographic and socioeconomic

© Burleigh Dodds Science Publishing Limited, 2024. All rights reserved.

characteristics and behavior/intention to comply. A characteristic of all types is that their attitude toward PBD is relatively low (indifferent or negative). The consumer type that has the most positive attitude compared to the other groups is the 'Convenient sustainables' who possess 'indifferent attitude'. The rest of the consumer types have a negative attitude. This attitude is expressed in behavior, as all groups are reluctant to adopt alternative meats. Thus, SFC issues related to meat reduction consumption seem to be the hardest issue to tackle. Attitudes toward OFC are not strongly positive (as expected due to the health benefits of their consumption) but range from negative to somewhat positive. This suggests trust issues relating to product certification.

The 'Half-way sustainables' have positive attitudes toward most of the dimensions of SFC but comply with one dimension of SFC (dominated by four of the five cohorts (BB, Generation X, Generation Y, and Generation Z). One reason for their attitudes and behavior could be that about one-half of the members of this group reside in rural and semirural areas (could be farmers as 'businessmen' or pensioners who came back to their village or town), thus adopting a more traditional lifestyle which brings them 'back' to the Mediterranean diet. Also, it is common for households living in provinces to have their own organic garden for household consumption or have direct access to produce (from friends or family). Moreover, living in rural and semi-rural areas, they have daily contact with nature and therefore are more likely to be aware of how important it is for the environment to be sustainably maintained. This group already demonstrates stronger SFC behavior (compared to the other types) and therefore an attractive target group for social marketing campaigns with the aim to maintain and extend their positive behavior. As it includes younger generational cohorts, this group may educate their children toward more support for SFC.

The 'Convenient sustainables' have positive and indifferent attitudes toward the dimensions of SFC and do not engage in SFC at the moment but are willing to think about it in the future (dominated by BB, Generation Y, and Generation Z and are 'city people'). This group cares about its convenience and will not engage in something that will need effort or extra resources (e.g. money to purchase organic when prices are high or search for a farmers or organic market to buy fresh or organic produce). Since it is not negatively positioned, this group can be targeted by social marketing campaigns and may produce favorable behavioral change in the long run. They are 'city people', campaigns need to focus on awareness of short-term direct benefits (because it consists of young generational cohorts who do not care for health benefits after 10 or 20 years) as well as how SFC may be convenient for those working many hours or are 'on the go'.

The third group, the 'Nihilists', are negatively positioned in attitudes, as well as their behavior/intention to comply. This consumer type is dominated

© Burleigh Dodds Science Publishing Limited, 2024. All rights reserved.

by young people (Generation Z and Y), who have the highest percentage of households with children (<18 years old), high education, and high-income families compared to the other consumer types. This result is of particular interest, as all these parameters lead to positive SFC attitude and behavior, and thus these findings do not reflect previous research (e.g. Annunziata et al., 2019; Vecchio et al., 2015). One plausible reason could be their distrust toward the government, authorities, and the food certification procedures and system for providing them with safe sustainable food. As Gutfreund (2016) states, Generation Z is 'Cynical, savvy and armed with information at their fingertips'. This group is resistant to change in the short run but needs consistent effort in order to change attitudes in the long run.

5.4 Policy implications

As this research suggests, whilst there are generable favorable attitudes to SFC, these still vary significantly across generational cohorts with some low rates of SFC behavior/intention to comply. This is consistent with the three types of consumers with distinctive attitudes and behavior. In particular, this research has highlighted mixed attitudes to SFC among younger generational cohorts compared to older cohorts. This suggests a need for a greater and more consistent focus on SFC through the educational system as a key component of behavioral change (Heimlich and Ardoin, 2008; Lindsey et al., 2014).

As regards Greece, there is a generally decreasing provision of time and resources devoted to environmental and sustainability topics from primary to secondary education, despite their importance in areas such as the environment, food security, and individual and public health. This lack of provision has been linked to a low level of interest in environmental and sustainability issues (Dabija et al., 2020) and will have a direct negative future impact on generational cohorts' SFC attitudes and behavior. Therefore, from the early years, SFC education is needed which should be continuous throughout the education system.

There is also a need for broader social marketing campaigns to raise awareness of SFC issues and target behavioral changes in food consumption patterns. These campaigns need significant resources and time to have an effect, particularly in the case of food purchasing and consumption patterns which often have deep cultural roots (Kotler et al., 2002; Bogueva et al., 2017). Campaigns will need to use the media most relevant to different cohorts combined with the consumer types. Therefore, they must include digital marketing campaigns involving social media to target younger cohorts (Kamenidou et al., 2020a). Additionally, attractive motives should be provided by the government to increase SFC behavior (such as coupons for organic foods), especially for those who reside in the city or have a low income.

© Burleigh Dodds Science Publishing Limited, 2024. All rights reserved.

Considering that distrust toward food certification is very common, governmental organizations and companies involved in product certification must regain consumers trust in certified food. This includes tackling problems such as high pricing of certified products, fake certification, or failures in certification, which need to be transparently reported and dealt with robustly. The most difficult consumer type to deal with is the 'Nihilists', where multiple long-run interventions should be implemented. The most demanding issue to deal with across types is associated with reducing meat consumption, which highlights the need for SFC education from the early age of consumers, i.e., from the pre-elementary and elementary (primary) school.

5.5 Limitations of the study

This research suffers from some unavoidable limitations which could be regarded as stimulus for future research. First, a non-probability sampling technique was used, so, generalization of the study's results to the overall population of Greek cohorts is not possible. The second issue refers to the use of cross-sectional data with the corresponding limitations as pointed out by Wong et al. (2021). Additionally, the sample size could be considered small compared to samples used in studies from other countries which yield the data from national databases. This limitation is also associated with the lockdowns and curfews implemented in 2020 by the government, and for so, a mixed method in data collection was required. A fourth limitation of the study is that it does not provide with data referring to citizens' engagement in environmental organizations which could shape their attitudes.

Lastly, there are additional variables that can be added referring to SFC, such as consuming insects for protein substitutes or food waste. The items used in this study were validated by the qualitative research employed as well as previous academic research. A development of a theoretical framework and a conceptual model that connects generational cohorts, cohort characteristics, and dimensions of SFC attitude and behavior would be of interest and a suggestion for further research.

6 Conclusion

This study had as its aim to explore cohort differences in SFC attitudes and behavior in Greece and to provide with an attitudinal and behavioral typology, offering a holistic approach to the issue to facilitate decision-making by government authorities and other stakeholders. It examined seven dimensions of SFC discussed in the literature as well as degrees of intention to comply with the SFC behavior of five generational cohorts. Results reveal that the three older cohorts, i.e. Generation X, Baby Boomers, and the Silent Generation, have

© Burleigh Dodds Science Publishing Limited, 2024. All rights reserved.

more positive attitudes toward SFC and comply more with SFC behavior than the Generation Z and the Generation Y cohort. Three consumer groups were identified in typology development: The 'Half-way sustainables', the 'Convenient sustainables', and the 'Nihilists', which have distinctive different characteristics, attitudes, and behavior.

Considering these cohort differences and consumer types, governments and other stakeholders can better plan and execute specific campaigns targeting each cohort and consumer types. Government interventions must operate at three levels. First, environmental education and education related to SFC are needed from the early years and should be continuous and not discontinued in adolescence (secondary education). Second, regaining trust in certified food is extremely important and leads to nihilism if not obtained. Third, digital social marketing campaigns are needed especially toward the younger generations and the consumer types that include young cohorts.

7 Where to look for further information

Further reading:

- Sustainable Diets and Biodiversity by Burlingame and Dernini (Eds), Nutrition and Consumer Protection Division, FAO.
- Food, Globalization and Sustainability by Oosterveer and Sonnenfeld.
- Sustainable food systems from agriculture to industry: Improving production and processing by Galanakis.
- Food Science, An Ecological Approach by Edelstein.
- Sustainable Diets: Food for Healthy People and a Healthy Planet: Workshop Summary.
- The 10 Principles of Food Industry Sustainability by Baldwin.
- Defining markets, defining moments: America's 7 generational cohorts, their shared experiences, and why businesses should care by Geoffrey, Schewe and Karlovich.

Key journals:

- *Journal of Cleaner Production*
- *Sustainability*
- *Food Policy*
- *Public Health Nutrition*
- *Sustainable Production and Consumption*
- *Journal of Nutrition Education and Behavior*
- *Food Quality and Preference*
- *Journal of Consumer Marketing*
- *Journal of Business Research*
- *Food Security*

© Burleigh Dodds Science Publishing Limited, 2024. All rights reserved.

- *Appetite*
- *Frontiers in Ecology and the Environment*
- *Nature Sustainability*

8 References

Adams, S. J. (2000). Generation X: how understanding this population leads to better safety programs. *Professional Safety* 45(1): 26.

Aguirre Sánchez, L., Roa-Díaz, Z. M., Gamba, M., Grisotto, G., Moreno Londoño, A. M., Mantilla-Uribe, B. P., Rincón Méndez, A. Y., Ballesteros, M., Kopp-Heim, D., Minder, B., Suggs, L. S. and Franco, O. H. (2021). What influences the sustainable food consumption behaviours of university students? A systematic review. *International Journal of Public Health* 66: 1604149.

Aleksandrowicz, L., Green, R., Joy, E. J. M., Smith, P. and Haines, A. (2016). The impacts of dietary change on greenhouse gas emissions, land use, water use, and health: A systematic review. *PLoS ONE* Wiley, A. S. (Ed.) 11(11): e0165797.

Allievi, F., Vinnari, M. and Luukkanen, J. (2015). Meat consumption and production-analysis of efficiency, sufficiency and consistency of global trends. *Journal of Cleaner Production* 92: 142–151.

Alvarado, R., Ortiz, C., Jiménez, N., Ochoa-Jiménez, D. and Tillaguango, B. (2021). Ecological footprint, air quality and research and development: the role of agriculture and international trade. *Journal of Cleaner Production* 288: 125589.

Amvrosiou, A., Theodoropoulou, E. and Mitoula, R. (2017). The study of the attitudes and perceptions of modern Greeks during the economic crisis on the consumption of organic products. *International Review of Social Sciences* 5(12).

Annunziata, A., Agovino, M. and Mariani, A. (2019). Sustainability of Italian families' food practices: Mediterranean diet adherence combined with organic and local food consumption. *Journal of Cleaner Production* 206: 86–96.

Aravossitas, T. and Sugiman, M. (2019). From brain drain to brain gain: New Greek migration to Canada and implications for the community. In: *'New' Migration of Families from Greece to Europe and Canada*. Springer. pp.33–56.

Arenas-Gaitán, J., Peral-Peral, B. and Reina-Arroyo, J. (2022). Food-related lifestyles across generations. *British Food Journal*. Available at: https://www.emerald.com/insight/content/doi/10.1108/BFJ-04-2021-0393/full/html. Accessed 9 March 2022 124(5): 1485–1501.

Azzurra, A., Massimiliano, A. and Angela, M. (2019). Measuring sustainable food consumption: A case study on organic food. *Sustainable Production and Consumption* 17: 95–107.

Bacon, L. and Krpan, D. (2018). (Not) Eating for the environment: the impact of restaurant menu design on vegetarian food choice. *Appetite* 125: 190–200.

Balderjahn, I., Peyer, M., Seegebarth, B., Wiedmann, K. P. and Weber, A. (2018). The many faces of sustainability-conscious consumers: A category-independent typology. *Journal of Business Research* 91: 83–93.

Balenović, I., Ignjatijević, S., Stojanović, G., Vapa Tankosić, J., Lekić, N., Milutinović, O., Gajić, A., Ivaniš, M., Bošković, J., Prodanović, R., Puvača, N. and Duđak, L. (2021). Factors influencing wine purchasing by generation Y and older cohorts on the Serbian wine market. *Agriculture* 11(11): 1054.

© Burleigh Dodds Science Publishing Limited, 2024. All rights reserved.

Baudry, J., Touvier, M., Allès, B., Péneau, S., Méjean, C., Galan, P., Hercberg, S., Lairon, D. and Kesse-Guyot, E. (2016). Typology of eaters based on conventional and organic food consumption: results from the NutriNet-Santé cohort study. *British Journal of Nutrition* 116(4): 700–709.

Bennett, J., Pitt, M. and Price, S. (2012). Understanding the impact of generational issues in the workplace. *Facilities* 30(7/8): 278–288.

Berdie, D. R., Anderson, J. F. and Niebuhr, M. A. (1986). *Questionnaires: Design and Use*, 2nd Edition. Metuchen, NJ: Scarecrow Press.

Beverland, M. B. (2014). Sustainable eating: mainstreaming plant-based diets in developed economies. *Journal of Macromarketing* 34(3): 369–382.

Bianchi, C. and Mortimer, G. (2015). Drivers of local food consumption: a comparative study. *British Food Journal* 117(9): 2282–2299.

Bogueva, D. and Marinova, D. (2020). Cultured meat and Australia's generation Z. *Frontiers in Nutrition* 7: 148.

Bogueva, D., Marinova, D. and Bryant, C. (2022). Meat me halfway: Sydney meat-loving men's restaurant experience with alternative plant-based proteins. *Sustainability* 14(3): 1290.

Bogueva, D., Marinova, D. and Raphaely, T. (2017). Reducing meat consumption: the case for social marketing. *Asia Pacific Journal of Marketing and Logistics* 29(3): 477-500.

Bollani, L., Bonadonna, A. and Peira, G. (2019). The millennials' concept of sustainability in the food sector. *Sustainability* 11(10): 2984.

Bonaccio, M., Bes-Rastrollo, M., de Gaetano, G. and Iacoviello, L. (2016). Challenges to the Mediterranean diet at a time of economic crisis. *Nutrition, Metabolism, and Cardiovascular Diseases* 26(12): 1057–1063.

Borusiak, B., Szymkowiak, A., Kucharska, B., Gálová, J. and Mravcová, A. (2022). Predictors of intention to reduce meat consumption due to environmental reasons–Results from Poland and Slovakia. *Meat Science* 184: 108674.

Calvo-Porral, C. and Pesqueira-Sanchez, R. (2019). Generational differences in technology behaviour: comparing millennials and Generation X. *Kybernetes* 49(11): 2755–2772.

Chabbi, A., Lehmann, J., Ciais, P., Loescher, H. W., Cotrufo, M. F., Don, A., SanClements, M., Schipper, L., Six, J., Smith, P. and Rumpel, C. (2017). Aligning agriculture and climate policy. *Nature Climate Change* 7(5): 307–309.

Chan, E. Y. and Zlatevska, N. (2019). Jerkies, tacos, and burgers: subjective socioeconomic status and meat preference. *Appetite* 132: 257–266.

Chon, K. S. (1990). Consumer satisfaction and dissatisfaction in tourism as related to destination image perception. Ph.D. Thesis. Blacksburg, VA: Virginia Polytechnic Institute and State University.

Collins, D. E. and Tilson, E. R. (2001). A new generation on the horizon. (Teaching Techniques). *Radiologic Technology* 73(2): 172–176.

Creswell, J. W. and Miller, D. L. (2000). Determining validity in qualitative inquiry. *Theory Into Practice* 39(3): 124–130.

Crippa, M., Solazzo, E., Guizzardi, D., Monforti-Ferrario, F., Tubiello, F. N. and Leip, A. (2021). Food systems are responsible for a third of global anthropogenic GHG emissions. *Nature Food* 2(3): 198–209.

da Silva, R., Bach-Faig, A., Raidó Quintana, B., Buckland, G., Vaz de Almeida, M. D. and Serra-Majem, L. (2009). Worldwide variation of adherence to the Mediterranean diet, in 1961–1965 and 2000–2003. *Public Health Nutrition* 12(9A): 1676–1684.

© Burleigh Dodds Science Publishing Limited, 2024. All rights reserved.

Dabija, D.-C., Bejan, B. M. and Puşcaş, C. (2020). A qualitative approach to the sustainable orientation of generation Z in retail: the case of Romania. *Journal of Risk and Financial Management* 13(7): 152.

Di Vita, G., Vecchio, R., Borrello, M., Zanchini, R., Maesano, G., Gulisano, G., Brun, F. and D'Amico, M. (2021a). Oh my darling clementine: heterogeneous preferences for sustainable citrus fruits. *Renewable Agriculture and Food Systems* 36(6): 557–568.

Di Vita, G., Zanchini, R., Falcone, G., D'Amico, M., Brun, F. and Gulisano, G. (2021b). Local, organic or protected? Detecting the role of different quality signals among Italian olive oil consumers through a hierarchical cluster analysis. *Journal of Cleaner Production* 290: 125795.

Diamantopoulos, A., Schlegelmilch, B. B., Sinkovics, R. R. and Bohlen, G. M. (2003). Can socio-demographics still play a role in profiling green consumers? A review of the evidence and an empirical investigation. *Journal of Business Research* 56(6): 465–480.

Dimock, M. (2019). Defining generations: Where millennials end and generation Z begins. *Pew Research Center*, 17(1). pp. 1-7.

Dobermann, D., Swift, J. A. and Field, L. M. (2017). Opportunities and hurdles of edible insects for food and feed. *Nutrition Bulletin* 42(4): 293–308.

Dudley, N. and Alexander, S. (2017). Agriculture and biodiversity: a review. *Biodiversity* 18(2–3): 45–49.

Eldesouky, A., Mesias, F. J. and Escribano, M. (2020). Consumer assessment of sustainability traits in meat production. A choice experiment study in Spain. *Sustainability* 12(10): 4093.

European Public Health Association (2017). Healthy and sustainable diets for European countries. Available at: https://eupha.org/repository/advocacy/EUPHA_report_on_healthy_and_sustainable_diets_20-05-2017.pdf.

FAO (2018). [Food and agriculture organization of the United Nations]. *Agriculture to Achieve the SDGs-20 Interconnected Actions to Guide Decision-Makers*. Available at: https://www.fao.org/3/I9900EN/i9900en.pdf. Rome, Italy: Food and Agriculture Organization of the United Nations.

FAO (2012). [Food and agriculture organization of the United Nations]. In: Sustainable Diets and Biodiversity, Directions and Solutions for Policy, Research and Action. Proceedings of the International Scientific Symposium "Biodiversity and Sustainable Diets United Against Hunger", 3–5 November 2010, FAO, Headquarters, Rome, in Burlingame, B. and Dernini S. (Eds), Nutrition and Consumer Protection Division, FAO, E-ISBN 978-92-5-107288-2.

Feil, A. A., Cyrne, C. C. da S., Sindelar, F. C. W., Barden, J. E. and Dalmoro, M. (2020). Profiles of sustainable food consumption: consumer behavior toward organic food in southern region of Brazil. *Journal of Cleaner Production* 258: 120690.

Fornell, C. and Larcker, D. F. (1981). Evaluating structural equation models with unobservable variables and measurement error. *Journal of Marketing Research* 18(1): 39–50.

Fotopoulos, C. and Krystallis, A. (2002). Organic product avoidance: reasons for rejection and potential buyers' identification in a countrywide survey. *British Food Journal* 104(3/4/5): 233–260.

Fountoulakis, K. N., Apostolidou, M. K., Atsiova, M. B., Filippidou, A. K., Florou, A. K., Gousiou, D. S., Katsara, A. R., Mantzari, S. N., Padouva-Markoulaki, M., Papatriantafyllou, E. I., Sacharidi, P. I., Tonia, A. I., Tsagalidou, E. G., Zymara, V. P., Prezerakos, P. E., Koupidis,

© Burleigh Dodds Science Publishing Limited, 2024. All rights reserved.

S. A., Fountoulakis, N. K. and Chrousos, G. P. (2021). Self-reported changes in anxiety, depression and suicidality during the COVID-19 lockdown in Greece. *Journal of Affective Disorders* 279: 624–629.

Funk, A., Sütterlin, B. and Siegrist, M. (2021). Consumer segmentation based on Stated environmentally-friendly behavior in the food domain. *Sustainable Production and Consumption* 25: 173–186.

Galli, A., Moreno Pires, S., Iha, K., Alves, A. A., Lin, D., Mancini, M. S. and Teles, F. (2020). Sustainable food transition in Portugal: assessing the Footprint of dietary choices and gaps in national and local food policies. *Science of the Total Environment* 749: 141307.

Garnett, T. (2006). Fruit and vegetables & UK greenhouse gas emissions: exploring the relationship. *UK: Food and Climate Research Network, University of Surrey (on-line).* Available at: https://tabledebates.org/sites/default/files/2020-10/fruitveg_paper_final_0.pdf.

Gazdecki, M., Goryńska-Goldmann, E., Kiss, M. and Szakály, Z. (2021). Segmentation of food consumers based on their sustainable attitude. *Energies* 14(11): 3179.

Geen, N. and Firth, C. (2006). The committed organic consumer. In: Proceedings of the Joint Organic Congress, Odense, Denmark.

Giosuè, A., Calabrese, I., Vitale, M., Riccardi, G. and Vaccaro, O. (2022). Consumption of dairy foods and cardiovascular disease: A systematic review. *Nutrients* 14(4): 831.

Grasso, A. C., Hung, Y., Olthof, M. R., Verbeke, W. and Brouwer, I. A. (2019). Older consumers' readiness to accept alternative, more sustainable protein sources in the European Union. *Nutrients* 11(8): 1904.

Green, R. F., Milner, J., Dangour, A. D., Haines, A., Chalabi, Z., Markandya, A., Spadaro, J. and Wilkinson, P. (2015). The potential to reduce greenhouse gas emissions in the UK through healthy and realistic dietary change. *Climatic Change* 129(1–2): 253–265.

Green, R. F., Joy, E. J. M., Harris, F., Agrawal, S., Aleksandrowicz, L., Hillier, J., Macdiarmid, J. I., Milner, J., Vetter, S. H., Smith, P., Haines, A. and Dangour, A. D. (2018). Greenhouse gas emissions and water footprints of typical dietary patterns in India. *Science of the Total Environment* 643: 1411–1418.

Greer, A. E., Davis, S., Sandolo, C., Gaudet, N. and Castrogivanni, B. (2018). Agricultural experiences are positively associated with high school students' fruit and vegetable perceptions and consumption. *Journal of Nutrition Education and Behavior* 50(2): 133–140.e1.

GSA (Greek Statistical Authority) (2020). Greek population (on-line). Available at: https://www.statistics.gr/documents/20181/972a68ea-7bc3-c764-94bd-2d989b25e6f6. Accessed 7 August 2020.

Gutfreund, J. (2016). Move over, Millennials: generation Z is changing the consumer landscape. *Journal of Brand Strategy* 5(3): 245–249.

Hair, B. J. F., Black, W. C. and Anderson, R. E. (2018). Multivariate data analysis. In: 8th, Illustr CENGAGE, IINDIA.

Halstead, P. (1996). Pastoralism or household herding? Problems of scale and specialization in early Greek animal husbandry. *World Archaeology* 28(1): 20–42.

Harguess, J. M., Crespo, N. C. and Hong, M. Y. (2020). Strategies to reduce meat consumption: A systematic literature review of experimental studies. *Appetite* 144: 104478.

Hauck, W. E. and Stanforth, N. (2007). Cohort perception of luxury goods and services. *Journal of Fashion Marketing and Management: an International Journal* 11(2): 175–188.

© Burleigh Dodds Science Publishing Limited, 2024. All rights reserved.

Heimlich, J. E. and Ardoin, N. M. (2008). Understanding behavior to understand behavior change: a literature review. *Environmental Education Research* 14(3): 215–237.

Held, B. and Haubach, C. (2017). The additional costs of organic food products - a basket of goods-based analysis differentiated by income. *Management Revu* 28(1): 6-61.

Hewlett, S. A., Sherbin, L. and Sumberg, K. (2009). How Gen Y and Boomers will reshape your agenda. *Harvard Business Review* 87(7–8): 71–153.

Hoek, A. C., Malekpour, S., Raven, R., Court, E. and Byrne, E. (2021). Towards environmentally sustainable food systems: decision-making factors in sustainable food production and consumption. *Sustainable Production and Consumption* 26: 610–626.

Iorgulescu, M.-C. (2016). Generation Z and its perception of work. *Cross-Cultural Management Journal* 18(1): 47–54.

Ivanova, O., Flores-Zamora, J., Khelladi, I. and Ivanaj, S. (2019). The generational cohort effect in the context of responsible consumption. *Management Decision* 57(5): 1162–1183.

Jackson, V., Stoel, L. and Brantley, A. (2011). Mall attributes and shopping value: differences by gender and generational cohort. *Journal of Retailing and Consumer Services* 18(1): 1–9.

Jorge, E., Lopez-Valeiras, E. and Gonzalez-Sanchez, M. B. (2020). The importance given to food naturalness attributes by millennial university students. *Sustainability* 12(2): 728.

Julianty, S. A., Kusdibyo, L. and Amalia, F. A. (2021). Predicting generation Z behavioral intention towards organic food consumptions. In: *Advances in Engineering Research*. Atlantis Press: 593–600.

Jurkiewicz, C. L. and Brown, R. G. (1998). Generational comparisons of public employee motivation. *Review of Public Personnel Administration* 18(4): 18–37.

Kaczorowska, J., Rejman, K. and Halicka, E. (2017). Purchasing behaviors and in-home food management of millennials in the context of sustainable food consumption. *Journal of Agribusiness and Rural Development* 16(3): 573–581.

Kamenidou, I., Mamalis, S., Mylona, I., Stavrianea, A. and Bara, E. Z. (2022). Perceptions of environmental benefits from sustainable food consumption patterns: evidence from the Generations Z and Y cohort. *International Journal of Agricultural Resources, Governance and Ecology* 18(1/2): 182–198.

Kamenidou, I., Mamalis, S. A., Pavlidis, S. and Bara, E. G. (2019). Segmenting the generation Z cohort university students based on sustainable food consumption behavior: A preliminary study. *Sustainability* 11(3): 837.

Kamenidou, I., Mamalis, S., Mylona, I. and Bara, E. Z. (2021). Comparing five generational cohorts on their sustainable food consumption patterns: recommendations for improvement through marketing communication. In: *Advances in Longitudinal Data Methods in Applied Economic Research* Tsounis, N., Vlachvei, A. (Eds). Cham: Springer International Publishing. pp.69–80.

Kamenidou, I., Rigas, K. and Priporas, C.-V. (2017). Household behavior on food security during an economic crisis. In: *Food Security and Sustainability* Mergos, G. and Papanastassiou, M. (Eds). Cham: Springer International Publishing. pp.243–261. Available at: http://link.springer.com/10.1007/978-3-319-40790-6_14. Accessed 23 March 2022.

Kamenidou, I., Stavrianea, A. and Bara, E.-Z. (2020a). Generational differences toward organic food behavior: insights from five generational cohorts. *Sustainability* 12(6): 2299.

© Burleigh Dodds Science Publishing Limited, 2024. All rights reserved.

Kamenidou, I., Stavrianea, A. and Liava, C. (2020b). Achieving a Covid-19 free country: citizens preventive measures and communication pathways. *International Journal of Environmental Research and Public Health* 17(13): 4633.

Kent, R. (1993). *Marketing Research in Action*. London: Routledge.

Kita, P., Žambochová, M., Strelinger, J. and Kitová Mazalánová, V. (2021). Consumer behaviour of Slovak households in the sphere of organic food in the context of sustainable consumption. *Central European Business Review* 10(1): 1-17.

Knaapila, A., Michel, F., Jouppila, K., Sontag-Strohm, T. and Piironen, V. (2022). Millennials' consumption of and attitudes toward meat and plant-based meat alternatives by consumer segment in Finland. *Foods* 11(3): 456.

Kneafsey, M., Venn, L., Schmutz, U., Balázs, B., Trenchard, L., Eyden-Wood, T., Bos, E., Sutton, G. and Blackett, M. (2013). Short food supply chains and local food systems in the EU. A state of play of their socio-economic characteristics. *JRC Scientific and Policy Reports* 123: 129.

Kol, O., Zimand-Sheiner, D. and Levy, S. (2023). A (local) apple a day: pandemic-induced changes in local food buying, a generational cohort perspective. *European Journal of International Management* 19(1): 1-26.

Kotler, P., Roberto, N. and Lee, N. (2002). *Social Marketing: Improving the Quality of Life* (2nd edn.). Thousand Oaks, CA: Sage Publications.

Kowalsky, T. O., Morilla Romero de la Osa, R. and Cerrillo, I. (2022). Sustainable diets as tools to harmonize the health of individuals, communities and the planet: A systematic review. *Nutrients* 14(5): 928.

Kristiansen, S., Painter, J. and Shea, M. (2021). Animal agriculture and climate change in the US and UK elite media: volume, responsibilities, causes and solutions. *Environmental Communication* 15(2): 153-172.

Kumar, A. and Smith, S. (2018). Understanding local food consumers: theory of planned behavior and segmentation approach. *Journal of Food Products Marketing* 24(2): 196-215.

Kymäläinen, T., Seisto, A. and Malila, R. (2021). Generation Z Food waste, diet and consumption habits: A Finnish social design study with future consumers. *Sustainability* 13(4): 2124.

Kyrkilis, D. and Simeon, S. (2015). Greek agriculture's failure. The other face of a failed industrialization. From accession to EU to the debt crisis. *Procedia Economics and Finance* 33: 64-77.

Kyroglou, G. and Henn, M. (2022). Young political consumers between the individual and the collective: evidence from the UK and Greece. *Journal of Youth Studies* 25(6): 833-853.

Lax, J. B. and Mertig, A. G. (2020). The perceived masculinity of meat: development and testing of a measure across social class and gender. *Food, Culture and Society* 23(3): 416-426.

Lazaridis, P. (2003). Household meat demand in Greece: A demand systems approach using microdata. *Agribusiness* 19(1): 43-59.

Lea, E. and Worsley, T. (2005). Australians' organic food beliefs, demographics and values. *British Food Journal* 107(11): 855-869.

Lee, J., Abell, N. and Holmes, J. L. (2017). Validation of measures of Cyberbullying perpetration and victimization in emerging adulthood. *Research on Social Work Practice* 27(4): 456-467.

© Burleigh Dodds Science Publishing Limited, 2024. All rights reserved.

Lindsey, R. V., Shroyer, J. D., Pashler, H. and Mozer, M. C. (2014). Improving students' long-term knowledge retention through personalized review. *Psychological Science* 25(3): 639–647.
Littrell, M. A., Jin Ma, Y. and Halepete, J. (2005). Generation x, Baby Boomers, and Swing: marketing fair trade apparel. *Journal of Fashion Marketing and Management: an International Journal* 9(4): 407–419.
Lorenz, B. A. and Langen, N. (2018). Determinants of how individuals choose, eat and waste: providing common ground to enhance sustainable food consumption out-of-home. *International Journal of Consumer Studies* 42(1): 35–75.
Lyons, S. T., Duxbury, L. and Higgins, C. (2007). An empirical assessment of generational differences in basic human values. *Psychological Reports* 101(2): 339–352.
Mamalis, S., Kamenidou, I., Gkitsas, S., Stavrianea, A., Gkagkani, D. and Pavlidis, S. (2022). Segmenting generation Z based on organic food decision-making: A preliminary study. In: International Conference on Applied Economics. Cham: Springer International Publishing. pp. 891–903.
Matacena, R., Zenga, M., D'Addario, M., Mari, S. and Labra, M. (2021). COVID-19 as an opportunity for a healthy-sustainable food transition. An analysis of dietary transformations during the first Italian lockdown. *Sustainability* 13(10): 5661.
McCrindle, M. and Wolfinger, E. (2009). *The ABC of XYZ: Understanding the Global Generations*. Sydney: University of New South Wales Press.
McKercher, B., Lai, B., Yang, L. and Wang, Y. (2020). Travel by Chinese: a generational cohort perspective. *Asia Pacific Journal of Tourism Research* 25(4): 341–354.
McMichael, A. J., Powles, J. W., Butler, C. D. and Uauy, R. (2007). Food, livestock production, energy, climate change, and health. *The Lancet* 370(9594): 1253–1263.
Mehrabi, S., Perez-Mesa, J. C. and Giagnocavo, C. (2022). The role of consumer-citizens and connectedness to nature in the sustainable transition to agroecological food systems: the mediation of innovative business models and a multi-level perspective. *Agriculture* 12(2): 203.
Mekonnen, M. M. and Hoekstra, A. Y. (2011). The green, blue and grey water footprint of crops and derived crop products. *Hydrology and Earth System Sciences* 15(5): 1577–1600.
Meredith, G. and Schewe, C. (1994). The power of cohorts: Americans who shared powerful experiences as young adults, such as the hardship of the great depression, fall into six cohorts. *American Demographics* 16: 22–22.
Molin, E., Martin, M. and Björklund, A. (2021). Addressing sustainability within public procurement of food: A systematic literature review. *Sustainability* 13(23): 13395.
Molinario, E., Lorenzi, C., Bartoccioni, F., Perucchini, P., Bobeth, S., Colléony, A., Diniz, R., Eklund, A., Jaeger, C., Kibbe, A., Richter, I., Ruepert, A., Sloot, D., Udall, A. M. and Bonaiuto, M. (2020). From childhood nature experiences to adult pro-environmental behaviors: an explanatory model of sustainable food consumption. *Environmental Education Research* 26(8): 1137–1163.
Morton, L. P. (2003). Targeting generation X. *Public Relations Quarterly* 48(4): 43.
Morton, L. P. (2002). Targeting generation Y. *Public Relations Quarterly* 47(2): 46.
Muposhi, A. and Dhurup, M. (2016). A qualitative inquiry of generation Y consumers' selection attributes in the case of organic products. *International Business and Economics Research Journal* 15(1): 1.

© Burleigh Dodds Science Publishing Limited, 2024. All rights reserved.

Nardi, V. A. M., Jardim, W. C., Ladeira, W. and Santini, F. (2019). Predicting food choice: a meta-analysis based on the theory of planned behavior. *British Food Journal* 121(10): 2250–2264.

Nguyen, P. H., Strizich, G., Lowe, A., Nguyen, H., Pham, H., Truong, T. V., Nguyen, S., Martorell, R. and Ramakrishnan, U. (2013). Food consumption patterns and associated factors among Vietnamese women of reproductive age. *Nutrition Journal* 12(1): 126.

Satinover Nichols, B. and Wehr Holt, J. (2023). A comparison of sustainability attitudes and intentions across generations and gender: a perspective from US consumers. *Cuadernos de Gestión* 23(1): 51–62.

Niederle, P. and Schubert, M. N. (2020). HOW does veganism contribute to shape sustainable food systems? Practices, meanings and identities of vegan restaurants in Porto Alegre, Brazil. *Journal of Rural Studies* 78: 304–313.

Niva, M., Mäkelä, J., Kahma, N. and Kjærnes, U. (2014). Eating sustainably? Practices and background factors of ecological food consumption in four Nordic countries. *Journal of Consumer Policy* 37(4): 465–484.

Norum, P. S. (2003). Examination of generational differences in household apparel expenditures. *Family and Consumer Sciences Research Journal* 32(1): 52–75.

Nosi, C., Zollo, L., Rialti, R. and Ciappei, C. (2020). Sustainable consumption in organic food buying behavior: the case of quinoa. *British Food Journal* 122(3): 976–994.

Ntanos, A., Skordoulis, M. and Ntanos, S. (2014). Millennial consumers' perceptions on the organic products. In: *Proceedings of the International Scientific Conference eRA-9, The SynEnergy Forum* ISSN-1791-1133. Piraeus, Greece. pp. 22–24.

Nuttavuthisit, K. and Thøgersen, J. (2017). The importance of consumer trust for the emergence of a market for green products: the case of organic food. *Journal of Business Ethics* 140(2): 323–337.

Oroian, C., Safirescu, C., Harun, R., Chiciudean, G., Arion, F., Muresan, I. and Bordeanu, B. (2017). Consumers' attitudes towards organic products and sustainable development: A case study of Romania. *Sustainability* 9(9): 1559.

Ostermann, C., Moyano, C. and Laufer, J. (2019). Brand positioning in higher education: the perception of Brazilian generational cohorts. *Review of Business Management* 20(3): 416–434.

Pairotti, M. B., Cerutti, A. K., Martini, F., Vesce, E., Padovan, D. and Beltramo, R. (2015). Energy consumption and GHG emission of the Mediterranean diet: a systemic assessment using a hybrid LCA-IO method. *Journal of Cleaner Production* 103: 507–516.

Paloviita, A. (2010). Consumers' sustainability perceptions of the supply chain of locally produced food. *Sustainability* 2(6): 1492–1509.

Pantano, E., Pizzi, G., Scarpi, D. and Dennis, C. (2020). Competing during a pandemic? Retailers' ups and downs during the COVID-19 outbreak. *Journal of Business Research* 116: 209–213.

Parry, E. and Urwin, P. (2011). Generational differences in work values: a review of theory and evidence: generational differences in work values. *International Journal of Management Reviews* 13(1): 79–96.

Perea, J. (2014). Organic beef farming in Spain: typology according to livestock management and economic variables. *Revista Cientifica-Facultad de Ciencias Veterinarias* 24(4).

Perry, B., Osbaldiston, R. and Henning, J. (2014). Tests of the validity and reliability of the community service attitudes scale. *Journal of College Student Development* 55(7): 726–731.

© Burleigh Dodds Science Publishing Limited, 2024. All rights reserved.

Pólya, É. and Máté, Z. (2021). The results of a primary research that investigated some aspects of conscious consumption among millennials and generation Z. *Academy of Strategic Management Journal* 20(6): 1–10.

Pomarici, E. and Vecchio, R. (2014). Millennial generation attitudes to sustainable wine: an exploratory study on Italian consumers. *Journal of Cleaner Production* 66: 537–545.

Popkin, B. M., Barclay, D. V. and Nielsen, S. J. (2005). Water and food consumption patterns of U.S. Adults from 1999 to 2001. *Obesity Research* 13(12): 2146–2152.

Popp, A., Lotze-Campen, H. and Bodirsky, B. (2010). Food consumption, diet shifts and associated non-CO2 greenhouse gases from agricultural production. *Global Environmental Change* 20(3): 451–462.

Pradhan, P., Kriewald, S., Costa, L., Rybski, D., Benton, T. G., Fischer, G. and Kropp, J. P. (2020). Urban food systems: how regionalization can contribute to climate change mitigation. *Environmental Science and Technology* 54(17): 10551–10560.

Priporas, C.-V., Kamenidou, I., Kapoulas, A. and Papadopoulou, F. M. (2015). Counterfeit purchase typologies during an economic crisis. *European Business Review* 27(1): 2–16.

Průša, P. and Sadílek, T. (2019). Green consumer behavior: the case of Czech consumers of generation Y. *Social Marketing Quarterly* 25(4): 243–255.

Reipurth, M. F. S., Hørby, L., Gregersen, C. G., Bonke, A. and Perez Cueto, F. J. A. (2019). Barriers and facilitators towards adopting a more plant-based diet in a sample of Danish consumers. *Food Quality and Preference* 73: 288–292.

Reisch, L. (2010). A definition of sustainable food consumption. In Policy Meets Research Workshop on Sustainable Food Consumption, Wien, Austria.

Reisch, L., Eberle, U. and Lorek, S. (2013). Sustainable food consumption: an overview of contemporary issues and policies. *Sustainability: Science, Practice and Policy* 9(2): 7–25.

Reisenwitz, T. H. and Iyer, R. (2009). Differences in generation X and generation Y: implications for the organization and marketers. *Marketing Management Journal* 19(2): 91–103.

Rödiger, M. and Hamm, U. (2015). How are organic food prices affecting consumer behaviour? A review. *Food Quality and Preference* 43: 10–20.

Rogler, L. H. (2002). Historical generations and psychology: the case of the Great Depression and World War II. *American Psychologist* 57(12): 1013–1023.

Román, S., Sánchez-Siles, L. M. and Siegrist, M. (2017). The importance of food naturalness for consumers: results of a systematic review. *Trends in Food Science and Technology* 67: 44–57.

Rosegrant, M. W., Ringler, C. and Zhu, T. (2009). Water for agriculture: maintaining food security under growing scarcity. *Annual Review of Environment and Resources* 34(1): 205–222.

Salas-Salvadó, J. and Papandreou, C. (2020). The Mediterranean diet: history, concepts and elements. In: *The Mediterranean Diet*. Available at: https://linkinghub.elsevier.com/retrieve/pii/B9780128186497000011. Accessed 23 March 2022. Elsevier. pp.3–11.

Sarti, S., Darnall, N. and Testa, F. (2018). Market segmentation of consumers based on their actual sustainability and health-related purchases. *Journal of Cleaner Production* 192: 270–280.

Scalvedi, M. L. and Saba, A. (2018). Exploring local and organic food consumption in a holistic sustainability view. *British Food Journal* 120(4): 749–762.

© Burleigh Dodds Science Publishing Limited, 2024. All rights reserved.

Schewe, C. D. and Meredith, G. (2004). Segmenting global markets by generational cohorts: determining motivations by age. *Journal of Consumer Behaviour* 4(1): 51–63.

Schewe, C. D., Meredith, G. E. and Noble, S. M. (2000). Defining moments: segmenting by cohorts. *Marketing Management* 9(3): 48–53.

Seconda, L., Baudry, J., Allès, B., Hamza, O., Boizot-Szantai, C., Soler, L. G., Galan, P., Hercberg, S., Lairon, D. and Kesse-Guyot, E. (2017). Assessment of the sustainability of the Mediterranean diet combined with organic food consumption: an individual behaviour approach. *Nutrients* 9(1): 61.

Serra-Majem, L., Ribas, L., Ngo, J., Ortega, R. M., García, A., Pérez-Rodrigo, C. and Aranceta, J. (2004). Food, youth and the Mediterranean diet in Spain. Development of KIDMED, Mediterranean Diet Quality Index in children and adolescents. *Public Health Nutrition* 7(7): 931–935.

Severo, E. A., Guimarães, J. C. F. de, Brito, L. M. P. and Dellarmelin, M. L. (2018). Environmental sustainability and sustainable consumption: the perception of baby Boomers, generation X and Y in Brazil. *Revista de Gestão Social e Ambiental* 11(3): 92–110.

Siatitsa, D. (2021). Youth housing in a context of socio-economic insecurity: the case of Greece. Κοινωνική Πολιτική 14: 149–162.

Siegrist, M., Visschers, V. H. M. and Hartmann, C. (2015). Factors influencing changes in sustainability perception of various food behaviors: results of a longitudinal study. *Food Quality and Preference* 46: 33–39.

Singh, A. P. and Dangmei, J. (2016). Understanding the generation Z: the future workforce. *South-Asian Journal of Multidisciplinary Studies* 3(3): 1–5.

Smith, S. and Paladino, A. (2010). Eating clean and green? Investigating consumer motivations towards the purchase of organic food. *Australasian Marketing Journal* 18(2): 93–104.

Spector, P. E. (1992). *Summated Rating Scale Construction: an Introduction Sage*. Newbury Park, CA: Sage Publications, Inc.

Stevanin, S., Voutilainen, A., Bressan, V., Vehviläinen-Julkunen, K., Rosolen, V. and Kvist, T. (2020). Nurses' generational differences related to workplace and leadership in two European countries. *Western Journal of Nursing Research* 42(1): 14–23.

Strauss, W. and Howe, N. (2020). Generation, Z (on-line). Available at: http://incomeresult.com/generation-z/#cite_note-McCrindleAU-27. Accessed 7 August 2020.

Strauss, W. and Howe, N. (1991). *Generations: the History of America's Future, 1584 to 2069*. New York: Quill.

Su, C. H., Tsai, C. H., Chen, M. H. and Lv, W. Q. (2019). US sustainable food market generation Z consumer segments. *Sustainability* 11(13): 3607.

Sütterlin, B., Brunner, T. A. and Siegrist, M. (2011). Who puts the most energy into energy conservation? A segmentation of energy consumers based on energy-related behavioral characteristics. *Energy Policy* 39(12): 8137–8152.

Szczebyło, A., Halicka, E., Rejman, K. and Kaczorowska, J. (2022). Is eating less meat possible? Exploring the willingness to reduce meat consumption among millennials working in polish cities. *Foods* 11(3): 358.

Tanner, C. and Wölfing Kast, S. (2003). Promoting sustainable consumption: determinants of green purchases by Swiss consumers: promoting sustainable consumption. *Psychology and Marketing* 20(10): 883–902.

Thøgersen, J. (2010). Country differences in sustainable consumption: the case of organic food. *Journal of Macromarketing* 30(2): 171–185.

© Burleigh Dodds Science Publishing Limited, 2024. All rights reserved.

Thøgersen, J. (2017). Sustainable food consumption in the nexus between national context and private lifestyle: A multi-level study. *Food Quality and Preference* 55: 16–25.

Tobarra, M. A., López, L. A., Cadarso, M. A., Gómez, N. and Cazcarro, I. (2018). Is seasonal households' consumption good for the nexus carbon/water footprint? The Spanish fruits and vegetables case. *Environmental Science and Technology* 52(21): 12066–12077.

Tobler, C., Visschers, V. H. M. and Siegrist, M. (2011). Eating green. Consumers' willingness to adopt ecological food consumption behaviors. *Appetite* 57(3): 674–682.

Torjusen, H., Lieblein, G., Wandel, M. and Francis, C. A. (2001). Food system orientation and quality perception among consumers and producers of organic food in Hedmark County, Norway. *Food Quality and Preference* 12(3): 207–216.

Trichopoulou, A., Costacou, T., Bamia, C. and Trichopoulos, D. (2003). Adherence to a Mediterranean diet and survival in a Greek population. *New England Journal of Medicine* 348(26): 2599–2608.

Trudel, R. (2019). Sustainable consumer behavior. *Consumer Psychology Review* 2(1): 85–96.

Tsakiridou, E., Mattas, K., Tsakiridou, H. and Tsiamparli, E. (2011). Purchasing fresh produce on the basis of food safety, origin, and traceability labels. *Journal of Food Products Marketing* 17(2–3): 211–226.

Tscharntke, T., Grass, I., Wanger, T. C., Westphal, C. and Batáry, P. (2021). Beyond organic farming: harnessing biodiversity-friendly landscapes. *Trends in Ecology and Evolution* 36(10): 919–930.

Twine, R. (2018). Materially constituting a sustainable food transition: the case of vegan eating practice. *Sociology* 52(1): 166–181.

UK Parliament (2011). Sustainable food (on-line). Available at: https://publications.parliament.uk/pa/cm201011/cmselect/cmenvaud/writev/food/m22.htm. Accessed 31 December 2021.

United Nations (2015). Transforming our world: the 2030 Agenda for Sustainable Development (on-line). Available at: https://sdgs.un.org/2030agenda.

United Nations (2019). *World Population Prospects Highlights, 2019 Revision Highlights, 2019 Revision*.

Van Loo, E. J., Hoefkens, C. and Verbeke, W. (2017). Healthy, sustainable and plant-based eating: perceived (mis)match and involvement-based consumer segments as targets for future policy. *Food Policy* 69: 46–57.

Vanham, D., Comero, S., Gawlik, B. M. and Bidoglio, G. (2018). The water footprint of different diets within European sub-national geographical entities. *Nature Sustainability* 1(9): 518–525.

Verain, M. C., Dagevos, H. and Antonides, G. (2015). Sustainable food consumption. Product choice or curtailment? *Appetite* 91: 375–384.

Verain, M. C. D., Snoek, H. M., Onwezen, M. C., Reinders, M. J. and Bouwman, E. P. (2021). Sustainable food choice motives: the development and cross-country validation of the Sustainable Food Choice Questionnaire (SUS-FCQ). *Food Quality and Preference* 93: 104267.

Vecchio, R., Annunziata, A., Krystallis, A. and Pomarici, E. (2015). Consumers' literacy and preferences for sustainability labels: an exploratory analysis on Italian young adults. *International Journal of Globalisation and Small Business* 7(3/4): 221–233.

© Burleigh Dodds Science Publishing Limited, 2024. All rights reserved.

Vilceanu, M. O., Grasso, O. and Johnson, K. (2019). Bridging the gap between public opinion research and consumer marketing research: insights into US shoppers of organic foods. In: Association of Marketing Theory and Practice Proceedings (vol. 2019). p. 7.

Wang, O. and Scrimgeour, F. (2021). Willingness to adopt a more plant-based diet in China and New Zealand: applying the theories of planned behaviour, meat attachment and food choice motives. *Food Quality and Preference* 93: 104294.

Wey Smola, K. and Sutton, C. D. (2002). Generational differences: revisiting generational work values for the new millennium. *Journal of Organizational Behavior* 23(4): 363–382.

Willett, W., Rockström, J., Loken, B., Springmann, M., Lang, T., Vermeulen, S., Garnett, T., Tilman, D., DeClerck, F., Wood, A., Jonell, M., Clark, M., Gordon, L. J., Fanzo, J., Hawkes, C., Zurayk, R., Rivera, J. A., De Vries, W., Majele Sibanda, L., Afshin, A., Chaudhary, A., Herrero, M., Agustina, R., Branca, F., Lartey, A., Fan, S., Crona, B., Fox, E., Bignet, V., Troell, M., Lindahl, T., Singh, S., Cornell, S. E., Srinath Reddy, K., Narain, S., Nishtar, S. and Murray, C. J. L. (2019). Food in the Anthropocene: the EAT–Lancet Commission on healthy diets from sustainable food systems. *The Lancet* 393(10170): 447–492.

Williams, K. C. and Page, R. A. (2011). Marketing to the generations. *Journal of Behavioral Studies in Business* 3(1): 37–53.

Wolburg, J. M. and Pokrywczynski, J. (2001). A psychographic analysis of generation Y College students. *Journal of Advertising Research* 41(5): 33+: 33–52.

Wong, K. T.-W., Zheng, V. and Wan, P.-S. (2021). Local versus national identity in Hong Kong, 1998–2017. *Journal of Contemporary Asia* 51(5): 803–827.

Wright, S. D., Caserta, M. and Lund, D. A. (2003). Older adults' attitudes, concerns, and support for environmental issues in the "new west". *International Journal of Aging and Human Development* 57(2): 151–179.

Wu, X., Xiong, J., Yan, J. and Wang, Y. (2021). Perceived quality of traceability information and its effect on purchase intention towards organic food. *Journal of Marketing Management* 37(13–14): 1267–1286.

Xiao, L., Liu, J. and Ge, J. (2021). Dynamic game in agriculture and industry cross-sectoral water pollution governance in developing countries. *Agricultural Water Management* 243: 106417.

Xie, X., Huang, L., Li, J. J., Zhu, H. (2020). Generational differences in perceptions of food health/risk and attitudes toward organic food and game meat: the case of the COVID-19 crisis in China. *International Journal of Environmental Research and Public Health* 17(9): 3148.

Xu, Z., Chen, X., Wu, S. R., Gong, M., Du, Y., Wang, J., Li, Y. and Liu, J. (2019). Spatial-temporal assessment of water footprint, water scarcity and crop water productivity in a major crop production region. *Journal of Cleaner Production* 224: 375–383.

Zagata, L. (2012). Consumers' beliefs and behavioural intentions towards organic food. Evidence from the Czech Republic. *Appetite* 59(1): 81–89.

Zhang, B., Mulhern, F. J., Wu, Y., Xu, M., Wang, W. and Gao, L. (2022). Thirty years and "I'm still Lovin' it!": brand perceptions of McDonald's among generation Y and generation Z consumers in China. *Asia Pacific Journal of Marketing and Logistics* 34(5): 906–921.

© Burleigh Dodds Science Publishing Limited, 2024. All rights reserved.

Chapter 5

Behavioural change towards sustainable food consumption

Réka Rozsnyói and Getachew Abate Kassa, Technical University of Munich, Germany

1 Introduction

The topic of sustainability has been growing in relevance over the past years across various industries such as construction, energy generation and mobility, and the food industry is no exception. Farmers, manufacturers, retailers and consumers are increasingly becoming more aware of the food industry's significant impact on the environment, producers, animals and personal and planetary health. For example, it is more and more known and acknowledged that agriculture is a significant contributor to a set of environmental problems such as soil degradation, water pollution, eutrophication and biodiversity loss (Reisch et al., 2013, p. 12; Willett et al., 2019). However, environmental impact differs widely between product categories. Meat and meat products, for instance, make up 9–14% of consumers' food-related emission footprint followed by a variety of dairy products. Vegetables, fruits and cereals, on the other hand, are associated with lower impacts (Reisch et al., 2013, p. 12). In Germany, food accounts for 16% of emissions equalling the emission footprint of national mobility, while in the UK, this number is as high as 18% (Reisch et al., 2013, p. 13). In general, 20–30% of European households' environmental impact can be attributed to food consumption (Verain et al., 2021, p. 1).

http://dx.doi.org/10.19103/AS.2023.0129.16
© Burleigh Dodds Science Publishing Limited, 2024. All rights reserved.

Food consumption is additionally associated with important health concerns like obesity, malnutrition and premature deaths, and ethical considerations ranging from food security, through animal welfare to working conditions (Reisch et al., 2013, 13). The unequal distribution of inputs and outputs also raises different priorities in different localities concerning health and ethics. That is, while in certain areas the main issue might be malnutrition and food insecurity, for instance, in other places working conditions might be a primary concern. With this in mind, there is an urgent need to better understand synergies between the different aspects of food consumption to enable the transition towards more healthy and sustainable production, distribution and diets that are also more equitable on a global scale. One way of doing so is by looking at the consumer, the end point of food production and retail, who can be considered as a 'key agent for change towards a more sustainable food system' (Vittersø and Tangeland, 2015, 91). As such, the consumer may become an integral part of the sustainable food transition. It is important to note, however, that consumption patterns, just like wealth, are unequally distributed among countries and different socio-economic groups, meaning that 23% of the world's population consume more than 77% of its resources, generating around 80% of all pollution in the meantime (Peano et al., 2019, p. 3). Hence, this chapter will focus more closely on consumers in Western countries, who currently have a much larger potential to reduce their environmental footprint through changing their consumption behaviour, more specifically their diets in this case.

Sustainable food consumption (SFC) is fundamental to the sustainability of human society and to the realization of the Sustainable Development Goals (SDGs), in particular, SDG-2 (no poverty), SDG-3 (good health and wellbeing), SDG-6 (clean water and sanitation) and SDG-12 (responsible consumption and production). Different factors affect the transition towards SFC: (1) consumer behaviour towards sustainability, (2) regulations and policies that support the transition towards sustainability and (3) network, collaborations and changes within the food value chain. The academic literature and practical application of the SFC concept are still at an infant stage. For example, the German Sustainable Development Strategy (Federal Statistical Office of Germany, 2021) does not have clear indicators and targets on SFC. Currently, despite some policy instruments and strategies towards organic agriculture, cultural landscape programmes, and related ecosystem services, there is no clear picture of sustainability-related regulations, policies and strategies that stipulate the effects of these actions towards SFC. Hence, it is of utmost importance to understand what influences and motivates consumers in making sustainable food choices, so that effective measures that promote such behaviour can be devised and implemented.

This chapter attempts to briefly highlight the factors that affect the transition towards SFC considering the multidimensionality of food (e.g. environmental,

© Burleigh Dodds Science Publishing Limited, 2024. All rights reserved.

social, economic, health and ethical effects) and taking into consideration consumer perceptions, attitudes and behaviours towards SFC. These would provide additional knowledge and information towards consumer choice and intentions in light of the different aspects of SFC. The chapter combines knowledge from different disciplines and frameworks to contribute to a comprehensive overview of consumers' motives and behaviours towards SFC, the barriers they encounter along the way and the sustainability impacts of various food groups and diets. This is done by juxtaposing and synthesizing articles and studies with relevance to sustainable consumption (motives), consumer segmentation and the environmental impacts of diets in general, and certain food products in particular.

2 Defining sustainable food consumption

Before any meaningful analysis of SFC can take place, it is necessary to find a common definition as to what SFC actually entails. Vittersø and Tangeland trace the concept back to Agenda 21, the main policy output of the UN Earth Summit in Rio in 1992, which advocates for consumption patterns within housing, mobility and food that 'reduce environmental stress and meet the basic needs of humanity' (2015, p. 92). Since then, multiple definitions have evolved, each with its own set of focus areas. Table 1 summarizes selected authors' and organizations' understanding of sustainable food and/or food consumption.

For the sake of this chapter, SFC will be defined broadly and will be understood as purchasing behaviour falling into any of the above categories. The reason for this is that the majority of this analysis focuses on consumer attitudes and behaviour instead of the actual impact of their beliefs. The gap between these two will also be discussed and bridged in Sections 4 and 5. Even though the sustainability benefits of organic food consumption are debated, this will be considered as a sustainable behaviour in this chapter, given that consumers do believe in the environmental benefits of organic production practices (Funk et al., 2021, p. 174).

3 Consumer food choices and motives towards sustainable food consumption

The Theory of Planned Behaviour (TPB) is one of the well-established models applied to understand and analyse food choice and consumption behaviours using constructs, including the concepts of perceived behavioural control (PBC), subjective norm and attitude (Ajzen, 1991; Bauer et al., 2018). This theory argues that behaviour is preceded by behavioural intentions, which are strongly dependent on one's attitudes and subjective norms. Additionally, behaviour is further influenced by PBC both directly and indirectly through its impact on behavioural intention (Madden et al., 1992). Yet, in practice attitudes

© Burleigh Dodds Science Publishing Limited, 2024. All rights reserved.

Table 1 Selected authors' and organizations' understanding of sustainable food and/or food consumption

Author	Definition or main aspects of sustainable consumption/food
UN Earth Summit in Vittersø and Tangeland (2015, p. 92)	'Reduce environmental stress and meet the basic needs of humanity'
UK Sustainable Development Commission in Reisch et al. (2013, p. 8)	'Safe, healthy and nutritious for consumers in shops, restaurants, schools, hospitals, and so forth; can meet the needs of the less well off at a global scale; provides a viable livelihood for farmers, processors, and retailers whose employees enjoy a safe and hygienic working environment; respects biophysical and environmental limits in its production and processing while reducing energy consumption and improving the wider environment; respects the highest standards of animal health and welfare compatible with the production of affordable food for all sectors of society; and supports rural economies and the diversity of rural culture, in particular by emphasizing local products that minimize food miles'
Lairon (2012, p. 32)	Based on 'agro-ecological staple food production including limited animal husbandry, short-distance production-consumption nets, minimal food processing and refining, important culinary skills, diet and nutrition education, and firm links to positive traits of ancestral local cultures as well as appropriate use of recent technology tools'
International Scientific Symposium on Biodiversity and Sustainable Diets United Against Hunger (2010, p. 10)	'Sustainable diets are those diets with low environmental impacts which contribute to food and nutrition security and to healthy life for present and future generations. Sustainable diets are protective and respectful of biodiversity and ecosystems, culturally acceptable, accessible, economically fair and affordable; nutritionally adequate, safe and healthy; while optimizing natural and human resources'
Eberle et al. in Reisch et al. (2013, p. 8)	Compatible with people's lifestyles by being 'feasible', available, affordable and accessible, while allowing for sociocultural diversity
Verain et al. (2021, p. 2)	'Produced, processed, packaged, transported and traded with respect for people, animals and the environment, without compromising future generations'

mostly fail to comprehensively explain behavioural intention, which also rarely leads to actual behaviour in a straightforward manner (Aertsens et al., 2009, p. 1142; Campbell-Arvai et al., 2014, p. 454; Dowd and Burke, 2013, p. 142; Terlau and Hirsch, 2015; Vermeir and Verbeke, 2006).

Various authors studied the gaps between these stages and found that individual values, perceived consumer effectiveness[1] and involvement are helpful mediators between attitudes, behavioural intention and behaviour (Aertsens et al., 2009; Dowd and Burke, 2013; Peeters et al., 2022; Terlau and

[1] "The extent to which the consumer believes that his personal efforts can contribute to the solution of a problem". (Vermeir and Verbeke, 2006, p. 175).

© Burleigh Dodds Science Publishing Limited, 2024. All rights reserved.

Hirsch, 2015; Vermeir and Verbeke, 2006). As a result, one set of values were identified to be connected to sustainable behaviour (universalism, benevolence, self-direction, honesty, idealism, equality, freedom and responsibility) and another set to less sustainable or conventional behaviour (power, hedonism, tradition, security, conformity, ambition) (de Jonge and van Trijp, 2013, p. 636; Vermeir and Verbeke, 2006, p. 174). At the same time, Aertsens et al. (2009) created an integrated framework based on TPB, emotions and an almost identical set of values compared to those listed earlier to analyse the relations between values, attitudes, societal and personal norms and sustainable purchasing behaviour. Similar to the former findings, they reserved conformity and power as indicators of conventional behaviour and self-direction as that of sustainable behaviour. However, they linked security (through health) and hedonism (through taste) as the strongest values to the purchasing of organic products, which stands in opposition to the finding of de Jonge and van Trijp (2013) and Vermeir and Verbeke (2006). These differences possibly give weight to the observation that the values connected with sustainable behaviour might vary among countries (Aertsens et al., 2009, p. 1145; Rahman and Luomala, 2021).

Beyond values, Aertsens et al. (2009, p. 1147) found that anticipated emotions can also indirectly affect behaviour through desires and intentions. Especially, anticipated guilt is considered to be an impactful 'pro-social' emotion that can induce conservatory behaviour and can exert a direct influence on moral norm, attitude and PBC. While personal norms exert a positive influence on behavioural intention, subjective norms are poor predictors and yet to be unanimously evaluated (Aertsens et al., 2009, p. 1148). Additionally, it might be useful to look at (positive) moral attitude and self-identity as further predictors of behavioural intention, independent of the above-mentioned factors (Dowd and Burke, 2013, p. 142).

Involvement was found to further influence behavioural intention and even actual behaviour by affecting the time and effort dedicated to information search and decision-making and therefore, impacting the formation of beliefs, attitudes and intentions, as well as the resulting behaviour (e.g. variety seeking, brand-switching, frequency of product use and shopping enjoyment) (Vermeir and Verbeke, 2006, p. 174, 184). Hence, if consumers can be convinced of the benefits of sustainable consumption, they might become more involved and their intention to buy sustainable products might increase. While a comparable relationship might be established for perceived consumer effectiveness, with consumers with higher PCE being more likely to buy sustainable food products, PCE seems to be resistant to manipulation and offer little help in influencing consumers' purchase intentions (Tobler et al., 2011, p. 679; Vermeir and Verbeke, 2006, p. 184).

© Burleigh Dodds Science Publishing Limited, 2024. All rights reserved.

The intention-behaviour gap can, in addition, be attributed to perceived and actual behavioural control (how easy or difficult it is to obtain the specific product), which is contingent on convenience, price and a product's availability, among other possible factors (Aertsens et al., 2009, p. 1150; Vermeir and Verbeke, 2006, p. 175). Disbelief, habits and situational factors such as promotions, time pressure and distractions tend to further inhibit intentions from translating into action (Campbell-Arvai et al., 2014, p. 466; Mäkiniemi and Vainio, 2014; Peeters et al., 2022; Vermeir and Verbeke, 2006, p. 173). In their study, Peeters et al. (2022) confirmed this and overcame the situational obstacles by successfully designing an environment where consumers were encouraged to reflect on their values during the decision-making process, which took place outside of the retail context. This resulted in a reduction of unsustainable meat choices from 18% within the control group to 1% among study participants (Peeters et al., 2022, p. 6).

Because of the various discrepancies pointed out earlier, further consideration of food choice motives is viewed as one of the critical factors in explaining SFC behaviours (Baudry, et al., 2017; Dowd and Burke, 2013). Environmental motives (e.g. avoidance for environmental reasons), taste, price, health and safety, convenience, quality, freshness (local and seasonality of products), natural content and ethical concerns have been identified as some of the relevant dimensions to measure food choice motives (Baudry, et al., 2017; Verain, et al., 2021). It is important to note here that whether a certain motive supports the purchase of sustainable food products or conventional ones depends on consumers' perceptions of the products. For example, while sustainable products are associated with better taste by one set of consumers, they are perceived as less tasty by other set of consumers (Verain et al., 2021, p. 2). Moreover, not all food choice motives are equally supported by each study, indicating that their relevance is highly dependent on target group or product category, for example (Dowd and Burke, 2013, p. 142).

Like values, food choice motives are also argued to be culture-dependent (Rahman and Luomala, 2021; Verain et al., 2021), making any generalizations across the globe challenging, especially since the majority of SFC-related studies have been conducted in Europe or in the United States. In Europe, Verain et al. found that consumers perceive different sustainability motives quite similarly in five countries, namely the Netherlands, Denmark, Czechia, France and Italy. They found that the consumers who studied in these countries generally observe two dimensions of sustainability: one pertaining to environmental, ethical and animal welfare concerns and the other to locality and seasonality. While environmental, ethical and animal welfare might be perceived as separate dimensions (e.g. in policy) and consumers might be able to differentiate between them on a theoretical level, this distinction often disappears in practice when the actual purchase decision is taking place. In

© Burleigh Dodds Science Publishing Limited, 2024. All rights reserved.

terms of local and seasonal, it could be argued that this dimension is redundant and instead could be part of the environmental dimension. However, there might be merit in making the separation, as focusing on local and seasonal can be a valuable tool to translate the abstract idea of sustainability into a more concrete, tangible and relatable concept through bringing it closer in time and distance to the consumer and their community (Verain et al., 2021).

A study by Chekima et al. focused on actual consumption of organic food instead of behavioural intentions and found that product-specific attitude, sensory appeal (e.g. taste) and health orientation are important considerations for Malaysian consumers (Chekima et al., 2017, p. 13). This is in line with the motivations of European consumers stated earlier. An interesting contribution of this study was to consider future orientation as a factor strengthening the effects of product-specific attitude and health orientation (Chekima et al., 2017, p. 14). Another study contrasting the motivations of Finnish consumers with those from Pakistan also found significant similarities between the motivations of consumers from these two distant localities. They were all motivated to purchase organic food by 'life goals' pertaining to environmental friendliness, health, naturalness, absence of chemicals, better life quality, security, fun, pleasure and enjoyment seeking, and to gain peace of mind, self-respect and self-fulfilment (Rahman and Luomala, 2021, p. 473). The authors concluded that although some of these motives can be linked to culturally typical life goals, they are mostly universal. The most important distinction lies rather between self-centred and other-centred reasons, neither of which can be associated more strongly with one country over another (Rahman and Luomala, 2021, p. 476-7).

Self-Determination Theory looks at the psychological needs (for competence, relatedness and autonomy) that affect the 'what' and 'why' behind people's goals (Deci and Ryan, 2000, p. 228). The theory establishes different motivational categories (e.g. intrinsic, extrinsic and amotivation) that have varying implications on consumers' (sustainable) food choices. For instance, consumers with intrinsic and internalized motivation in Schösler et al.'s study were more likely to choose plant-based protein products, and those with internalized motivation, even if intrinsic motivation was low, purchased organic or free-range meet more frequently (2014, p. 66). Extrinsically motivated or amotivated consumers were mostly younger, less educated males with larger families who preferred instant products and meals and did not reflect often on their food choices (Schösler et al., 2014, p. 67). In contrary, some extrinsically motivated participants in Sahelices-Pinto et al.'s study were characterized by the strongest motivations to purchase organic products, even if these did not derive from their own convictions (2021, p. 454). It must be emphasized, therefore, that extrinsic motivation is not inherently contrary to SFC. If external expectations guide the individual

© Burleigh Dodds Science Publishing Limited, 2024. All rights reserved.

towards sustainability, extrinsic motivational structures have the potential to significantly promote SFC (Sahelices-Pinto et al., 2021, p. 454; Schösler et al., 2014, p. 67).

4 Sustainable production and health impact of dietary composition

During the past decades, consumption patterns and nutrition issues have gained prominence in the discussion on sustainable development as food insecurity, malnutrition and micro-nutrient deficiencies are among the key challenges at the global level with more than 820 million people not having enough food, resulting in protein-energy undernutrition as well as micronutrient deficiencies (FAO et al., 2019). At the same time, even more people consume too much food of poor quality, resulting in overweight and obesity. Often these forms of malnutrition coexist in the so-called double burden of disease, particularly in societies in the nutrition transition. The Global Burden of Disease Study 2017 (Lancet, 2018) hence concluded that unhealthy diets pose a greater risk to morbidity and mortality. In the past few years, responses to the double burden nutrition challenges included attempts to increase public awareness on healthy diet, interventions to change the food environment, quality standards, generic healthy eating campaigns and nutrition education. In particular, pursuance of nutrition-sensitive agriculture (NSA) has been viewed as an effective approach to deal with nutrition issues, especially in developing countries (Ruel, et al., 2018). Although many publications have highlighted the potential of food system to influence nutrition on a large scale, methodologies and tools for understanding these linkages are limited.

Fortunately, sustainable diets are one way in which individuals and communities can translate awareness of the problem of endangered health/nutrition and environment into action. Therefore, another big part of the literature has focused on the environmental impacts of different diets varying in their composition of ingredients (Foresight, 2011; Garnett, 2011; Jungbluth et al., 2000). For example, there is a broad consensus about the environmental benefits of diets comprising less animal-based products (meat and dairy) and more plant-based products (Baroni et al., 2007; Garnett, 2011; Jungbluth et al., 2000; Willett et al., 2019). Many studies have found that consuming beef has the single largest environmental footprint among all food categories, especially in terms of greenhouse gas emissions, followed by other animal-source foods like different meat products and dairy – together these products are responsible for three-quarter of climate change effects observed today (Baroni et al., 2007, p. 283; Willett et al., 2019, p. 471). In contrast, cereals/grains, fruits and vegetables are generally more environment friendly, regardless of whether measured per unit weight,

© Burleigh Dodds Science Publishing Limited, 2024. All rights reserved.

serving, unit of energy or protein weight (Reisch et al., 2013, p. 12; Willett et al., 2019, p. 470). However, there are a few exceptions: rice, for example, is associated with high methane emissions and staple crops like wheat, white rice and certain cereals are argued to be responsible for a third to a half of pressures on specific environmental domains (Foresight, 2011; Willett et al., 2019, p. 471). The impacts of seafood vary significantly depending on production method (Willett et al., 2019, p. 470). Even though some product categories inherently produce more emissions, it is important to highlight that production and transportation methods have a major role to play in the final footprint of any given product. The region of production and the production and distribution system are of significance here as land-use change, growing produce under heated glass, air-freighting and refrigeration, for instance, are all associated with outstanding emission factors (Foresight, 2011, p. 134). Hence, it is not sufficient to only monitor what one consumes, but also the source of the product and circumstances under which it reached their plate must be considered (Jungbluth et al., 2000).

This creates further complications for SFC. Although reasons and motivations for SFC were examined earlier, it was not discussed whether the belief of consumers in the sustainability of specific product categories is justified. For example, organic is widely accepted by consumers as a sustainable alternative to conventional food production, yet research has been ambiguous about this topic. While there seems to be a consensus that organic production is more conscious of environmental resources, the use of fertilizer and animal welfare, it is at the same time argued to be an inefficient production method with lower yields and higher spatial requirements compared to conventional agriculture (Antonides, 2022, p. 171; Rahman and Luomala, 2021, p. 465; Reisch et al., 2013, p. 12). Moreover, organic certification does not extend beyond the farm gate and, thereby, fails to account for the environmental impacts of the transportation of the certified product (Jungbluth et al., 2000; Vittersø and Tangeland, 2015, p. 97).

Finally, it is essential to examine the health impacts of different food categories in parallel with their environmental impacts. For example, beef is not only the most emission-intensive food product, but red meat in general is also associated with strong adverse health effects and is documented to be a relevant diet-related cause of premature deaths in Europe, North America, East and South-East Asia (Malley et al., 2021, p. 18). In comparison, when it comes to fruits and vegetables, the lack of these foods in one's diet is what poses a significant health risk in South Asia and sub-Saharan Africa (Malley et al., 2021, p. 18). Recommendations stemming from these facts might prescribe a reduction in red meat consumption and an increase of fruits and vegetables in the diets of affected groups, which are also key levers for climate change mitigation (Garnett, 2011; Willett et al., 2019).

© Burleigh Dodds Science Publishing Limited, 2024. All rights reserved.

5 Diversity and heterogeneity within sustainable food consumption

As highlighted in previous sections, social influence, habit formation, individual self, feelings and cognition and tangibility are some of the factors to consider in understanding and differentiating sustainable consumer behaviours (White et al., 2019, p, 23.). Similarly, intrinsic (e.g. environmental responsibilities, gaining knowledge and willingness to resource conservation and reducing the environmental impact) and extrinsic factors (e.g. product pricing, quality performance, safety and promotion) are factors that affect consumers' sustainable food purchase decisions (Nekmahmud and Fekete-Farkas, 2020). Verain et al. (2012) indicate that the consumer segment for local and seasonal products could be different from the segment that generally focuses on environmental and social sustainability of food products. In a study that assesses consumer preferences for ethically produced chicken meat products, Escobedo del Bosque et al. (2021) identified three consumer segments: (1) price-sensitive consumers, (2) price-sensitive and origin-oriented consumers and (3) origin-oriented consumers. This suggests the need for consideration of heterogeneity in consumer behaviour (e.g. Dolnicar and Grün, 2009) in the importance they attach to specific sustainability-related components and in the type of sustainable behaviours they perform. Identifying and understanding consumer segments with similar intentions and behaviours would thus help strategize and positioning of appropriate sustainable products in a given local and regional market.

Generally, researchers tend to group consumers across studies into segments similar to the following: (1) consumers with the most frequent and pronounced sustainable behaviour (e.g. 'sustainable activists', 'doers' and 'consequent pro-environmental'), (2) consumers that care about the environment and sustainability to a given degree, but do not necessarily act on it as strongly as the first group for a variety of reasons (e.g. 'sustainable believers', 'conscious' and 'origin-focused food savers') and (3) consumers that do not care much about the environment or even have a negative attitude towards the concept of sustainability (e.g. 'moderates', 'reluctant' and 'renouncement-aversive') (Funk et al., 2021; Gazdecki et al., 2021; Su et al., 2019). Out of these three groups, the second (or middle) one is usually the largest and most heterogeneous and can sometimes be subdivided into additional segments (de Jonge and van Trijp, 2013; Funk et al., 2021; Verain et al., 2012).

The most significant segmentation variables are usually centred around attitudes (and values), lifestyle and behaviour (involvement) and comprise environmental concern or attitude towards the environment and sustainability, knowledge about and interest in sustainability and sustainable consumption and purchase frequency of products deemed sustainable (and/or frequency

© Burleigh Dodds Science Publishing Limited, 2024. All rights reserved.

of meat consumption) (Gazdecki et al., 2021, pp. 22–23; Verain et al., 2012, p. 127). Members of each segment share many characteristics in terms of their attitudes towards and frequency of sustainable purchasing behaviour. However, these similarities might be targeted towards a variety of different sustainability practices. For example, Funk et al. (2021), studying different types of environment-friendly consumers considering food waste management, identified four in-between segments that could all be placed under the umbrella of the second segment above, while having distinctive motives and preferences regarding sustainability. As such, two groups attributed great importance to minimizing food waste, but each of them followed different behaviours to act on their conviction. 'Origin-focused food savers' aimed to minimize food waste by using the food themselves and avoiding leftovers, while 'food waste reducing sharers' tackled the same issue by purchasing suboptimal products and products close to or past the expiration date and by sharing food with their environment – a practice strongly rejected by origin-focused food savers (Funk et al., 2021). In the same way, these two groups had different food motivations as well. The former placed outstanding importance on origin, transportation and seasonality, while the latter was most concerned about taste (Funk et al., 2021, p. 183)

Another possible segmentation might be made based on price: between price-sensitive and non-price sensitive consumers as well as between price-sensitive consumers that care about the environment and those that do not (de Jonge and van Trijp, 2013; Verain et al., 2012, p. 130). Although demographics are regularly considered when analysing consumer segments, a correlation between such data and actual consumption behaviour is contested (Schösler et al., 2014, p. 65; Verain et al., 2012, p. 126). For instance, while some studies find that younger people are more likely to be environmentally conscious (Funk et al., 2021; Gazdecki et al., 2021), in others they were largely immune to the issue of sustainability (Vanhonacker et al., 2013). Again, others found no significant impact of age on sustainable behaviour (Lea and Worsley, 2005; Schösler et al., 2014, p. 65). Similarly, higher income does not seem to have an unequivocal positive or negative relationship with one's sustainable behaviour (Aertsens et al., 2009, p. 1157; Schösler et al., 2014, p. 65); although it was often associated with higher levels of environmental concern (Gazdecki et al., 2021), it was also characteristic of ambiguous and frequent meat consumers (Funk et al., 2021). In the same manner, education does not appear to have a straightforward correlation with environmental behaviour (Aertsens et al., 2009, p. 1152; Funk et al., 2021; Gazdecki et al., 2021; Schösler et al., 2014, p. 65; Vanhonacker et al., 2013). Nonetheless, gender seems to have a correlation with sustainable behaviour, with women being more likely to purchase organic and other climate-friendly food compared to men (Aertsens et al., 2009; Hughner et al., 2007; Mäkiniemi and Vainio, 2014), more likely to care about

© Burleigh Dodds Science Publishing Limited, 2024. All rights reserved.

the environment and sustainability (Su et al., 2019, p. 9) and more likely to reduce meat consumption or become vegetarian or vegan (Campbell-Arvai et al., 2014, p. 462; Haas et al., 2019, p. 5; Tobler et al., 2011, p. 678). Overall, there can be multiple different directions consumers take in their attempt to be more sustainable. Some might buy organic products, others might focus on the regionality of products or food certifications and still others might reduce their meat and dairy consumption. Because of this, it is essential that each segment is encouraged and supported in its SFC through adequate accessibility and availably of the desired sustainable products and practices.

6 Conclusion

People's behaviour and SFC are highly complex phenomena that can only be analysed by involving a multiplicity of factors. Therefore, this chapter aimed at highlighting the most influential components of individual's attitudes, motives and behaviours towards SFC, and the key environmental and health impacts of dietary compositions, as well as the various consumer segments that make up the totality (non-)sustainable consumers. In general, sustainable consumers seem to respond to similar values and food choice motives and experience similar barriers to SFC. They are typically driven by health motives, environmental motives, naturalness, taste, quality, freshness and hindered by prices, appearance and conservation. Perceived low availability and, therefore, lack of time to seek out sustainable products, as well as limited knowledge and awareness about sustainable food can further deter SFC. While pro-environmental consumers commonly hold values such as universalism, benevolence, self-direction, honesty, idealism, equality, freedom and responsibility, their non-sustainable counterparts are more likely to have a high regard for power, hedonism, tradition, security, conformity and ambition. It must be remembered, however, that this is not always the case and, just like with values, the results might vary between different countries, cultures or even product categories.

In terms of environmental impact, diets with less (or no) meat and dairy products tend to be more environment friendly and are often associated with better health outcomes as well. Beyond food groups, the origin, production method and transportation of food products, among other factors, must also be accounted for in order to create a thorough analysis of the actual footprint of a given product. Additionally, while organic foods are often regarded as more sustainable by consumers, further considerations need to be made to evaluate these claims.

Finally, based on their values, motivations and behaviours, consumers can be grouped into different segments that can be loosely described as one with the most outstanding sustainability performance, one with the least (or

© Burleigh Dodds Science Publishing Limited, 2024. All rights reserved.

lack of) sustainable behaviour and one (or many) in between. While socio-demographic factors might not be decisive segmentation criteria, gender seems to be a strong predictor of SFC, with women being more likely to follow environment-friendly diets.

7 Where to look for further information

- Rocha, A. and Neto, B. (2022). *Sustainable Food Production and Consumption*. https://doi.org/10.3390/books978-3-0365-4854-8.
- Meixner, O., Riefler, P. and Schanes, K. (2021). *Sustainable Consumer Behavior and Food Marketing*. https://doi.org/10.3390/books978-3-0365-2595-2.

8 References

Aertsens, J., Verbeke, W., Mondelaers, K. and Van Huylenbroeck, G. (2009). Personal determinants of organic food consumption: A review. *British Food Journal* 111(10), 1140–1167. https://doi.org/10.1108/00070700910992961.

Ajzen, I. (1991). The theory of planned behavior. *Organizational Behavior and Human Decision Processes* 50(2), 179–211.

Baudry, J., Péneau, S., Allès, B., Touvier, M., Hercberg, S., Galan, P., Amiot, M. J., Lairon, D., Méjean, C. and Kesse-Guyot, E. (2017). Food choice motives when purchasing in organic and conventional consumer clusters: Focus on sustainable concerns (the NutriNet-Santé cohort Study). *Nutrients* 9(2), 88. https://doi.org/10.3390/nu9020088.

Bauer, D., Arnold, J. and Kremer, K. (2018). Consumption-intention formation in education for sustainable development: An adapted model based on the theory of planned behavior. *Sustainability* 10(10), 3455. https://doi.org/10.3390/su10103455.

Baroni, L., Cenci, L., Tettamanti, M. and Berati, M. (2007). Evaluating the environmental impact of various dietary patterns combined with different food production systems. *European Journal of Clinical Nutrition* 61(2), Article 2. https://doi.org/10.1038/sj.ejcn.1602522.

Campbell-Arvai, V., Arvai, J. and Kalof, L. (2014). Motivating sustainable food choices: The role of nudges, value orientation, and information provision. *Environment and Behavior* 46(4), 453–475. https://doi.org/10.1177/0013916512469099.

Chekima, B. and Oswald, A. I., Wafa, S. A. W. S. K. and Chekima, K. (2017). Narrowing the gap: Factors driving organic food consumption. *Journal of Cleaner Production*, 166, 1438–1447. https://doi.org/10.1016/j.jclepro.2017.08.086.

de Jonge, J. and van Trijp, H. C. M. (2013). Meeting heterogeneity in consumer demand for animal welfare: A reflection on existing knowledge and implications for the meat sector. *Journal of Agricultural and Environmental Ethics*, 26(3), 629–661. https://doi.org/10.1007/s10806-012-9426-7.

Deci, E. L. and Ryan, R. M. (2000). The "what" and "why" of goal pursuits: Human needs and the self-determination of behavior. *Psychological Inquiry* 11(4), 227–268. https://doi.org/10.1207/S15327965PLI1104_01.

© Burleigh Dodds Science Publishing Limited, 2024. All rights reserved.

Dolnicar, S. and Grün, B. (2009). Environmentally friendly behavior: Can heterogeneity among individuals and contexts/environments be harvested for improved sustainable management? *Environment and Behavior* 41(5), 693–714. https://doi.org/10.1177/0013916508319448.

Dowd, K. and Burke, K. J. (2013). The influence of ethical values and food choice motivations on intentions to purchase sustainably sourced foods. *Appetite* 69, 137–144. https://doi.org/10.1016/j.appet.2013.05.024.

Escobedo del Bosque, C. I., Spiller, A. and Risius, A. (2021). Who wants chicken? Uncovering consumer preferences for produce of alternative chicken product methods. *Sustainability* 13, 2440. https://doi.org/10.3390/su13052440.

FAO, IFAD, UNICEF, WFP and WHO (2019). *The State of Food Security and Nutrition in the World 2019*. Safeguarding against economic slowdowns and downturns. FAO, Rome.

Federal Statistical Office of Germany (2021). *Sustainable Development in Germany*. Indicator Report (vol. 2021).

Foresight (2011). *The Future of Food and Farming: Challenges and Choices for Global Sustainability*. The Government Office for Science. Available at: https://www.gov.uk/government/publications/future-of-food-and-farming.

Funk, A., Sütterlin, B. and Siegrist, M. (2021). Consumer segmentation based on Stated environmentally-friendly behavior in the food domain. *Sustainable Production and Consumption* 25, 173–186. https://doi.org/10.1016/j.spc.2020.08.010.

Garnett, T. (2011). Where are the best opportunities for reducing greenhouse gas emissions in the food system (including the food chain)? *Food Policy* 36, S23–S32. https://doi.org/10.1016/j.foodpol.2010.10.010.

Gazdecki, M., Goryńska-Goldmann, E., Kiss, M. and Szakály, Z. (2021). Segmentation of food consumers based on their sustainable attitude. *Energies* 14(11), Article 11. https://doi.org/10.3390/en14113179.

Haas, R., Schnepps, A., Pichler, A. and Meixner, O. (2019). Cow milk versus plant-based milk substitutes: A comparison of product image and motivational structure of consumption. *Sustainability* 11(18), Article 18. https://doi.org/10.3390/su11185046.

Hughner, R. S., McDonagh, P., Prothero, A., Shultz, I. I., C. J. and Stanton, J. (2007). Who are organic food consumers? A compilation and review of why people purchase organic food. *Journal of Consumer Behaviour* 6(2–3), 94–110. https://doi.org/10.1002/cb.210.

International Scientific Symposium on Biodiversity and Sustainable Diets United Against Hunger (2010). FAO Headquarters, Rome.

Jungbluth, N., Tietje, O. and Scholz, R. W. (2000). Food purchases: Impacts from the consumers' point of view investigated with a modular LCA. *The International Journal of Life Cycle Assessment* 5(3), 134–142. https://doi.org/10.1007/BF02978609.

Lairon, D. (2012). Biodiversity and sustainable nutrition with a food-based approach. In: Burlingame, B. and Dernini, S. (Eds), *Sustainable Diets and Biodiversity. Directions and Solutions for Policy, Research and Action*, 30–35. Food and Agriculture Organization. Available at: https://www.fao.org/policy-support/tools-and-publications/resources-details/en/c/522860/.

Lancet (2018). *Changes in Health in the Countries of the UK and 150 English Local Authority Areas 1990–2016: A Systematic Analysis for the Global Burden of Disease Study 2016*, vol 392. www.thelancet.com.

© Burleigh Dodds Science Publishing Limited, 2024. All rights reserved.

Madden, T. J., Ellen, P. S. and Ajzen, I. (1992). A comparison of the theory of planned behavior and the theory of reasoned action. *Personality and Social Psychology Bulletin* 18(1), 3–9. https://doi.org/10.1177/0146167292181001.

Malley, C. S., Hicks, W. K., Kulyenstierna, J. C. I., Michalopoulou, E., Molotoks, A., Slater, J., Heaps, C. G., Ulloa, S., Veysey, J., Shindell, D. T., Henze, D. K., Nawaz, O., Anenberg, S. C., Mantlana, B. and Robinson, T. P. (2021). Integrated assessment of global climate, air pollution, and dietary, malnutrition and obesity health impacts of food production and consumption between 2014 and 2018. *Environmental Research Communications* 3(7), 075001. https://doi.org/10.1088/2515-7620/ac0af9.

Nekmahmud, M. and Fekete-Farkas, M. (2020). Why not green marketing? Determinates of consumers' intention to green purchase decision in a new developing nation. *Sustainability* 12(19), 7880; https://doi.org/10.3390/su12197880.

Peano, C., Merlino, V. M., Sottile, F., Borra, D. and Massaglia, S. (2019). Sustainability for food consumers: Which perception? *Sustainability* 11(21), Article 21. https://doi.org/10.3390/su11215955.

Peeters, A.-L., van der Werff, E. and Tromp, N. (2022). Designing for value-behaviour consistency: Ethical choice architecture to stimulate sustainable meat purchase. *Cleaner and Responsible Consumption* 5, 100067. https://doi.org/10.1016/j.clrc.2022.100067.

Rahman, S. U. and Luomala, H. (2021). A comparison of motivational patterns in sustainable food consumption between Pakistan and Finland: Duties or self-reliance? *Journal of International Food and Agribusiness Marketing* 33(5), 459–486. https://doi.org/10.1080/08974438.2020.1816243.

Reisch, L., Eberle, U. and Lorek, S. (2013). Sustainable food consumption: An overview of contemporary issues and policies. *Sustainability: Science, Practice and Policy* 9(2), 7–25. https://doi.org/10.1080/15487733.2013.11908111.

Ruel, M. T., Quisumbinga, A. R. and Balagamwala, M. (2018). Nutrition-sensitive agriculture: What have we learned so far? *Global Food Security* 17, 128–153.

Schösler, H., de Boer, J. and Boersema, J. J. (2014). Fostering more sustainable food choices: Can self-determination theory help? *Food Quality and Preference* 35, 59-69. https://doi.org/10.1016/j.foodqual.2014.01.008.

Terlau, W. and Hirsch, D. (2015). Sustainable consumption and the attitude-behaviour-gap phenomenon–Causes and measurements towards a sustainable development. *International Journal on Food System Dynamics* 6(3), Article 3. https://doi.org/10.18461/ijfsd.v6i3.634.

Vanhonacker, F., Van Loo, E. J., Gellynck, X. and Verbeke, W. (2013). Flemish consumer attitudes towards more sustainable food choices. *Appetite* 62, 7–16. https://doi.org/10.1016/j.appet.2012.11.003.

Verain, M. C. D., Bartels, J., Dagevos, H., Sijtsema, S. J., Onwezen, M. C. and Antonides, G. (2012). Segments of sustainable food consumers: A literature review. *International Journal of Consumer Studies* 36(2), 123–132. https://doi.org/10.1111/j.1470-6431.2011.01082.x.

Verain, M. C. D., Snoek, H. M., Onwezen, M. C., Reinders, M. J. and Bouwman, E. P. (2021). Sustainable food choice motives: The development and cross-country validation of the Sustainable Food Choice Questionnaire (SUS-FCQ). *Food Quality and Preference* 93, 104267. https://doi.org/10.1016/j.foodqual.2021.104267.

© Burleigh Dodds Science Publishing Limited, 2024. All rights reserved.

Vermeir, I. and Verbeke, W. (2006). Sustainable Food Consumption: Exploring the Consumer "Attitude-Behavioral Intention" Gap. *Journal of Agricultural and Environmental Ethics* 19, 169–194. https://doi.org/10.1007/s10806-005-5485-3.

Vittersø, G. and Tangeland, T. (2015). The role of consumers in transitions towards sustainable food consumption. The case of organic food in Norway. *Journal of Cleaner Production* 92, 91–99. https://doi.org/10.1016/j.jclepro.2014.12.055.

White, K., Habib, R. and Hardisty, D. J. (2019). How to SHIFT consumer behaviors to be more sustainable: A literature review and guiding framework. *Journal of Marketing* 83(3), 22–49.

Willett, W., Rockström, J., Loken, B., Springmann, M., Lang, T., Vermeulen, S., Garnett, T., Tilman, D., DeClerck, F., Wood, A., Jonell, M., Clark, M., Gordon, L. J., Fanzo, J., Hawkes, C., Zurayk, R., Rivera, J. A., De Vries, W., Majele Sibanda, L., Afshin, A., Chaudhary, A., Herrero, M., Agustina, R., Branca, F., Lartey, A., Fan, S., Crona, B., Fox, E., Bignet, V., Troell, M., Lindahl, T., Singh, S., Cornell, S. E., Srinath Reddy, K., Narain, S., Nishtar, S. and Murray, C. J. L. (2019). Food in the anthropocene: The EAT–Lancet Commission on healthy diets from sustainable food systems. *The Lancet* 393(10170), 447–492. https://doi.org/10.1016/S0140-6736(18)31788-4.

© Burleigh Dodds Science Publishing Limited, 2024. All rights reserved.

Part 2

Product attributes

Chapter 6

Understanding consumer perceptions and attitudes towards nutrition labels and health claims in food

Joe Bogue and Lana Repar, University College Cork, Ireland

1 Introduction

Making healthy food choices has an important role to play in one's personal health and well-being and in the prevention of chronic diseases, such as heart disease, obesity, cancers and type 2 diabetes. Consumers' food choices and behaviours, in addition to sustainable choices, are extremely complex phenomena and there are many factors, related to the person themselves (age, gender, and education), their environment (family life stage, degree of urbanization, and size of household) and the food (taste, texture, and

http://dx.doi.org/10.19103/AS.2023.0129.09
© Burleigh Dodds Science Publishing Limited, 2024. All rights reserved.

appearance) that influence those choices. For packaged foods, what is on the label and label design has a significant influence on consumer purchase decisions and repeat purchases, and this can, in turn, influence the healthy or unhealthy choices made by consumers. Nutrition labels and health claims (NLHCs) on packaged foods are important instruments used by food and health policymakers in seeking to promote healthy food choices and reduce the consumption of unhealthy foods that influence the levels of obesity and occurrence of non-communicable diseases (NCDs) in society.

This chapter examines current research around NLHCs and how they influence the consumer now, and in the future, and also shape consumer behaviour, government and food and health policymakers' guidelines and product reformulations by food and beverage companies. The role of NLHCs in informing and influencing consumers' dietary behaviours will be examined as well as the effectiveness of labels in influencing purchase decisions and how consumers assess health claims. To identify contemporary issues in consumer perceptions and attitudes towards NLHCs, a systematic literature review (SLR) was conducted using a PRISMA approach, which focused on articles published between 2015 and 2022, from 20 leading journals, to synthesize the key research themes and findings. This SLR provides a roadmap and state of the art in relation to the existing body of research work on understanding consumer perceptions and attitudes towards NLHCs on packaged foods.

2 Overview of nutrition labelling and health claims

Nutrition-related NCDs are on the rise due to an increase in the consumption of cheap, energy dense and poor nutrient foods, coupled with more sedentary modern lifestyles (Cavaliere et al., 2015; Mandle et al., 2015). Therefore, the communication of relevant, and clear information, through NLHCs is an important strategy by food and health policymakers in terms of encouraging consumers' healthy food choices and positive dietary behaviours. Nutrition information on food labels is a significant source of information for consumers, though the importance and usage vary by consumers and their backgrounds, e.g. the extent of their nutrition knowledge, education, interest in healthy eating behaviours, willingness to pay for packaged foods with health labels and their socio-economic characteristics. Consumers' perceptions, understanding and attitudes towards NLHCs will influence the usage of NLHCs and purchase behaviour in relation to healthy foods. The effectiveness of NLHCs will also depend on optimization of label design and phrasing of claims. Figure 1 depicts a conceptual framework for a multidisciplinary approach to NLHCs research.

The modern food consumer is exposed to an information overload and is bombarded with many nutrition and health claims messages in relation to

© Burleigh Dodds Science Publishing Limited, 2024. All rights reserved.

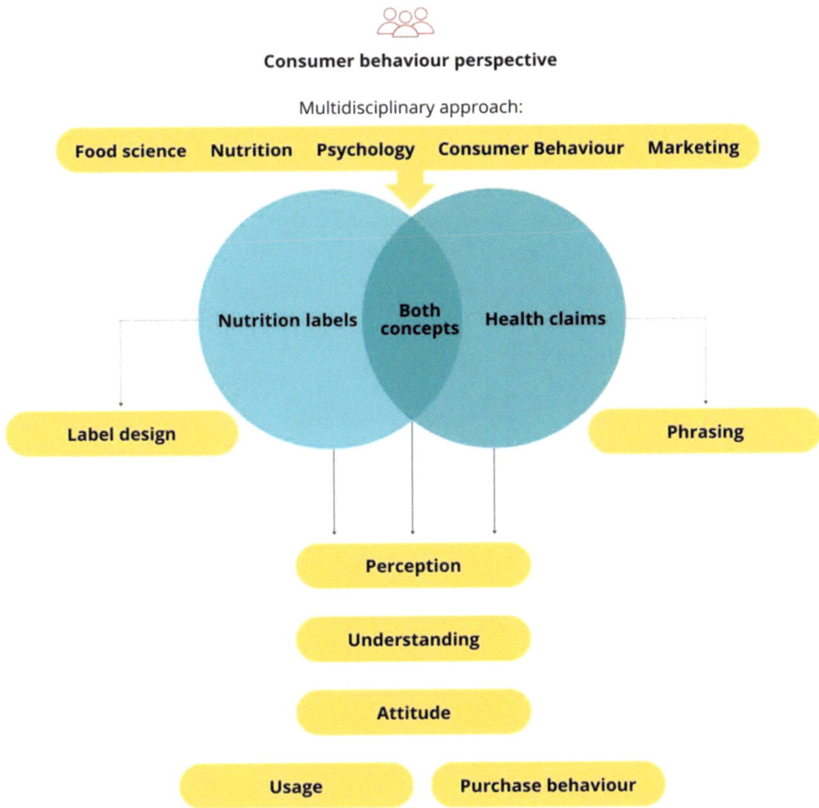

Figure 1 Conceptual framework for nutrition labels and health claims research.

packaged foods. However, NLHCs, with clear substantiated information, can help food and beverage companies distinguish their packaged foods in the marketplace and give them a competitive advantage, as they promote the health benefits of packaged foods, and also novel foods such as functional foods. Food labels can also provide information on food allergies that affect a significant number of children and adults globally and they can provide important warnings about the possible presence of proteins causing the allergies in foodstuffs or processing facilities (Saab, 2021).

Nutrition labelling can fall into two categories: back-of pack (BOP) and front-of-pack (FOP) labels (Mandle et al., 2015). According to Benson et al. (2019, p.1) *'NHCs are text or images on food or drinks which highlight nutritional or energy content and (for health claims) their impact on health'*. They go on to explain further that nutritional claims indicate that a food has a beneficial nutritional property, while a health claim suggests or implies a relationship exists between

© Burleigh Dodds Science Publishing Limited, 2024. All rights reserved.

food and health. According to the US Food and Drug Administration (2022), *'Health claims describe a relationship between a food substance (a food, food component, or dietary supplement ingredient), and reduced risk of a disease or health-related condition'*.

In the European Union, NLHCs on packaged foods are legislated by Commission Regulation (EC) 1924/2006(1) and Statutory Instrument No. 11 of 2014(2). These pieces of legislation ensure that any claim made on a food label is clear, accurate and substantiated, allowing consumers to make informed decisions (Donovan et al., 2020). NLHCs should not be false or misleading, or encourage excessive consumption of foods, and there is a list of the permitted nutrition claims with their conditions of use in the Annex of Regulation 1924/2006 (FSAI, 2021). Nutrition labelling among many requirements must be expressed per 100 g or per 100 ml and must be presented in tabular form, in a legible way, and in a conspicuous location (FSAI, 2021). In the United States, NLHCs are regulated by the FDA and there are three categories of claims that are defined by statute and/or FDA regulations for use on food and dietary supplement labels: health claims, nutrient content claims and structure/function claims (FDA, 2022). In this chapter, we refer to nutrition labels as different ways of providing nutrition information (e.g. mandatory BOP nutrition facts or voluntary FOP traffic light

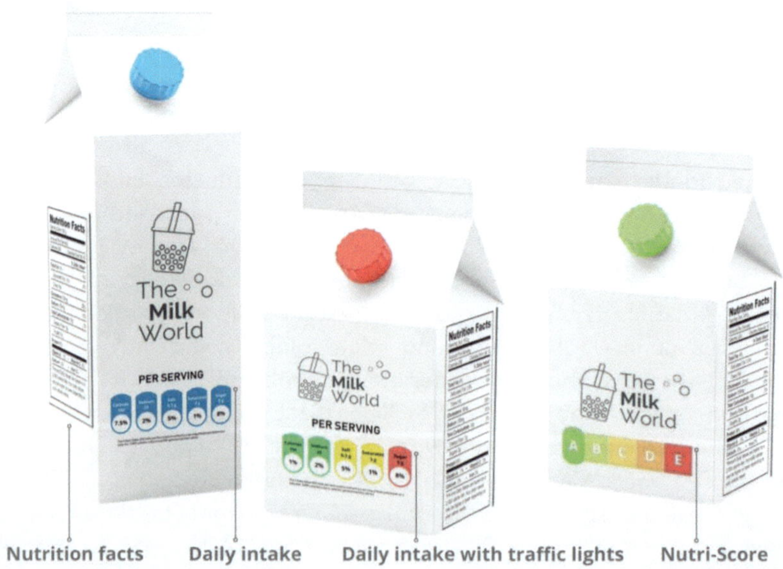

Figure 2 An example of different types of nutrition label designs.

© Burleigh Dodds Science Publishing Limited, 2024. All rights reserved.

labels) while health claims relate to any link to health benefits. Figure 2 depicts some of the designs used to communicate nutrition information to consumers. The impact of NLHCs on consumers is further explored in this chapter.

3 Developing a systematic literature review of nutrition labels and health claims research

To get an objective and holistic overview of consumer perceptions and attitudes towards NLHCs on packaged foods and to unlock current research, a systematic literature review (SLR) was undertaken by the authors. SLRs are becoming increasingly significant outside medical sciences, where they were initially developed, and today have a wide variety of different approaches and protocols. An SLR is a systematic, explicit and reproducible method for identifying, evaluating and synthesizing the existing body of completed and recorded work on a specific topic, produced by researchers (Fink, 2014). High-quality SLRs help researchers to objectively synthesize the literature under review and give direction for policymakers to make better evidence-based decisions and are a powerful form of research (Kraus et al., 2020). More robust SLRs use PRISMA (Preferred Reporting Items for Systematic Reviews and Meta-Analyses) protocols but there are also adjusted, less time-consuming SLR versions that are tailored according to research needs. Some SLRs take into consideration a specified time frame and all relevant papers published within this period, with the application of inclusion and exclusion criteria, while in other cases specific journals might be selected for a review.

The argument for selecting specific journals for an SLR is that such journals are highly specialized and high ranking and therefore trustworthy and generally believed to accept rigorously peer-reviewed papers. They feature papers that are widely cited, timely, relevant and add a significant contribution to the body of knowledge (Kraus et al., 2020; Betini et al., 2014). The selection of such journals serves as an additional filter for ensuring that an SLR is capturing state-of-the-art and most important recent developments on a selected research topic. In addition, there are relatively less duplicates, depending on the number of search strings, which can make the review quicker without compromising on rigour. However, it can be challenging to collect a meaningful number of relevant papers from selected journals, leading to a need to either increase the number of journals or expand the time frame, or both.

In this chapter on NLHCs, the aim was to answer the following research questions: (1) *What are the key research themes that emerged between 2015 and 2022 related to NLHCs from a consumer's perspective?* (2) *What are the key findings and how do they relate to consumer usage of NLHCs and ultimately change dietary behaviour?* (3) *What methodologies have been used to address*

© Burleigh Dodds Science Publishing Limited, 2024. All rights reserved.

Table 1 Search strategy for a systematic literature review of nutrition labels and health claims

Criteria	Inclusion criteria	Exclusion criteria	Search strings
Journals	Highly ranked in the area of food science and technology	Outside of the identified 20 journals; no peer-review, no empirical study; not deemed relevant; full-text not available	TITLE/KEYWORDS/ABSTRACT **For nutrition labels:** • (nutrition* OR food*) AND label* • package* • ('front-of-pack' OR 'front of pack' OR 'front of package' OR 'front-of-package' OR 'front of packaging' OR 'FOP') • symbol* • sign* • (guide* OR guideline*) • fact* **For health claims:** • food* AND health* AND • claim* • statement* • symbol* • sign*
'Other methods' search	Highly relevant peer-reviewed papers and journals		
Time frame	2015–22	Earlier than 2015	
Type	Original primary research articles, with quantitative and/or qualitative study completed	Reviews, meta-analyses, reports, books, editorials, comments	
Language	English	Non-English	
Area	Food and consumers	Not including food and consumers explicitly	
Category	Nutrition labels and health claims	Not including either nutrition labels or health claims	

the research on NLHCs? (4) *How will future trends impact on NLHCs and the food consumer?*

To answer the research questions, an SLR search strategy was devised for NLHCs (Table 1). When selecting journals to be included in this SLR on NLHCs, the following criteria were taken into account. First, three websites with journal rankings were consulted, focusing in general on the 'food' area: (1) Journal Citation Reports for the category 'food science and technology' (looking at a new metric Journal Citation Indicator) which provides a single journal-level metric normalized for different fields of research and their publications); (2) Scimago Journal & Country Rank for the category 'food science' (metrics based on Scopus® data as of April 2022); and (3) Google Scholar for the category 'life sciences & earth sciences' and subcategory 'food science & technology'. The identified journals were listed and discussed based on their rank (e.g. top three ranked journals), appearance across different lists and the journal's relevance for the NLHCs topic. Second, two non-exhaustive exploratory searches with keywords *nutrition label* and *health claim*, with a time frame set to 2010–22, were conducted in Google Scholar, Web of Science, Science Direct, JStor, Scopus, EBSCO and PubMed databases to capture the key journals. Finally, the

© Burleigh Dodds Science Publishing Limited, 2024. All rights reserved.

authors discussed which potential journals were considered relevant from a consumer behaviour perspective.

Initially, 20 journals were selected from the lists based on three websites, two database searches and authors' discussions. An exploratory search of the 20 journals' volumes was undertaken using *nutrition AND label AND food* and *health AND claim AND food* keywords for the time frame 2015–22 and for 'research article' type to gauge the scope of available papers. In the final step, 20 journals were confirmed based on the availability of papers, ranking, relevance for the NLHCs topic and authors' perceived importance of the journals in the context of consumer behaviour (see Fig. 3 for selected journals). The selected journals had a minimum of 10 returned records (two search strings combined) and a total of 16 475 records were returned in the exploratory search, indicating solid potential for contribution to the SLR. A multidisciplinary approach to this SLR is reflected in a diversity of disciplines covered by the 20 journals selected: food science, nutrition, consumer behaviour, marketing and psychology. To ensure that the SLR scope was inclusive enough and to avoid too limited recall,

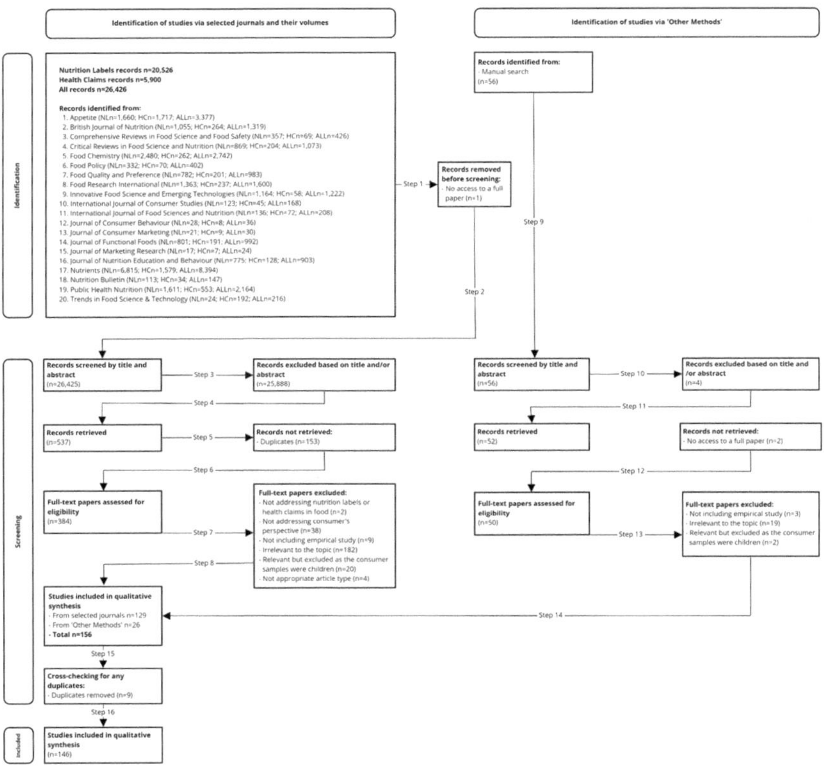

Figure 3 PRISMA diagram for search articles.

© Burleigh Dodds Science Publishing Limited, 2024. All rights reserved.

several journals and papers that were deemed relevant were included in the SLR as 'Other methods' (Snyder, 2019). The records were retrieved by the 10th of July 2022.

The main tool for the literature search and management was EndNote 20, while the Excel (Version, 2108), MIRO and SankeyMATIC software were used to support data visualization. The search strings included truncation, Boolean separators and keywords that were identified by examining the titles and keywords in a few dozen articles related to the topic of interest. Search strings were typed in a search box of EndNote 20 online search option in Web of Science Core Collection database for each journal separately, using *Title/Keywords/Abstract* search field descriptor (the exceptions were *Innovative Food Science & Emerging Technologies*, *Journal of Nutrition Education and Behaviour* and *Trends in Food Science & Technology* journals which were searched partially through their webpage). Search strings for the nutrition labels were typed first (seven different strings) and then the health claims strings (four different strings).

The PRISMA diagram captures the number of records found for each search in detail (Fig. 3). This SLR specifically included research articles that addressed issues in adult or adolescent consumers' perceptions and attitudes towards NLHCs. From the 26426 records identified through the selected journals, the majority (77.7%) addressed nutrition labels. The lower number of articles addressing health claims possibly indicates scope for more research that could explore consumers' understanding and acceptance of health claims on packaged foods.

One hundred and twenty-nine studies from the 20 journals selected and 26 studies identified through a manual search ('Other Methods') were checked for duplicates with a total of 146 studies included in the final qualitative analysis. Other studies were excluded as they were not relevant to the topic, e.g. the focus was too product-specific, or studies included a particular group of participants with social or health conditions (n = 201), did not address consumers' perspectives (n = 38), included children as participants, which was not the focus of this SLR (n = 22), or did not include an empirical study (n = 12).

The Sankey diagram follows the selected studies and summarizes the key characteristics and research trends in NLHCs based on the SLR, as defined by the parameters outlined earlier in this chapter (Fig. 4). The methodologies used by researchers in this SLR often included an experiment or a set of experiments with stimuli, which varied from physical food products with differentiated (and very creative and realistic) mock packaging to images displayed on a computer and a virtual grocery store experience. Eye-tracking was noted in several studies as a tool that was particularly useful to record consumers' gaze, duration and fixation when it came to BOP and FOP labels. Another design that was utilized were within- and between-subject design experiments. The selected studies

© Burleigh Dodds Science Publishing Limited, 2024. All rights reserved.

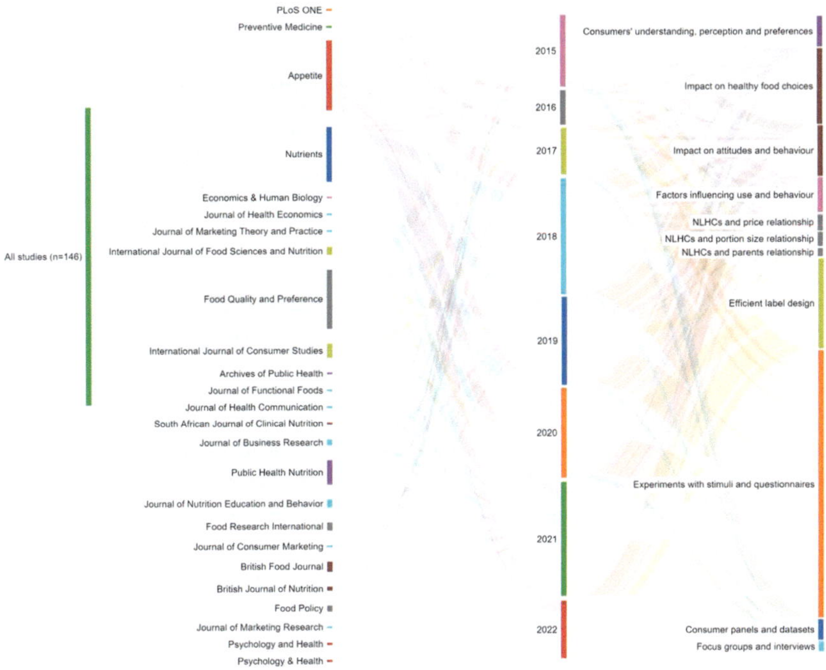

Figure 4 Sankey diagram summarizing research trends and methodologies in nutrition labels and health claims.

mostly employed a quantitative survey in the form of a questionnaire with choice and ranking tasks, in-person and in controlled laboratory conditions or online, while qualitative methods such as focus groups and intercept interviews were less common. The profile of participants in the studies included consumers in general, and in some cases parents with kids and college students were also targeted. Several studies used consumer panel data sets with recorded purchases. Sampling techniques were usually through a convenience sampling or highly structured recruitment, supported by an external marketing agency with set parameters. Sample sizes in the quantitative studies varied, from n = 24 to n = 535 000, depending on whether the participants were individuals, households or consumer panels and on the type of experiment or survey conducted. The selected studies either focused on one country or several countries and NLHCs were researched in 41 different countries in this SLR, with the United States, United Kingdom, Australia, Uruguay, Italy and Canada featuring most frequently. The top journals where the studies were published were *Appetite* (n = 34), *Food Quality and Preference* (n = 29), *Nutrients* (n = 27), *Public Health Nutrition* (n = 12) and *International Journal of Consumer Studies* (n = 7). Chronologically, the highest number of studies were published in 2018 (n = 27) and 2021 (n = 27), followed by a growth during 2019 (n = 21) and 2020

© Burleigh Dodds Science Publishing Limited, 2024. All rights reserved.

(n = 22), while by July 2022 there were 14 published studies included in this SLR. In the earlier years (2015 n = 16; 2016 n = 8; 2017 n = 11), less studies were recorded, which indicates an increased interest over the following years in the NLHCs' research topic.

The selected studies are further synthesized and presented by themes and the full list of studies and their details is located in the Appendix. The following sections address the main research issues and findings on NLHCs.

4 Key research areas: consumers' understanding, perceptions and preferences for nutrition labels and health claims

How consumers understand, perceive and assess NLHCs is an important research area in terms of how they influence usage and purchase decisions in relation to packaged foods. This is particularly true in relation to the development and commercialization of new foods with innovative functional ingredients. More effective NLHCs can increase usage by consumers and help reduce the economic and social costs linked to diet-related illnesses. It is generally accepted that NLHCs influence consumer perceptions of healthy packaged foods and can increase the healthiness perceptions of specific packaged foods. According to Benson et al. (2019), NLHCs have an influence on perceptions of food and consumer behaviour and Leathwood et al. (2007) note they are potentially powerful tools to convey information on food characteristics to consumers and facilitate well-formed food choices.

Selected research studies have explored consumers' understanding, perceptions and preferences for NLHCs and, to various degrees, they helped consumers change their eating habits and impacted their product selection. Hassan and Dimassi (2017) found that the food label knowledge average score was 63.1% in their sample of Lebanese shoppers, prompting suggestions that the low knowledge score necessitates nutrition education on how to read and use labels. Koen et al. (2018a) recorded even lower average food and nutrition label knowledge average score (44.4%) for consumers in South Africa. For Chinese consumers, Liu et al. (2015) reported a moderate degree of subjective understanding and a low degree of objective understanding of food labels and 70% of the participants stated they rarely, or never, use nutrition labels when food shopping. Hutton and Gresse (2022) assessed the objective understanding of FOP labels in South Africa and concluded that the labels significantly improved participants' ability to identify healthier food products. When it comes to understanding health claims on extra-virgin olive oil, Lombardi et al. (2021) highlighted that only 36% of Italian respondents understood the meaning of the authorized health claims.

© Burleigh Dodds Science Publishing Limited, 2024. All rights reserved.

Benson et al. (2019) further uncovered that the knowledge of consumers in Northern Ireland and Republic of Ireland when it came to particular claims (such as 'low in fat') was basic and superficial, and participants cited nutrition and health claims together, as well as official and unregulated claims (such as 'gluten free' and 'organic'), when asked for examples for both, which suggested that the poor knowledge might be linked to feelings of confusion or information overload. In addition, as suggested by Mazzu et al. (2021a), who looked at the effect of labels across seven European countries, there was a significant effect in the interaction between country and FOP labels regarding comprehensibility and liking, meaning food and health policymakers should take into consideration comparative studies when promoting food education and FOP labels.

As noted by van Buul and Brouns (2015), nutrition and health claims are mostly only perceived positive by a specific target market to help them achieve their health goals. Some studies demonstrated that consumers had positive perceptions and showed an interest in including nutritional information and health-warning labels on food products, and they assigned greater importance to health-warning labels, citing health as a crucial driver for taking warning labels into consideration (Ares et al., 2018b; Annunziata et al., 2016). In Slovenia, research revealed that consumers were moderately doubtful of NLHCs, but when they chose a category, such as breakfast cereals, the NLHCs were more important than whether visual images were present (Klopčič et al., 2020). Other perceptions from consumers' side included a belief that certain types of individuals or groups (on a diet or managing their weight, eating healthily, with specific illnesses or conditions, parents with children) were most likely to benefit from the use of NLHCs; labels supported purchase decisions and contributed to the definition of healthier choices when they were perceived as useful and easy to use, yet the conflicting impact on purchase intentions was also noted (labels influence/do not influence purchase); and views that products with certain claims likely had ingredients being substituted (e.g. reduction in fat led to an increase in sugar) and might be bland or poor in taste (Mazzu et al., 2022; Benson et al., 2019). In terms of trust, studies found that consumers were strongly aware that labels were controlled by law and therefore a general belief was they must be true. However, negative perceptions were identified in cases where consumers who did not trust the political or economic system also did not accept foods with NLHCs, or the labels were seen as a form of marketing and consumers were unable to verify their accuracy (Klopčič et al., 2020; Benson et al., 2019; Menger-Ogle and Graham, 2018). Across different countries, beliefs in NLHCs varied and influenced consumers' attitudes to foods that carried NLHCs.

© Burleigh Dodds Science Publishing Limited, 2024. All rights reserved.

The NLHC uptake and efficiency will ultimately depend on consumers' acceptance. Consumers preferred claims that showed a reduced presence of unhealthy substances, rather than those that highlighted the presence of healthy ones, and the highest part-worth utility was observed for the claim that directly communicated health benefits ('Protects your health') (Miklavec et al., 2016; Gracia and Barreiro-Hurle, 2019). Consumers in the study by Koen et al. (2018a) preferred less complex terminology, more pictures/colour, a single health endorsement logo and bigger font sizes. Pichierri et al. (2021) further emphasized the importance of text clarity, as they found that clearer health claims were perceived as more useful and could increase product attractiveness and consumers' purchase intentions.

Factors impacting consumers' understanding and use of the NLHCs were identified broadly as socio-economic and lifestyle variables and nutrition knowledge levels. Consumers paying attention to, and using nutrition labels, and with higher capacity to identify healthier products were females, young adults, with tertiary education, singles, employed, physically active and non-smokers (Packer et al., 2021; Egnell et al., 2018a; Cheah et al., 2015). In contrast, Hassan and Dimassi (2017) stated that older, obese shoppers having kids, suffering from chronic illness or allergies, following a specific diet and residing in big cities, scored significantly higher on food label knowledge. Food labelling was more useful for specific consumer segments such as athletes, consumers with health conditions, those concerned about healthy lifestyles and weight, and those with high compensatory beliefs (Tonnesen et al., 2022; Moreira et al., 2019). Similarly, Cavaliere et al. (2015) in their study of Italian consumers found two different profiles of consumers who were interested in nutrition or health-related claims on packaged foods. Nutrition claims were of interest to those who focused on household well-being such as females, families with young children and those with a higher nutritional knowledge. Consumers who were older, with limited income, and with health conditions were more interested in health claims.

Some studies interestingly concluded that socio-economic variables were not significant predictors of consumers reading labels, self-reported use or probability of purchasing labelled foods (Bryła, 2020; Smed et al., 2017; Liu et al., 2015). Instead, nutrition knowledge was found to play a key role in accurate label reading and impacting how much consumers believed in claims (Andrews et al., 2021; Benson et al., 2019; Soederberg Miller and Cassady, 2015). Nevertheless, Steinhauser et al. (2019a) noted that consumers with higher nutrition knowledge and/or higher health motivations looked longer at claims, but this did not show an effect on the purchase decision.

© Burleigh Dodds Science Publishing Limited, 2024. All rights reserved.

5 Key research areas: nutrition labels and health claims and how they influence healthy choices

One of the key issues in relation to NLHCs is what information is to be communicated to consumers and how it influences consumers' healthy eating choices. NLHCs can help consumers identify healthier food choices and can have an influence on consumption patterns (Packer et al., 2021; Benson et al., 2018). Irish research indicated that consumers were more heavily influenced by health claims compared to nutrition claims, with 78% of participants stating that health claims influenced purchases (Donovan et al., 2020).

Studies that were selected for the SLR showed mixed evidence of the NLHCs' impact on consumers' healthy choices and unearthed some mechanisms that might be hindering the selection of healthier foods. Nutrition information and warnings showed the potential to encourage consumers to make healthier choices and boosted the intention to avoid processed foods (Adasme-Berríos et al., 2022; Packer et al., 2021; Centurión et al., 2019; Ares et al., 2018a; Machín et al., 2018a; Ni Mhurchu et al., 2018). Follett et al. (2022) suggested that nutritional score programmes had a positive effect on probability of rejecting less healthy items, both in the short and long term. Health warnings and FOP labels reduced perceived healthfulness and selection of sugar-sweetened beverages in choice scenarios and consumers were less motivated and less likely to select an energy-dense snack, especially when labels included an image and text (Jáuregui et al., 2022; Miller et al., 2022; Ventsel et al., 2022; Schubert et al., 2021; Clarke et al., 2020; Billich et al., 2018). Other factors that positively contributed to encouraging consumers' healthier choices and eliciting dietary self-control included familiarity with a nutrition/health symbol, negatively framed graphic health warnings, being exposed to health rather than commercial messages, willingness to compromise pleasure for health and higher general health consciousness and interest (Choi et al., 2021; De Gennaro et al., 2021; Nobrega et al., 2020; Kergoat et al., 2019; Rosenblatt et al., 2018; Miklavec et al., 2016).

In some cases, NLHCs had a limited effect helping consumers select healthier products (Siegrist et al., 2019). Subjective food attributes such as taste, convenience and affordability outweighed the benefits of label information about healthier food products and claims, and had detrimental effects on sensory expectations and purchase intentions (Anders and Schroeter, 2017; Orquin and Scholderer, 2015). Thiene et al. (2018) suggested in their Northern Irish research that individuals tend to give up palatability to obtain healthier food options as a result of their understanding of nutritional level information on the FOP label and those with a weight problem (objectively measured or perceived) seem to pay attention to the

© Burleigh Dodds Science Publishing Limited, 2024. All rights reserved.

labels and use them for healthier choices. They found that those with higher education levels had a tendency to rank packaged foods more accurately and that FOP labels were effective at improving participants' ability to correctly rank packaged foods' healthiness. Velasco Vizcaíno and Velasco (2019) found that brand familiarity and brand trust shielded a product from being evaluated as unhealthy and decreased consumers' alertness, even when warning labels were present. In relation to packaged foods with health benefits, such as functional foods, the brand itself often has a strong influence on consumer acceptance in terms of whether it is a relevant and credible brand (Wennström and Mellentin, 2003).

Income also had an effect on healthy choices, as lower-income consumers perceived ultra-processed products to be significantly more healthful and were more likely to select sugar-sweetened beverages compared to those of higher-income consumers (Mantzari et al., 2020; Machín et al., 2017). In addition, the so-called halo effect, i.e. a food product is perceived healthier than it actually is, was recorded in several studies citing various factors contributing to this effect. For example, Dang and Nichols (2022) noted that the inclusion of a positive nutrient on the label led to greater purchase interest and an unhealthy food was perceived as healthier. Similarly, the inclusion of health-related imagery increased the perceived benefits of consumption of a food and indicated that consumers could infer additional health properties from non-diagnostic information featured on the packaging (Delivett et al., 2020). Three other studies highlighted the effect of different claims on the 'halo' effect. Drinks that displayed a 'disease risk reduction' claim were perceived as healthier than those without, regardless of products' healthfulness (Franco-Arellano et al., 2020). Fibre-related claims resulted in positive ratings of a product, which extended beyond perceptions of only the fibre content, to perceptions of overall healthiness, naturalness, quality, vitamin content and intentions to purchase and recommend the product (Mediano Stoltze et al., 2021). Despite the fact that products labelled with natural claims contained the same number of calories as their counterparts, consumers evaluated them as more healthful (Skubisz, 2017).

6 Key research areas: the impact of the nutrition labels and health claims on consumers' attitudes and behaviour

How effective are NLHCs in influencing consumers' decisions in terms of altering behaviour and impacting purchase decisions? European regulations enforce that only substantiated and approved statements can be used as NLHCs and it is essential to understand consumer perceptions of these claims, to understand purchase intent and consumer choices (van Buul and Brouns, 2015). Labels on

© Burleigh Dodds Science Publishing Limited, 2024. All rights reserved.

packaged foods with health claims, nutrition claims and nutrition information are ubiquitous, but often consumers find these labels too difficult to understand or time-consuming to read and interpret the information. This is a salient point in terms of developing new packaged foods and, particularly so, in relation to developing new functional foods and related brands.

In selected studies, the impact that NLHCs had on consumers' attitudes and purchase behaviour varied. The presence of NLHCs increased visual attention, raised the likelihood of affecting the final purchase and encouraged more healthful and dietary quality food choices (Rønnow, 2020; Ballco et al., 2019; Machín et al., 2019). It also resulted in a reduction of snack intake and encouraged consumers to consume healthier snacks (Baum et al., 2017). When it came to households' responses, Cawley et al. (2015) found that nutrition labelling led households to buy a more nutritious mix of products by reducing the purchase of less nutritious foods, but without increasing purchases of nutritious foods. Smed et al. (2019) noted that labels had a positive effect on the volume share of purchased products by households and Fichera and von Hinke (2020) reported that households reduced the quantity of purchased labelled foods, improved their nutritional composition and reduced the total monthly calories from labelled products. In addition to labels, the following factors encouraged changes in consumers' behaviour: visual attention and consumers' gaze duration at claims, differences in perceptions and importance of health and taste features of food, and presence of detailed nutrition information (Pabst et al., 2021; Ballco et al., 2020a; Steinhauser et al., 2019b; Fisher, 2018).

There were studies that suggested only limited or no meaningful change occurred in consumers' explicit attitudes, evaluation processes, food and beverage choices, product perceptions or purchase intentions due to NLHCs (Asbridge et al., 2021; Neuhofer et al., 2020; Menger-Ogle and Graham, 2018; Hamlin et al., 2015). Instead, NLHCs competed for consumers' attention and purchase intentions with non-nutrition elements such as the type of product, brand or product name, imagery, pricing, purchasing habit and 'clean' labels related to how foods were processed (Rybak et al., 2021; Tortora et al., 2019; Bartels et al., 2018; Boztuğ et al., 2015).

Information on the FOP labels can influence the portion size consumed. In some cases, it has been hypothesized that adding more nutrition information on product packaging could potentially reduce the calorie intake and/or consumption of sugary packaged foods. However, the results from such interventions have been mixed. Different FOP labels can have different effects on getting consumers to decrease their portion size for less healthy products (Egnell et al., 2018b; Talati et al., 2018). Some studies found that while the nutrition labels with more information about sugar caused consumers to be more visually attentive to nutrition information, they did not affect their choice of beverage nor improved the effectiveness of the dietary information on

© Burleigh Dodds Science Publishing Limited, 2024. All rights reserved.

subsequent choices (Neuhofer et al., 2020). On the other hand, better labelling can lead to a reduction in total monthly calories and the quantity of purchased foods and other studies found that NLHCs can help individuals meet their recommended nutritional daily guidelines (Benson et al., 2022; Fichera and von Hinke, 2020). Familiarity with foods carrying claims and belief in claims were seen by Benson et al. (2018) as the most consistent predictors of potion size selection.

Food and beverage companies use many ways to market their products using FOP label visual and written claims to capture parents' attention and influence them in their food choices. In many cases, the foods and beverages are of poor nutritional content. FOP labels have an important role to play in parents' healthy food choices. So how can policymakers reduce the purchase of nutrient poor foods or sugary drinks and snacks? Velázquez et al. (2021) in their research with Uruguayan mothers found that nutrient claims increased perceived healthiness and encouraged purchase. For sugary snacks, text and pictorial warning labels do have a role in reducing parents' likelihood of purchase for their young children (Taillie et al., 2022). However, in many cases, parents are easily convinced that a product is healthier than it actually is based on visuals with fruit, health claims and visuals suggesting the product is more natural (Abrams et al., 2015). In Australian research, star ratings and visuals were found to be the most significant attributes driving choice (Russell et al., 2017). For time-pressed parents who want to make better choices for their young children, clearer communication on FOP labels is required to help them make healthy food choices.

Willingness to pay a premium price is another important indicator of consumer choice for marketers. The relationship between price and labelling provides a measure of whether consumers are willing to pay for foods with health labelling. In fact, labelling is very effective in terms of providing information that influences consumers to pay a premium price. However, for certain consumers, price may be the only consideration, regardless of nutritional value or quality (Koen et al., 2018b). Bernard et al. (2019) in their research found that labels can influence willingness to pay and shape perceptions beyond the attributes they make a claim regarding on a pack. However, willingness to pay may be dependent on the claims, such as for fat-related health claims but not for fibre claims, while a high sugar label may reduce the likelihood for selecting a sugary drink (Ballco et al., 2019; Acton et al., 2018). Cooper et al. (2020), in their research on Australian consumers' willingness to pay for food products with a Health Star Rating label, found that two-thirds of consumers were willing to pay more for a product with such a label. Wilson and Lusk (2020) surveyed 1122 US consumers using three packaged foods with redundant labels (non-GMO sea salt, gluten-free orange juice and no-hormone-added chicken breast) to

© Burleigh Dodds Science Publishing Limited, 2024. All rights reserved.

determine if greater knowledge of the claims would decrease willingness to pay. They found that 41–58% of consumers were willing to pay a premium price for redundantly labelled foods, and people with farm work experience and with higher scientific literacy were less willing to pay for such foods.

7 Key research areas: label design of nutrition labels and health claims

One stream of NLHCs research has been dedicated to assessing the optimal label design that will enable consumers to quickly and easily understand the nutritional value of packaged foods and encourage healthier food choices. Studies compared several labels, such as BOP labels, Nutrition Facts Panel (NFP) and FOP labels, both monochrome and in colour, including nutritional/health warnings and symbols, Reference Intakes (RI), Guideline Daily Amounts (GDA), Multiple Traffic Lights (MTL), Health Star Rating (HSR), NutrInform Battery (NIB), Nutri-Score (NS), Physical Activity Equivalent (PAE) and emoticons, and their impact on consumers' understanding of a food's nutritional quality, perception of healthiness and purchase intentions. The results regarding the efficiency of different labels have been mixed. However, studies demonstrated that, in many cases, consumers improved their ability to classify foods based on their nutritional value, thanks to the labels. Still, some studies found that labelling has only a limited role in increasing judgement accuracy, consistency of nutritional information usage, quality food choices and calorie intake (Finklestein et al., 2021; Carter and Gonzalez-Vallejo, 2018).

Becker et al. (2015) stated that the FOP types of labels were attended more often and earlier than the NFPs and suggested that this is due to their placement on the front of the packaging and design characteristics. Nutritional/health warning labels and symbols performed better than other labels in terms of consumer behaviour towards unhealthful product categories and when identifying foods with high content of nutrients associated with non-communicative diseases (Deliza et al., 2020; Ares et al., 2018c; Goodman et al., 2018; Khandpur et al., 2018; Arrúa et al., 2017). MTL labels significantly increased the probability of avoiding the least healthy options and choosing the healthiest items among the alternatives and were found to be a consumer-friendly way of communicating nutrition information (Defago et al., 2020; Siegrist et al., 2015). Gabor et al. (2022) reported that the most accurate estimates on nutritional quality perception came from products with GDA label. While the labels often impacted consumers' nutritional value and healthiness understanding, as well as products' ratings, some studies suggested that purchase intention is not moderated accordingly, which indicated a potential mismatch with consumers' actions and this needs to

© Burleigh Dodds Science Publishing Limited, 2024. All rights reserved.

be accounted for in NLHCs' research and applications (Medina-Molina and Pérez-González, 2021; Lundeberg et al., 2018).

Two voluntary FOP labels that captured particular attention of researchers were Nutri-Score and Health Star Rating. Originating from the French public health agency Santé publique France, which developed the label following a request from the French Ministry of Health, the NS can be found in several EU countries, while the HSR is an Australian and New Zealand Governmental initiative. Several studies highlighted NS as the preferred label by consumers with positive impact on healthfulness and nutritional quality estimation accuracy, associated with higher objective understanding than monochrome and nutrient-specific labels and it offered the potential to boost sales of healthy products without affecting sales of the unhealthy ones (Kontopoulou et al., 2022; Packer et al., 2022; van den Akker et al., 2022; Aguenaou et al., 2021; Bossuyt et al., 2021; De Temmerman et al., 2021; Goiana-da-Silva et al., 2021; Fialon et al., 2020; Egnell et al., 2018c). Nevertheless, Finkelstein et al. (2019) found that NS did not result in the reduction of calories purchased or sugar from beverages and concluded that while NS might be preferred if the goal was to improve overall diet quality, other labels such as MTL might perform better in reducing total energy intake. In some cases, NS had no effect on consumers' attitudes, taste perceptions and purchase intentions, and hence it did not modify cognitive responses of consumers towards tested foods, and instead, another label (NutrInform Battery) was more informative and helpful (Folkvord et al., 2021; Mazzu et al., 2021b).

When it came to the HSR label's effectiveness, consumers perceived it as more useful and easier to understand than other labels and consumers' ability to select healthier products was heightened when products displayed the HSR label regardless of consulting BOP labels (Anderson and O'Connor, 2019; Neal et al., 2017). On the other hand, the HRS label had a significant depressive effect on consumer preference, and it performed worse than the other labels in capturing attention and altering perceived healthfulness and purchase intention (Ares et al., 2018c; Hamlin and McNeill, 2016). Pelly et al. (2020) noted that consumers saw benefits in the HSR as being simple, uncluttered and useful for quick comparison across products, but were sceptical about how the ratings were derived and therefore more transparency on the development of the label and education on its application could improve consumers' perceptions of this label.

In addition to design, colours displayed on labels played a role in consumers' perceptions of the healthiness of packaged foods. Studies exploring the effect of colours found that colour had the largest contribution to perceived healthfulness, compared to packaging shape and textual information, while the use of colour in FOP labels increased attention to the label, and consumers opted to purchase more green-labelled products compared to red-labelled

© Burleigh Dodds Science Publishing Limited, 2024. All rights reserved.

ones (Bromberg et al., 2022; Cabrera et al., 2017; Becker et al., 2015). Regarding the importance of congruency in packaging design, consumers were slower to identify a colour-flavour incongruent target compared to a colour-flavour congruent one and switched to a word-based search if the target was not found (Huang et al., 2021). A salient, red traffic light label potentially reduced unhealthy food-related decision-making and activated areas in the frontal lobe implicated in the execution of responses and motor inhibition (Zhang et al., 2020). However, the correct interpretation of nutrition information through colour coding can vary and reliance on colours might not necessarily support healthier food choices in all situations and other clues, such as emoticon labels, can yield stronger effects on perceptions of taste and healthiness than colour labels (Nyilasy et al., 2016; Vasiljevic et al., 2015). In addition, country-specific formative testing might be needed as interpretations and preferences for colours and textual information differ on the FOP label (Pettigrew et al., 2021).

Furthermore, in their systematic review on colour-coded and warning nutrition schemes, Song et al. (2021) found that all colour-coded and warning labels had beneficial effects through encouraging the purchase of healthier packaged foods and reducing the purchase of less healthful options. They also suggested that the labels improved consumers' understanding of nutrition information. Using scanned data from over 30 million transactions related to yoghurt and ready meals purchase from a major retailer in the UK over a three-year period, Boztuğ et al. (2015) suggested that the introduction of the monochrome Guideline Daily Amount (GDA) label reduced the attraction of unhealthier packaged foods in terms of market share, however, it did not affect product choice behaviour. Price and habit had a greater impact on consumers' purchase behaviour and product choice, compared to the GDA label introduction. How then to increase consumers' usage of NLHCs and make them more effective in their contribution to consumption of healthier diets? Soederberg Miller et al. (2019) researched the impact of label training and found that a nutrition e-training tool was an effective way to boast students' label-reading skills and health food choices. In the future, it is likely that new technologies and the metaverse will have an important role to play in consumers' usage of NLHCs and colour and design will become more significant in signalling healthfulness.

8 Developing more market-oriented nutrition labels and health claims

Adopting a market-oriented approach to developing new packaged foods suggests early engagement with the consumer to develop successful new packaged foods. Market orientation is linked to more successful new product

© Burleigh Dodds Science Publishing Limited, 2024. All rights reserved.

development (NPD) and more closely identifying consumers' needs. Narver and Slater (1990) suggested that market orientation helped create the necessary behaviours within an organization that offered superior value for consumers and sustained business performance. NLHCs can be very effective marketing tools for food and beverage companies but require in-depth consumer insights to design more effective and optimum labels targeted at specific consumer segments, e.g. teenagers or elderly consumers. Ma and Zhuang (2021) in their review of eye-tracking studies found that although updated designs of nutrition labels improved attention capture, this did not necessarily lead to healthier food choices. Thus updated, or newly designed NLHCs on packaged foods, need consumer input to ensure they both grab attention, initiate purchase and thus lead to healthier food choices.

Furthermore, developing more effective labels targeted at certain segments, such as elderly consumers, who may need information presented in a different way, or format, than other segments, based on their specific needs is required. It may also be about a more in-depth understanding of whether consumers, across different segments and socio-demographic groups, understand the information on labels and what format elicits the best response in terms of healthy eating behaviours by using methodologies, such as ethnography or semiotics. In addition, measuring NLHCs choices in a laboratory setting or online potentially changes consumers' behaviour as they might guess a study's purpose and hence give appropriate (and not accurate) answers. Thus, a key factor in developing new NLHCs is to use the consumer to co-design labels with food and health policymakers and food and beverage companies in a market-oriented approach to communicate the health benefits of specific foods. There are a range of methodologies that can be better used to develop more market-oriented packaging for consumers, including semiotics, sensory and conjoint analysis, ethnography, advanced eye tracking technologies and focus group discussions. This helps to understand how the target market perceive the NLHCs, their likely usage and how it will influence the purchase decision. The integration of the consumer as co-designer in the NPD process is a more market-oriented approach and has been shown to lead to more successful NPD performance in firms.

Box 1: Semiotics for nutrition labelling and health claims

Semiotics concerns the study of meaning in terms of signs and symbols and has a key role to play in labelling and packaging design. The key issue here is the balance between the design of a product label and the information that is needed on the label, so that consumers

© Burleigh Dodds Science Publishing Limited, 2024. All rights reserved.

make an informed decision during the purchase event. Semiotics is an investigation into how meaning is created and communicated and Bignell (2002, p. 1) defines semiotics as '*a way of analysing meanings by looking at the signs (like words, for instance, but also pictures, symbols etc.) which communicate meanings*'. Packaging is a key component to communicate information on any packaged food and provides many purchase cues in terms of the value proposition, provenance and all the meanings that it communicates. Cues on food packaging in addition to NLHCs, which interact with those NLHCs, include language on the packs, materials, colours, logos, crests, fonts, serving size information and symbols (gluten free, etc.). Ares et al. (2011) found in their study of consumers' perceptions of yogurt labels in Spain and Uruguay that due to cultural differences, consumers perceived signs differently and this shows the significance of understanding how consumers perceive signs on labels. Do consumers really understand the messages that food and beverage companies or food and health policymakers are trying to convey through signs, symbols and labels on food packaging? How do all these pieces of information combine on food packaging to offer meaning to the consumer?

9 Future trends in nutrition labelling and health claims in food

Certain trends in the future will impact on how consumers perceive and use NLHCs on packaged foods. Three trends that will impact on the development and design of NLHCs are outlined in the following sections.

9.1 Trend 1: The metaverse is here!

Thanks to the rapid development of technology, consumers will soon be faced with a parallel reality called the metaverse, which will undoubtedly shape many aspects of their lives, from how they communicate with each other, entertain, provide education and receive knowledge, and also how they buy, sell and experience foods on a daily basis. This virtual space promises to bring together people, food and beverage companies and products in both the digital and physical environments to interact, create and collaborate (Hackl et al., 2022). According to Terry and Keeney (2022), by 2032, consumers will have an option to build their own metaverses, as part of a shared metaverse, which will be filled with their personal and professional tools. This poses both opportunities and challenges for all stakeholders across the agri-food system. The metaverse will open a wide array of platforms where producers, food and beverage companies, academics, researchers, food and health policymakers and NGOs will be able to create content and directly communicate with consumers without limitations in terms of time or location, which will allow for faster information sharing and close relationship building.

© Burleigh Dodds Science Publishing Limited, 2024. All rights reserved.

Users' experience is a central part of the metaverse, and text, photos, sounds, videos and tactile simulations will be easily transformed into electronic and virtual shapes and communicated to the user. However, the food experience itself will likely require a more complex approach when it comes to transforming it into something meaningful for consumers. For example, virtual restaurants, events and supermarkets might offer a variety of Non-fungible Tokens (NFTs) in food and virtual products which will then either become part of consumers' virtual assets or will be redeemed in 'real' life. So, how will this impact on NLHCs? It is likely that NLHCs will be equally, if not more, important in a metaverse as in our current reality, with the possibility to create even more user-friendly and useful information and suggestions that will help consumers better understand and use them for their benefit and make healthy food choices. They will be an intricate part of virtual foods and will need to be adjusted together with virtual food packaging. The research into NLHCs in the future might be easier with utilization of eye-tracking tools, such as Augmented Reality glasses (a combination of real and virtual worlds) that will become more mainstream in the metaverse and will provide feedback, or feedback in 'real time', on which part of NLHCs capture the consumer's attention and for how long. In addition, the standardization of nutritional labels across the EU, or globally, might be easier in the metaverse as, for example, nutritional labels for foods in virtual retailers could be easily available to consumers in their own language and customizable for different cultural situations, so that consumers readily understand the information.

How will consumers interact with NLHCs in the metaverse and is there an opportunity for food and beverage companies to design market-oriented information and labels to guide consumers in terms of healthy eating behaviours? As consumers shop in a virtual world, there is an opportunity for food and beverage companies and food and health policymakers to more deeply, and effectively, interact and connect with consumers with NLHCs on food packaging. In the virtual world, every consumer will likely have an avatar, which is their digital representative, to represent them as they shop, socialize, learn and work in the digital sphere (Marr, 2022). Research on the effective design and use of NLHCs in the metaverse will be important in years to come, as consumers adapt to this dual reality. Interactivity with NLHCs will produce opportunities for new way of communicating in the 3D realm with consumers to promote healthy dietary behaviours.

9.2 Trend 2: A different kind of consumer?

Another aspect that will direct NLHC research in the next 10 years is that Generation Z (individuals born roughly between mid and late 1990s and 2010) will soon mature and take over in terms of purchasing power. In addition,

© Burleigh Dodds Science Publishing Limited, 2024. All rights reserved.

Generation Alpha (individuals born roughly between mid-2010s and mid-2020s) will be well capable of navigating the digital world and will want an immersive experience (van Rijmenam, 2022). This means serving consumers with different needs, aspirations and ways of purchasing and it will become the norm. The new consumer will be tech savvy, demanding different experiences and also seeking virtual assets and consumer-generated content. So, what will this mean for NLHCs on packaged foods? It will mean the use of new technologies on packaging with more user-friendly and verified information. This will optimize new scanning devices, such as phones and watches, and a new advanced QR code type technology to deliver clear NLHC information. In addition, the increased impact of social media in food choices and in terms of raising awareness of NLHCs will be important, such as the role of influencers in promoting healthy lifestyles and consumption of healthy foods. The development of new apps that can be used to interpret dietary guidelines and healthy eating strategies will provide this future consumer with more information to make informed decisions.

While this chapter focused on packaged foods only, in the future, for food service providers, it is likely that there will be more information required on menus, including both calorie and nutritional information which will be communicated in an interactive way through new technologies. In addition, in their research, Bablani et al. (2020) found that FOP labels, such as the Australasian Health Star Rating scheme, had a modest role in driving healthier product reformulation, but that reformulation was higher for those packaged foods that were least healthy. So, the introduction of more market-oriented labelling schemes can have a positive role in encouraging food and beverage companies to reformulate unhealthy packaged foods for future consumers.

9.3 Trend 3: New foods, new segments and new nutritional labels and health claims

Another key trend will be the development of new foods, such as new functional foods as new functional ingredients are commercialized by food and beverage companies. Considering the future consumer will likely live longer, this will result in the increased importance of functional foods and the related NLHCs targeted at the elderly consumer segment. There are few food products, if any, in global markets specifically marketed towards the elderly market segment. These foods may be consumed for ongoing health conditions, to prevent a specific illness or in terms of their affective values. Globally, 16% of the population will be a member of this extremely heterogeneous group aged 65 or over by 2050, with more complicated buying behaviours than other generations (United Nations, 2020). It is likely for the elderly consumer

© Burleigh Dodds Science Publishing Limited, 2024. All rights reserved.

segment that packaged foods will need to be adapted in many ways, such as smaller portion size, ease of pack opening, enhanced sensory profiles and familiar brands. To communicate directly to this segment, food and beverage companies and food and health policymakers will need information that is clearly communicated in terms of nutritional labels (e.g. protein, calories, nutrients) with health claims that are endorsed by the scientific community. Other issues, such as font size and colours used in packaged foods, will need to be adjusted for this target segment. It is likely also that there will be more nutrition information for all consumers, such as calories, fat and protein content, next to food meals when ordering at food service outlets.

Sustainability and environmental labelling will become more influential in the purchase decision for consumers in the future. Can these labels be co-labelled with NLHCs to strengthen the combined message to consumers? There will be a demand for new types of labels such as environmental labels as food and beverage companies attempt to satisfy consumers' increasing desire for information. The French Government has enacted legislation to develop and implement an environmental label system (Hélias et al., 2022). What will environmentally sustainable labels look like and can they be combined with NLHCs to provide information on the carbon footprint of the product, linked to nutritional and health information, to allow consumers make a really informed purchase decision? Gröppel-Klein et al. (2020) found that combining a health claim with a sustainability label increased the credibility of the health claim, which then helped convey the product's health message. They found that this only applied to co-labelling with an organic label. Interestingly, they found that an organic label benefited from co-labelling with the health claim, and itself then became more credible to consumers. What information will these co-labels provide? How will consumers interact with them and how will they be designed to provide clear usable and actionable information? This is a possible future research area.

10 Conclusion

In this chapter, a systematic literature review (SLR) was used with a PRISMA approach to identify the contemporary issues in understanding consumer perceptions and attitudes towards NLHCs on packaged foods. Key research trends were identified and future trends were discussed that may impact how consumers perceive and understand NLHCs and that will need to be addressed to ensure consumers make healthy food choices. These results can guide future research in consumer perceptions and attitudes towards NLHCs and play a role in healthy food choices and ultimately improving the health of populations. There was mixed evidence on the design efficiency of NLHCs to alter dietary behaviour and impact on healthy choices. There is a strong need for a more

© Burleigh Dodds Science Publishing Limited, 2024. All rights reserved.

market-oriented approach to the design of NLHCs, using the consumer as an *active* co-designer, which would integrate the consumer into the design process at an early stage and to also devise education intervention strategies that are more effective in terms of consumer understanding and usage of labels.

11 Where to look for further information

The following book can provide a gentle introduction to the subject of nutrition labels and health claims for non-specialist audience:

- Bobo, J. A. (2021). *Why Smart People Make Bad Food Choices: The Invisible Influences that Guide Our Thinking*, Mango Media, United States.

Recommended books on nutrition, nutritional labels and health claims:

- Hammond, G., O'Connor, C., Piché, L. A., Noss Whitney, E. and Rady Rolfe, S. (2023). *Understanding Nutrition* (3rd edn). Cengage, Canada.
- Ghosh, D., Bogueva, D. and Smarta, R. (Eds.) (2023). *Nutrition Science, Marketing Nutrition, Health Claims, and Public Policy* (1st edn). Academic Press by Elsevier Inc., United States.
- Astley, S. (Ed.) (2019). *Health Claims and Food Labelling*. Royal Society of Chemistry, United Kingdom.

The Nutrition Society and American Society for Nutrition are useful sources for more advanced information about nutrition, and the following links contain suggested reading lists and journals for those looking to dive deeper into the area of nutrition and labelling:

The Nutrition Society:

- https://www.nutritionsociety.org/publications/textbooks.

American Society for Nutrition:

- https://nutrition.org/publications/.

Examples of relevant organisations can be found here:

European Commission:

- https://food.ec.europa.eu/safety/labelling-and-nutrition/nutrition-and-health-claims_en.

European Food Safety Authority (EFSA):

- https://www.efsa.europa.eu/en/applications/nutrition/regulations andguidance.

© Burleigh Dodds Science Publishing Limited, 2024. All rights reserved.

US Food and Drug Administration:

- https://www.fda.gov/food/nutrition-education-resources-materials/new-nutrition-facts-label.

Academy of Nutrition and Dietetics:

- https://www.eatright.org/.

The Association of UK Dietitians:

- https://www.bda.uk.com/resource/food-labelling-nutrition-information.html.

Suggested websites that include more details on nutrition labels and health claims:

UK Food Labelling and Packaging Legislation:

- https://www.gov.uk/food-labelling-and-packaging/food-labelling-what-you-must-show.

UK Legislation – Nutrition and Health Claims Made on Food (Section 4):

- https://www.gov.uk/government/publications/nutrition-legislation-information-sources/nutrition-legislation-information-sheet--2.

US Food and Drug Administration – Food Labelling and Nutrition:

- https://www.fda.gov/food/food-labeling-nutrition.

US Food and Drug Administration – How to Understand and Use the Nutrition Facts Label:

- https://www.fda.gov/food/new-nutrition-facts-label/how-understand-and-use-nutrition-facts-label.

US Food and Drug Administration – Label Claims for Conventional Foods and Dietary Supplements:

- https://www.fda.gov/food/food-labeling-nutrition/label-claims-conventional-foods-and-dietary-supplements.

US Food and Drug Administration – Questions and Answers on Health Claims in Food Labelling:

- https://www.fda.gov/food/food-labeling-nutrition/questions-and-answers-health-claims-food-labeling.

© Burleigh Dodds Science Publishing Limited, 2024. All rights reserved.

European Commission Food Safety – Labelling and Nutrition:

- https://ec.europa.eu/food/safety/labelling-and-nutrition_en.

European Commission - Health Claims:

- https://food.ec.europa.eu/safety/labelling-and-nutrition/nutrition-and-health-claims/health-claims_en.

National Health Service (NHS) - Food Labels:

- https://www.nhs.uk/live-well/eat-well/food-guidelines-and-food-labels/how-to-read-food-labels/.

Safefood - How to Read Nutrition Labels:

- https://www.safefood.net/how-to/nutrition-labels?gad=1&gclid=CjwKCAjwt52mBhB5EiwA05YKo3NFc2l21mSitDnG6oDydUAO9zmdE0HjixDuFGHqKib7jV4eFDvyFhoCCdgQAvD_BwE.

Another useful resource is the EU register of health claims:

- https://food.ec.europa.eu/safety/labelling-and-nutrition/nutrition-and-health-claims/eu-register-health-claims_en.

Selected country-specific sources of information for nutrition labelling and health claims:

United Kingdom:

- https://www.gov.uk/government/publications/nutrition-and-health-claims-guidance-to-compliance-with-regulation-ec-1924-2006-on-nutrition-and-health-claims-made-on-foods/nutrition-and-health-claims-guidance-to-compliance-with-regulation-ec-19242006.

Canada:

- https://www.canada.ca/en/health-canada/services/understanding-food-labels/nutrition-claims.html.

Australia and New Zealand:

- https://www.foodstandards.gov.au/consumer/labelling/nutrition/Pages/default.aspx.

Singapore:

- https://www.hpb.gov.sg/docs/default-source/newsroom/a-handbook-on-nutrition-labelling-(singapore)-_june-2020.pdf?sfvrsn=b232e5ec_2.

© Burleigh Dodds Science Publishing Limited, 2024. All rights reserved.

Mexico:

- https://www.gob.mx/cofepris/acciones-y-programas/manual-de-la-modificacion-a-la-norma-oficial-mexicana-nom-051-scfi-ssa1-2010-272744?state=published.

Conference of interest:

- The Academy of Nutrition and Dietetics sponsors the Food & Nutrition Conference & Expo® to address key issues affecting the health of Americans:
- https://eatrightfnce.org/.

Selected universities and centres of interest based on the research and teaching excellence in nutrition and food science:

University of California - Davis (USA):

- https://www.hpb.gov.sg/docs/default-source/newsroom/a-handbook-on-nutrition-labelling-(singapore)-_june-2020.pdf?sfvrsn=b232e5ec_2.

Cornell University (USA):

- https://www.human.cornell.edu/dns.

University of Illinois at Urbana - Champaign (USA):

- https://nutrsci.illinois.edu/.

Wageningen University (The Netherlands):

- https://www.wur.nl/en/research-results/chair-groups/agrotechnology-and-food-sciences/human-nutrition-and-health.htm.

King's College London (United Kingdom):

- https://www.kcl.ac.uk/slcps/our-departments/department-of-nutritional-sciences.

© Burleigh Dodds Science Publishing Limited, 2024. All rights reserved.

12 Appendix: Studies (n = 146) on nutrition labels and health claims included in the qualitative analysis and their characteristics

No.	Authors	Year	Journal	Methodology	Country	Participants	Key findings
1	Abrams et al.	2015	*Appetite*	90-min focus groups (4)	USA	30 consumers (parents)	Parents were easily led to believe the product was healthier based on visuals of fruit, more realistic pictures, health claims, cross-branding with healthier foods and visuals suggesting the product is more natural
2	Acton and Hammond	2018	*Appetite*	Experimental marketplace study	Canada	675 consumers	With the price increase, participants were significantly less likely to select a sugary drink and opted for drinks with fewer calories and less sugar
3	Acton et al.	2018	*Preventive Medicine*	Experimental marketplace study	Canada	675 consumers	Consumers indicated almost unanimous support for implementing FOP nutritional labelling and the influence of FOP labels may vary based on nutritional quality of food products
4	Adasme-Berríos et al.	2022	*Nutrients*	Questionnaire	Chile	807 consumers	The direct effect of nutritional warnings raised the intention to avoid processed foods, while eating motivation was negative in its direct effect on the same avoidance intention
5	Aguenaou et al.	2021	*Archives of Public Health*	Questionnaire (online and in-person)	Morocco	814 consumers	Nutri-Score was associated with the highest improvement in the ability to correctly classify foods according to their nutritional quality
6	Anders and Schroeter	2017	*International Journal of Consumer Studies*	Data set from a national survey	Canada	8114 consumers	Taste, convenience and affordability may outweigh the benefits of information on healthier, alternative food choices

© Burleigh Dodds Science Publishing Limited, 2024. All rights reserved.

No.	Authors	Year	Journal	Methodology	Country	Participants	Key findings
7	Anderson and O'Connor	2019	*Food Quality and Preference*	Questionnaire and stimuli	Australia	253 consumers	Consumers' ability to choose healthier products was increased when products displayed a HSR label
8	Andrews et al.	2021	*Journal of Business Research*	Experiments with stimuli	USA	2000 consumers	Consumers with greater objective nutrition knowledge performed better on nutrition use accuracy
9	Annunziata et al.	2016	*Appetite*	Conjoint analysis	Italy	300 consumers	Consumers assigned greater utility to health warnings, followed by nutrition information
11	Ares et al.	2018a	*Public Health Nutrition*	Online questionnaire	Uruguay	1416 consumers	Consumers showed a positive attitude towards nutritional warnings and health motivation was a crucial driver for taking nutrition warnings into consideration
10	Ares et al.	2018b	*Food Quality and Preference*	Repeated purchase simulation	Uruguay	395 consumers	Warnings modified the choices of around half of the participants and consumers who reacted more strongly to the warnings were characterized by greater health motivation
12	Ares et al.	2018c	*Food Quality and Preference*	Visual search task and online questionnaire	Uruguay	1004 consumers	HSR label performed worse than the other two schemes in capturing attention and altering perceived healthfulness and purchase intention
13	Arrúa et al.	2017	*Public Health Nutrition*	Visual search task and questionnaire	Uruguay	387 consumers	Nutrition warnings improved consumers' ability to correctly identify a food product with high content of a key nutrient
14	Asbridge et al.	2021	*Appetite*	Online experimental study	UK	1185 consumers	Health warning labels did not significantly change consumers' food choice nor explicit attitudes

© Burleigh Dodds Science Publishing Limited, 2024. All rights reserved.

15	Aschemann-Witzel and Grunert	2015	*Food Quality and Preference*	Experiments	USA and Denmark	626 consumers	Soft information positively influenced health inferences and attitudes in Denmark. Scientific information positively influenced health inferences but not the attitudes in the United States
16	Ballco et al.	2019	*Food Research International*	Experiments with eye-tracking	Spain	100 consumers	The presence of nutritional claims generally increased visual attention regarding fixation count and may be linked to the final purchase decision
17	Ballco et al.	2020a	*Food Quality and Preference*	Experiments with eye-tracking	Spain	218 consumers	There was a relationship between the most highly valued NLHCs and visual attention, so product selection is also based on the visual attention that consumers pay to the packaging
18	Ballco et al.	2020b	*Food Quality and Preference*	Experiment with economic incentives	Spain	245 consumers	Consumers were willing to pay a premium for the fat-related health claims in addition to the nutrition claims but not for the claims related to the fibre content
19	Bartels et al.	2018	*International Journal of Consumer Studies*	Cross-sectional survey	USA	60 consumers	Nutrition information received strong competition for participants' attention from other non-nutrition elements, such as the brand, imagery and price
20	Baum et al.	2017	*Appetite*	Randomized crossover design	USA	33 consumers	Protein source did not affect appetite response or snack intake, but educational nutrition messaging could help in reducing snack intake between meals
21	Becker et al.	2015	*Food Policy*	Experiments with change detection methodology	USA	125 consumers	FOP labels were attended more often and earlier than mandated Nutrition Facts Panel due to their placement on the front of the package and design characteristics

© Burleigh Dodds Science Publishing Limited, 2024. All rights reserved.

No.	Authors	Year	Journal	Methodology	Country	Participants	Key findings
22	Benson et al.	2018	*Nutrients*	Cross-sectional study	Northern Ireland and Republic of Ireland	1039 consumers	Claims influenced consumers' fillingness perceptions of some foods, but there was little influence on tastiness, healthiness perceptions or the portion size
23	Benson et al.	2019	*Nutrients*	Focus groups (10)	Northern Ireland and Republic of Ireland	78 consumers	Knowledge was a key factor influencing how much consumers believed the claims and their perceptions of products displaying claims
24	Benson et al.	2022	*Appetite*	Within-subject experiments	Northern Ireland	115 consumers	The presence of claims significantly reduced the amount of energy, fat, saturated fat, sugar, carbohydrates and sodium in the first study but no impact on nutritional composition was noted in subsequent lunch servings
25	Bernard et al.	2019	*Food Quality and Preference*	Field experiments	USA	328 consumers	Labels have capabilities to shape consumers' perceptions and influence their willingness to pay beyond those attributes they make a claim regarding
26	Bialkova et al.	2016	*Appetite*	Taste experiment	Germany	240 consumers	The message displayed on the FOP labels altered consumers' evaluation and choice
27	Billich et al.	2018	*Appetite*	Online choice experiment	Australia	994 consumers	Warnings and labels significantly reduced selection of a sugar-sweetened beverage
28	Bossuyt et al.	2021	*Nutrients*	Experiment with eye-tracking	Belgium	398 consumers	Nutri-Score had a positive impact on healthfulness estimation accuracy while Nutrition Facts Panel had no effect or negative effect on healthfulness accuracy
29	Boztuğ et al.	2015	*Food Policy*	Purchase transaction, product and consumer data from a retailer	UK	Around 50millionpurchase transactions from consumers	The GDA label reduced attraction of unhealthier products in terms of market share but did not affect product choice behaviour

© Burleigh Dodds Science Publishing Limited, 2024. All rights reserved.

30	Bromberg et al.	2022	*International Journal of Food Sciences and Nutrition*	Cross-sectional telephone survey (questionnaire)	Israel	1042 consumers	Most consumers reported they frequently used the Nutrition Facts Tables and intended to purchase fewer relabelled and more green-labelled products
31	Bryła	2020	*Nutrients*	Online questionnaire using Computer-Assisted Web Interviews	Poland	1051 consumers	The importance of the information about the fat content and health effects of consuming a food product were significant predictors of label use, unlike the demographic or socio-economic variables
32	Cabrera et al.	2017	*Public Health Nutrition*	Multiple studies: online questionnaire, choice conjoint analysis, associations test, visual search	Uruguay	496 consumers	Perceived healthfulness was significantly shaped by shape, colour and textual information, with colour having the largest contribution to perceived healthfulness
33	Carbonneau et al.	2015	*British Journal of Nutrition*	Experiment with ratings on a Likert scale	Canada	141 consumers	Labelling did not influence energy intake, but a low-fat label might increase users' fasting hunger perceptions compared to energy label or no label
34	Carter and Gonzalez-Vallejo	2018	*Appetite*	Questionnaire with stimuli (online and in-person)	USA	297 students (18+)	FOP labelling had a limited role in increasing judgement accuracy, consistency of nutritional information use and food choice
35	Cavaliere et al.	2015	*Food Quality and Preference*	In-person questionnaire	Italy	240 consumers	Females, families with young children and consumers with a higher nutritional knowledge paid attention to nutrition claims. Older consumers with limited income and a health condition were more interested in health claims

© Burleigh Dodds Science Publishing Limited, 2024. All rights reserved.

No.	Authors	Year	Journal	Methodology	Country	Participants	Key findings
36	Cavaliere et al.	2016	*Food Quality and Preference*	In-person questionnaire	Italy	300 consumers	Highly health-oriented consumers were more likely to refer to the extensive information reported on Nutrition Facts panel. Claims were of more interest to consumers with low health-orientation
37	Cawley et al.	2015	*Public Health Nutrition*	Aggregate sales data for 102 categories of food	USA	Data on consumers' purchase from 168 supermarkets	After the introduction of the nutrition ratings, sales of less nutritious foods fell. Ratings led consumers to buy a more nutritious mix of products
38	Centurión et al.	2019	*Journal of Nutrition Education and Behavior*	Experiment with eye-tracking	Uruguay	100 consumers	Nutritional warnings caught consumers' attention and reduced the amount of visual attention needed to evaluate healthfulness
39	Cheah et al.	2015	*British Food Journal*	Data set from a national survey	Malaysia	30 992 consumers	Females, young adults, Malays, tertiary educated, singles, employed, physically active and non-smokers were significantly associated with increased odds of nutrition label use
40	Choi et al.	2021	*Journal of Health Communication*	Between-subject design experiments with stimuli	USA	166 consumers	Health consciousness positively influenced evaluative responses to unhealthy food messaging while health literacy negatively influenced perceived healthiness and purchase intention when the nutrition label communicated a higher serving size
41	Clarke et al.	2020	*Appetite*	Between-subject design experiments with stimuli	UK	4134 consumers	Consumers were least likely to select an energy-dense snack in the image-and-text health warning labels group

© Burleigh Dodds Science Publishing Limited, 2024. All rights reserved.

42	Cooper et al.	2020	*Nutrients*	Online questionnaire	Australia	1024 consumers	Almost two-thirds of participating household grocery shoppers were willing to pay more (up to an additional 3.7% of the price) for a product with HSR label
43	Dallas et al.	2015	*Appetite*	Multiple studies: online questionnaire, in-person taste test, between-subject experimental design	USA	273 consumers	Participants incorrectly interpreted the serving sizes, served more cookies for themselves and more cheese crackers to another person, bought more lasagnas for others and divided lasagna into larger slices
44	Dang and Nichols	2022	*Journal of Marketing Theory and Practice*	Between-subject design experiments with eye-tracking	USA	722 consumers (including undergraduate students)	When a positive nutrient was added to the label, unhealthy food was perceived as healthier and healthy food was unaffected
45	De Gennaro et al.	2021	*Journal of Functional Foods*	Online questionnaire	Italy	1030 consumers	Health claims were of interest, albeit marginal, to consumers and the attributes 'geographic indication organic' were ranked the highest
46	De Temmerman et al.	2021	*Appetite*	Online between-subject and mixed-design experiments	Belgium	707 consumers	The presence of Nutri-Score enabled consumers to assess the healthiness of products better
47	Defago et al.	2020	*International Journal of Consumer Studies*	In-person choice experiment	Peru	100 college students	Exposure to MTL labels significantly increased the probability of avoiding the least healthy options and of choosing the healthiest product

© Burleigh Dodds Science Publishing Limited, 2024. All rights reserved.

No.	Authors	Year	Journal	Methodology	Country	Participants	Key findings
48	Delivett et al.	2020	*Appetite*	Randomized experiments with stimuli	UK, Italy, Romania	1016 consumers	The inclusion of a health-related images increased consumers' perceived benefits of consuming the product, with minimal effect on the perceived risks
49	Deliza et al.	2020	*Food Quality and Preference*	Visual search and online questionnaire	Brazil	1994 consumers	Nutritional warnings demonstrated advantage over the GDA and MTL labels in facilitating the identification of products with high content of nutrients associated with non-communicable diseases
50	Devia et al.	2021	*Food Quality and Preference*	Choice experiment	Uruguay	790 consumers	References to home-made and images of natural foods on the labels had a positive effect on consumers' purchase intention and tended to create a belief that products were healthful
51	Egnell et al.	2018a	*PLoS ONE*	Online questionnaire	France	3751 consumers	Women, younger people, non-smokers, individuals with higher education levels and with children had a higher capacity to identify healthier products
52	Egnell et al.	2018b	*Nutrients*	Online questionnaire	France	25 772 consumers	Nutri-Score consistently lowered the portion sizes selected by consumers followed by MTL, compared to no label

© Burleigh Dodds Science Publishing Limited, 2024. All rights reserved.

53	Egnell et al.	2018c	*Nutrients*	Online questionnaire with stimuli	Argentina, Australia, Bulgaria, Canada, Denmark, France, Germany, Mexico, Singapore, Spain, UK and USA	12 015 consumers	Nutri-Score performed the best in assisting consumers to rank the products according to their nutritional quality, followed by MTL, HSR, warning symbol and RIs
54	Egnell et al.	2019	*Nutrients*	Online questionnaire	The Netherlands	1032 consumers	Nutri-Score demonstrated the highest overall performance in assisting consumers rank the products according to their nutritional quality
55	Embling et al.	2019	*Food Quality and Preference*	Online questionnaire and *ad libitum* experiment in the lab	UK	393 consumers	There was no effect of variety labelling on ideal portion size, expected fullness or food intake
56	Fenko et al.	2018	*Food Quality and Preference*	Field experiment with eye-tracking	The Netherlands	48 consumers	Consumers fixated longer and more often on the MTL labels than choices logos. Participants in the time constraint condition demonstrated less visual attention towards health labels
57	Fialon et al.	2020	*Nutrients*	Online randomized controlled trial (questionnaire)	Italy	1032 consumers	Nutri-Score showed the highest overall performance in assisting consumers to correctly rank the products according to their nutritional quality compared to RIs

© Burleigh Dodds Science Publishing Limited, 2024. All rights reserved.

No.	Authors	Year	Journal	Methodology	Country	Participants	Key findings
58	Fichera and von Hinke	2020	*Journal of Health Economics*	Data set from a rolling panel of households	UK	360 921 observations for 20 707 households	Labelling resulted in a reduction in the quantity purchased of labelled store-brands foods and improvement in their nutritional composition
59	Finkelstein et al.	2019	*Nutrients*	Within-subject crossover design using online experimental grocery store	Singapore	154 consumers	Nutri-Score performed statistically better than MTL and control but, unlike MTL, it did not statistically reduce calories or sugar from beverages
60	Finkelstein et al.	2021	*Appetite*	Within-subject crossover design using online experimental grocery store	Singapore	117 consumers	Neither Singapore's Healthier Choice Symbol (HCS)-only condition nor HCS paired with PAE condition led to a change in the number of calories per serving purchased
61	Fisher	2018	*Appetite*	Lab experiments with stimuli	USA	87 consumers (college students)	The presence of a nutrition label reduced valuations by an average of 25 cents
62	Folkvord et al.	2021	*Food Quality and Preference*	Between-subject design via online questionnaire	The Netherlands	192 consumers	No effect of the Nutri-Score label was found on consumers' attitudes, taste perceptions or purchase intention
63	Follett et al.	2022	*Appetite*	DMARC database on food items selected by consumers, with demographics	USA	3498 observations across 403 consumers	Nutritional-Score programme had a positive effect on the probability of rejecting less healthy foods in the short and long term

© Burleigh Dodds Science Publishing Limited, 2024. All rights reserved.

64	Franco-Arellano et al.	2020	*Appetite*	Online questionnaire	Canada	1997 consumers	Less healthy drinks displaying any kind of FOP labelling were perceived as less healthy compared to the control group
65	Gabor et al.	2020	*Food Quality and Preference*	Experiment with eye-tracking	Croatia	76 consumers	Nutri-Score required least visual attention from consumers regarding dwell-time, number of fixations and average fixation duration, followed by MTL and GDA. The most accurate estimates came from products carrying GDA label
66	Goiana-da-Silva et al.	2021	*Nutrients*	Questionnaire	Portugal	1059 consumers	All FOP labels led to a higher percentage of correct responses on the ranking task compared to no label condition
67	González-Vallejo et al.	2016	*Appetite*	Online questionnaire	USA	196 consumers	Judgement consistency was higher for more educated consumers and those who reported having regular meals had models that were in greater agreement with the expert's model
68	Goodman et al.	2018	*Nutrients*	Between-groups design experiment (online questionnaire)	UK, Canada, USA and Australia	11 617 consumers	FOP labels that contained 'high in' descriptions, red colour and intuitive warning symbols were more effective at communicating high levels of nutrients of public health concern
69	Gracia and Barreiro-Hurle	2019	*Nutrients*	Direct ranking preference method (online questionnaire)	Spain	400 consumers	Consumers preferred claims that showed a reduced presence of unhealthy substances above those that highlighted the presence of healthy substances
70	Hagmann and Siegrist	2020	*Food Quality and Preference*	Online experiment (questionnaire)	Switzerland	1313 consumers	Nutri-Score label led to the greatest accuracy in identifying the healthier of two snacks but with a minimal effect when only some of the products were labelled

© Burleigh Dodds Science Publishing Limited, 2024. All rights reserved.

No.	Authors	Year	Journal	Methodology	Country	Participants	Key findings
71	Hamlin et al.	2015	*Public Health Nutrition*	In-person experiment	New Zealand	250 consumers (college students)	The presence of FOP labels led to significant and positive changes in consumers' purchase intentions, but these changes were not affected by the nature of FOP labels used, their size, or the product's nutritional status
72	Hamlin and McNeill	2016	Nutrients	Intercept survey with factorial experiment and response form	New Zealand	1200 consumers	The presence of the HSR label had a significant depressive effect on consumers' preferences but this impact was not moderated in any way by the nutritional status expressed by the label
73	Hamlin and McNeill	2018	*Nutrients*	Longitudinal study; intercept survey with factorial experiment and response form	New Zealand	2600 consumers	The HSR label has potentially started to influence consumers' choice, but the impact was still very small and statistically sub-significant relative to other consumer decision inputs shown on the package
74	Hassan and Dimassi	2017	*International Journal of Consumer Studies*	Cross-sectional study with a pre-coded structured questionnaire	Lebanon	748 consumers	The average food label knowledge score was 63.1% and older, obese consumers with kids, suffering from chronic illness or allergies, following a specific diet and residing in big cities scored significantly higher
75	Huang et al.	2021	*Food Quality and Preference*	Lab experiment with eye-tracking	China	24 consumers	Consumers were slower to find a colour-flavour incongruent target than a colour-flavour congruent one

© Burleigh Dodds Science Publishing Limited, 2024. All rights reserved.

76	Hung and Verbeke	2019	*Food Quality and Preference*	Large-scale online survey (questionnaire)	UK, Germany, the Netherlands, Spain, Slovenia, Czechia, France, Denmark, Greece and Lithuania	5337 consumers	The interaction between active ingredients and claimed health benefits might influenced health claim's ratings. The link between health claim use and health relevance was inconsistent
77	Hutton and Gresse	2022	*South African Journal of Clinical Nutrition*	Cross-sectional study; interviewer-administered questionnaire	South Africa	359 consumers	All the evaluative FOP labels significantly improved consumers' ability to identify healthier food products compared to the control group. The reductive RIs showed no significant impact on consumers' understanding
78	Jáuregui et al.	2022	*Public Health Nutrition*	Randomized controlled experiment with online questionnaire	UK, USA, Canada, Australia and Mexico	22 140 consumers	'High in' warning labels had the greatest impact in lowering perceived healthfulness across all countries, compared to the control group. The GDA did not reduce perceived healthfulness in most surveyed countries
79	Jo and Lusk	2018	*Food Quality and Preference*	Online questionnaire	USA, China and Korea	580 consumers	When neutral or negative exogeneous information about healthfulness of food items was provided, US and Korean consumers tended to consider healthier foods more expensive, but this was not the case with Chinese consumers
80	Kergoat et al.	2019	*Journal of Consumer Marketing*	Between-subject design lab experiments	France	254 consumers (undergraduate students)	Consumers exposed to the commercial message ate more sweets than consumers exposed to the health message

© Burleigh Dodds Science Publishing Limited, 2024. All rights reserved.

No.	Authors	Year	Journal	Methodology	Country	Participants	Key findings
81	Khandpur et al.	2018	*Nutrients*	Online randomized controlled experiment	Brazil	1607 consumers	Warning labels assisted consumers improve their understanding of excess nutrient content, ability to identify the healthier product, decrease perceptions of product healthfulness and correctly identify healthier products relative to the control
82	Klopčič et al.	2020	*Food Quality and Preference*	Conjoint analysis and focus group	Slovenia	249 consumers	Slovenian consumers were moderately doubtful of NLHCs. The NLHCs were more important to them when they were choosing their breakfast cereal compared to whether visual images were presented
83	Koen et al.	2018a	*International Journal of Consumer Studies*	Cross-sectional study with interviewer-administered electronic questionnaire	South Africa	960 consumers	36% of consumers frequently/always read nutrition information on food labels. Buying the same type of products all the time (34.28%) and not interested in the nutrition information (31.44%) were the main reasons for not reading the nutrition information
84	Koen et al.	2018b	*International Journal of Consumer Studies*	Focus group (9)	South Africa	67 consumers	Price was sometimes the only consideration when consumers selected foods, irrespective of quality and nutritional value
85	Kontopoulou et al.	2022	*Nutrients*	Online consumer questionnaire	Greece	1278 consumers	Nutri-Score label showed greater improvements compared to the GDA label when classifying products according to their nutritional quality

© Burleigh Dodds Science Publishing Limited, 2024. All rights reserved.

86	Kumar and Kapoor	2017	*British Food Journal*	Questionnaire	India	300 consumers	Consumers attached considerable importance to the labels and read them before making a final purchase decision
87	Liu et al.	2015	*Food Quality and Preference*	Questionnaire	China	660 consumers	70% of consumers rarely, or never, used nutrition labels when shopping for food.
88	Lombardi et al.	2021	*Food Research International*	Questionnaire	Italy	1030 consumers	36% of consumers understood the meaning of the authorized health claims on extra-virgin olive oil
89	Lundeberg et al.	2018	*Appetite*	Online questionnaire	USA	306 consumers (undergraduate students)	Consumers who viewed the star labels rated products with the lowest healthfulness as significantly less healthful and rated products with the highest healthfulness as significantly more healthful compared to consumers who viewed the same products with MTL labels
90	Machín et al.	2017	*Journal of Nutrition Education and Behaviour*	Between-subject design experiment	Uruguay	300 consumers	Low-income consumers perceived ultra-processed products to be significantly more healthful than did middle- and high-income consumers
91	Machín et al.	2018a	*Appetite*	Between-subject design experiment	Uruguay	1182 consumers	The FOP nutrition information effectively improved the average healthfulness of consumers' choices compared to the control
92	Machín et al.	2018b	*Journal of Nutrition Education and Behaviour*	Within- and between-subject design experiments (online questionnaires)	Uruguay	1228 consumers	The inclusion of information about low nutrient content in the MTL label statistically significantly increased the perception of the healthfulness of products with high nutrient content
93	Machín et al.	2019	*Food Quality and Preference*	Lab experiment with eye-tracking	Uruguay	199 consumers	When the warnings were present, consumers selected products with fewer warnings and lower average sodium, saturated fat and sugar content

© Burleigh Dodds Science Publishing Limited, 2024. All rights reserved.

No.	Authors	Year	Journal	Methodology	Country	Participants	Key findings
94	Mantzari et al.	2020	*Appetite*	Lab experiment	UK	401 consumers	Consumers with lower socio-economic position (45%) were more likely to select a sugar-sweetened beverage compared to those with higher socio-economic position but socio-economic position did not moderate the impact of labels on drink selection
95	Mazzu et al.	2021a	*International Journal of Food Sciences and Nutrition*	Cross-country experimental study	France, Germany, Greece, Italy, Portugal, Romania and Spain	2776 consumers	NutrInform Battery label obtained the highest performance across countries and showed limited impact of sociocultural differences
96	Mazzu et al.	2021b	*International Journal of Food Sciences and Nutrition*	Between-subject design experiment	Italy	200 consumers	NutrInform Battery label was found more informative and helpful than Nutri-Score in terms of consumers' understanding of the product composition
97	Mazzu et al.	2022	*International Journal of Food Sciences and Nutrition*	Questionnaires	UK, Italy and France	2812 consumers	A structural link between the usefulness and ease of use of the labels and the way consumers formed their attitudes and intentions towards buying healthier products was suggested

© Burleigh Dodds Science Publishing Limited, 2024. All rights reserved.

	2020	*Food Quality and Preference*	Lab experiments	Singapore	258 consumers	Consumers were willing to pay more for products with added labels and selected larger portions of the 'healthier choice' products	
99	Mediano Stoltze et al.	2021	*Appetite*	Online questionnaire	Chile	497 consumers	Fiber-related claims had the main effect which led to more positive ratings of the products, compared to having no nutrient content claims or fat-related claims
100	Medina-Molina and Pérez-González	2021	*British Food Journal*	Questionnaire	Spain	301 consumers (college students)	The different formats of the interpretative food labels did not moderate the relationship between perceived healthiness and consumers' purchase intention
101	Menger-Ogle and Graham	2018	*Food Quality and Preference*	Interview with stimuli and rating tasks	Nepal	239 consumers	FOP nutrition claims were largely described as useful, despite their inconsistent influence on consumers' perceptions
102	Miklavec et al.	2015	*Food Quality and Preference*	Conjoint and cluster analysis (questionnaire)	Slovenia	371 consumers	Consumers had generally positive preference for a familiar probiotic claim and negative preference for a non-familiar fat metabolism claim
103	Miklavec et al.	2016	*Food Research International*	Conjoint analysis with word-association tasks (online questionnaire)	Slovenia	1050 consumers	Worded claims had less relative importance (29.5%) than the symbols (70.5%), it was shown that careful choice of the wording can affect consumers' preferences considerably

© Burleigh Dodds Science Publishing Limited, 2024. All rights reserved.

No.	Authors	Year	Journal	Methodology	Country	Participants	Key findings
104	Miller et al.	2022	*Appetite*	Experiment with online store and an app	Australia	511 consumers	Warning labels reduced young adults' selection of sugar-sweetened beverages and instead promoted substitution with water
105	Moreira et al.	2019	*International Journal of Consumer Studies*	Online questionnaire	Portugal	308 consumers	Consumers usually did not read food labels due to lack of time and excessive information
106	Neal et al.	2017	*Nutrients*	Randomized controlled trial	Australia	1578 consumers	When compared to the control, there was no difference in the mean healthiness of purchases for HSR, MTL or daily intake but warning labels resulted in healthier food purchases
107	Neuhofer et al.	2020	*Economics & Human Biology*	Randomized controlled trial with eye-tracking	USA	164 consumer	Consumers were more visually attentive to nutrition information displayed by the updated nutrition facts label, but viewing the updated label did not affect choice of beverages nor did it improve the effectiveness of the dietary information on consumers' subsequent choices
108	Ni Mhurchu et al.	2018	*Appetite*	Randomized controlled trial with an app	New Zealand	1255 consumers	Consumers were most likely to view labels for convenience foods, cereals, snacks, bread, bakery and oils and less likely for sugar, honey products, eggs, fish, fruit, vegetables and meat

© Burleigh Dodds Science Publishing Limited, 2024. All rights reserved.

109	Nikolova and Inman	2015	*Journal of Marketing Research*	Large-scale quasi-experiment with shopper panel data set	USA	Data of weekly purchases from over 535 000 members of a grocery chain's frequent shopper programme	The point-of-sale nutrition scoring system assisted consumers in making healthier food choices, such that they switched to higher-scoring products in the post-rollout period
110	Nobrega et al.	2020	*Food Quality and Preference*	Online questionnaire	Brazil	820 consumers	Claims had a significant and positive effect on the perceived healthfulness of all categories. Nutritional warnings had the largest relative importance and significantly decreased perceived healthfulness
111	Nyilasy et al.	2016	*Public Health Nutrition*	Between-subject design experiment	USA	196 consumers	For healthy foods, the nutritional colouring schemes reduced perceived healthiness, irrespective of which nutrients were coloured red or green, and for unhealthy foods there was no significant difference in perceptions of food healthiness
112	Oh et al.	2021	*Psychology & Health*	Scenario-based experiments (online questionnaires)	USA	952 consumers	Nutrition information did not generally affect consumers' choices between low-calorie and high-calorie options, irrespective of dietary restraints
113	Orquin and Scholderer	2015	*Food Policy*	Experiments (online and in-person)	Denmark	810 consumers	The studied claims had little effect on consumer judgements of food healthfulness but had detrimental effects on sensory expectations and purchase intentions
114	Pabst et al.	2021	*Appetite*	Cross-country discrete choice experiment	Australia, Italy and Germany	2176 consumers	Consumers across all countries had a significant positive utility for detailed nutrition information. Ingredient information received positive utility only in Italy

© Burleigh Dodds Science Publishing Limited, 2024. All rights reserved.

No.	Authors	Year	Journal	Methodology	Country	Participants	Key findings
115	Packer et al.	2021	*Nutrients*	Online randomized controlled experiment	UK	4504 consumers	The probability of correct ranking at follow-up was significantly greater for the Nutri-Score, MTL and warning labels compared to control. The largest effects were noted for Nutri-Score and then MTL
116	Packer et al.	2022	*Nutrients*	Online randomized controlled experiment	UK	4530 consumers	FOP labels were perceived favourably, in particular Nutri-Score and MTL
117	Pelly et al.	2020	*Nutrients*	Focus groups (4)	Australia	15 consumers	The nutrition information was viewed positively but consumers showed little confidence in the HSR due to a perceived lack of transparency in the criteria used to determine the number of stars
118	Pettigrew et al.	2021	*Public Health Nutrition*	Online choice tasks (questionnaire)	UK, USA, Canada, Australia, New Zealand, China and India	7545 consumers	Coloured versions of labels, and in particular those with just a summary indicator, outperformed the monochrome version that included nutrient-specific information
119	Pichierri et al.	2021	*British Food Journal*	Online questionnaire	Italy	185 consumers	Clearer health claims were perceived as more useful and may increase product attractiveness and purchase intention
120	Rønnow	2020	*Nutrients*	Home-scan panel, nutritional data, questionnaire	Denmark	1210 household panel	The use of FOP labels increased overall dietary quality, which was driven by reduced intake of added sugar and increased intake of fibre

© Burleigh Dodds Science Publishing Limited, 2024. All rights reserved.

121	Rosenblatt et al.	2018	*Appetite*	Mixed experimental design	Australia	96 consumers	Health warnings primed healthier dietary decision-making behaviour and negatively framed graphic health warnings were the most effective
122	Russell et al.	2017	*Appetite*	Questionnaire	Australia	520 consumers (parents)	Product visuals, followed by star ratings, were the most significant attributes in driving choice. Written claims and other visuals were the least significant
123	Rybak et al.	2021	*Journal of Business Research*	Between-subject design experiments (questionnaire)	USA	1601 consumers	Results indicated significant indirect effects on purchase intentions for processing claims and ingredient lists through the mediators of clean label and nutrition evaluations (in parallel), and perceived healthfulness (in serial)
124	Sarda et al.	2020	*Nutrients*	Online questionnaire	France	4006 consumers	The impact of the Nutri-Score on purchasing behaviours was greater over time on younger populations and frequent label readers
125	Schubert et al.	2021	*Appetite*	Computer-based tasks and EEG	Australia	63 consumers	Both general and specific health warning labels decreased willingness to consume sugar-sweetened beverages perceived as tasty, compared to the control condition
126	Siegrist et al.	2015	*Food Quality and Preference*	Experiment with eye-tracking	Switzerland	98 consumers	Consumers needed more time to process the GDA format compared to the MTL and nutrition table format
127	Siegrist et al.	2019	*British Journal of Nutrition*	Online experiment (questionnaire)	Switzerland	780 consumers	The median accuracy in selecting healthier cereals was only marginally lower in the control condition (91%) compared with the nutrition table (94%), the label (94%) and the combined conditions (97%)

© Burleigh Dodds Science Publishing Limited, 2024. All rights reserved.

No.	Authors	Year	Journal	Methodology	Country	Participants	Key findings
128	Skubisz	2017	*Appetite*	Experiment (online questionnaire)	USA	578 consumers	Consumers estimated that products with the 'natural' claim contained 18% fewer calories despite the fact that such products contained the same number of calories as their regular counterparts
129	Smed et al.	2017	*British Food Journal*	Data set from a consumer panel in two countries	Denmark and the Netherlands	Scanner data on purchase from ~6300 Danish and Dutch households	Households with children tended to have a lower probability of purchasing labelled products compared to other household types. Urbanity increased the probability
130	Smed et al.	2019	*Public Health Nutrition*	Data set from a consumer panel	The Netherlands	Scanner data from 831 to 7216 households	An increase in the market share of products that displayed the label led to an increase in the volume share purchased of products eligible for the label for dairy, yoghurts and sauces
131	Steinhauser et al.	2019a	*Nutrients*	Purchase simulation with eye-tracking	Germany	156 consumers	The longer a consumer looked at a specific claim, the more likely a consumer was to purchase the respective product
132	Steinhauser et al.	2019b	*Appetite*	Purchase simulation with eye-tracking	Germany	156 consumers	Each claim was noticed by at least 85% of the consumers. Consumers looked at health claims longer than nutrition or taste claims
133	Taillie et al.	2022	*Appetite*	Randomized trial in a virtual convenience store	USA	2219 consumers (parents)	Consumers exposed to the text or pictorial sugar warning labels were less likely to select the labelled snack compared to those in the barcode control group
134	Talati et al.	2018	*Public Health Nutrition*	Questionnaire	Australia	1505 consumers	The presence of the HSR label resulted in a small, but significant reduction in the portion size selected as appropriate for consumption of pizzas and cornflakes, compared to no FOP label

© Burleigh Dodds Science Publishing Limited, 2024. All rights reserved.

135	Talati et al.	2019	*Nutrients*	Online questionnaire	UK, USA, Australia, Canada, Argentina, Bulgaria, Denmark, France, Germany, Mexico, Singapore and Spain	12 015 consumers	Consumers who viewed the MTL label provided the most favourable ratings. Perceptions of the other FOP labels were either mixed or neutral
136	Temple et al.	2016	*Journal of Nutrition Education and Behavior*	Lab-based convenience store experiment	USA	36 consumers	Increasing the price of energy drinks reduced purchases and increased purchasing of other caffeinated beverages. Labels affected sales in adolescents
137	Tonnesen et al.	2022	*Food Quality and Preference*	Online choice and portion size experiments (questionnaire)	Denmark	1494 consumers	The choice of products carrying a claim was linked to an increased perceived healthiness, health interest in food and a sense of reward associated with NLHCs and negatively linked to subjective nutritional knowledge
138	Tortora et al.	2019	*Food Research International*	Choice-conjoint analysis with eye-tracking	Uruguay	124 consumers	Nutritional warnings were efficient in attracting consumers' attention and required less time and fewer fixations to process compared to the facts up front panel
139	Van den Akker et al.	2022	*Appetite*	Lab-in-field experiment	The Netherlands	300 consumers	Nutri-Score promoted choice of the healthiest cereal and labels did not affect consumers' serving size selection
140	Vanderlee et al.	2021	*Public Health Nutrition*	Online randomized control trial	Canada	1997 consumers (online commercial panel)	All FOP labels decreased consumers' perceived healthiness of less healthy products

© Burleigh Dodds Science Publishing Limited, 2024. All rights reserved.

No.	Authors	Year	Journal	Methodology	Country	Participants	Key findings
141	Vasiljevic et al.	2015	*Appetite*	Between-subject design experiments	UK	955 consumers	Consumers rated the chocolate bar as tastier and more desirable when compared to the cereal bar, and the cereal bar as healthier than the chocolate bar, regardless of label
142	Velasco Vizcaíno and Velasco	2019	*Journal of Business Research*	Experiments (including online questionnaires)	UK, Ecuador and Chile	1352 consumers (including college students)	Brand familiarity and brand trust together created a shield that protected the product from being evaluated as unhealthy, even when the MTL labels were present
143	Velázquez et al.	2021	*Food Quality and Preference*	Online choice-based conjoint questionnaire	Uruguay	1213 consumers (parents)	Nutrient claim had the strongest effect regardless of the product and increased healthiness perceptions and encouraged consumers' choice
144	Ventsel et al.	2022	*Appetite*	Between-subject design experiments	UK	1472 consumers	Health warning labels and irrelevant aversive labels both reduced motivation towards snack foods, but only when the labels were physically present
145	Vidal et al.	2019	*Public Health Nutrition*	Between-subject design experiment with eye-tracking	Uruguay	201 consumers	The average percentage of consumers that fixated their gaze on the nutritional warnings during the choice task was slightly but significantly higher for consumers who attended to nutrition messages compared to the control group
146	Zhang et al.	2020	*Nutrients*	Food choice task with functional MR imaging	China	44 consumers	Compared to the GDA only, a salient red traffic-light label potentially reduced unhealthy food-related decision-making and activated the superior medial frontal gyrus and the supplementary motor area that are implicated in the execution of responses and monitor inhibition

© Burleigh Dodds Science Publishing Limited, 2024. All rights reserved.

13 References

Abrams, K. M., Evans, C. and Duff, B. R. L. (2015). Ignorance is bliss. How parents of preschool children make sense of front-of-package visuals and claims on food, *Appetite* 87: 20–29. https://doi.org/10.1016/j.appet.2014.12.100.

Acton, R. B. and Hammond, D. (2018). Do manufacturer 'nutrient claims' influence the efficacy of mandated front-of-package labels?, *Public Health Nutrition* 21(18): 3354–3359. https://doi.org/10.1017/S1368980018002550.

Acton, R. B., Vanderlee, L. and Hammond, D. (2018). Influence of front-of-package nutrition labels on beverage healthiness perceptions: results from a randomized experiment, *Preventive Medicine* 115: 83–89. https://doi.org/10.1016/j.ypmed.2018.08.022.

Adasme-Berríos, C., Aliaga-Ortega, L., Schnettler, B., Parada, M., Andaur, Y., Carreño, C., Lobos, G., Jara-Rojas, R. and Valdes, R. (2022). Effect of warning labels on consumer motivation and intention to avoid consuming processed foods, *Nutrients* 14(8): 1547. https://doi.org/10.3390/nu14081547.

Aguenaou, H., El Ammari, L., Bigdeli, M., El Hajjab, A., Lahmam, H., Labzizi, S., Gamih, H., Talouizte, A., Serbouti, C., El Kari., K., Benkirane, H., El Berri, H., AL-Jawaldeh, A. and Yahyane, A. (2021). Comparison of appropriateness of Nutri-Score and other front-of-pack nutrition labels across a group of Moroccan consumers: Awareness, understanding and food choices, *Archives of Public Health* 79(1): 71. https://doi.org/10.1186/s13690-021-00595-3.

Anders, S. and Schroeter, C. (2017). The impact of nutritional supplement intake on diet behavior and obesity outcomes, *PLoS ONE* 12(10): e0185258. https://doi.org/10.1371/journal.pone.0185258.

Anderson, C. L. and O'Connor, E. L. (2019). The effect of the health star rating on consumer decision-making, *Food Quality and Preference* 73: 215–225. https://doi.org/10.1016/j.foodqual.2018.11.005.

Andrews, J. C., Netemeyer, R., Burton, S. and Kees, J. (2021). What consumers actually know: the role of objective nutrition knowledge in processing stop sign and traffic light front-of-pack nutrition labels, *Journal of Business Research* 128: 140–155. https://doi.org/10.1016/j.jbusres.2021.01.036.

Annunziata, A., Pomarici, E., Vecchio, R. and Mariani, A. (2016). Nutritional information and health warnings on wine labels: exploring consumer interest and preferences, *Appetite* 106: 58–69. https://doi.org/10.1016/j.appet.2016.02.152.

Ares, G., Piqueras-Fiszman, B., Varela, P., Marco, R. M., López, A. M. and Fiszman, S. (2011). Food labels: do consumers perceive what semiotics want to convey?, *Food Quality and Preference* 22(7): 689–698. https://doi.org/10.1016/j.foodqual.2011.05.006.

Ares, G., Aschemann-Witzel, J., Curutchet, M. R., Antúnez, L., Moratorio, X. and Bove, I. (2018a). A citizen perspective on nutritional warnings as front-of-pack labels: Insights for the design of accompanying policy measures, *Public Health Nutrition* 21(18): 3450–3461. https://doi.org/10.1017/S1368980018002045.

Ares, G., Aschemann-Witzel, J., Curutchet, M. R., Antúnez, L., Machín, L. and Vidal, L., Martinez, J. and Gimenez, A. (2018b). Nutritional warnings and product substitution or abandonment, *Policy*.

Ares, G., Varela, F., Machín, L., Antúnez, L., Gimenez, A., Curutchet, M. R. and Aschemann-Witzel, J. (2018c). Comparative performance of three interpretative front-of-pack

© Burleigh Dodds Science Publishing Limited, 2024. All rights reserved.

nutrition labelling schemes: insights for policy making, *Food Quality and Preference* 68: 215–225. https://doi.org/10.1016/j.foodqual.2018.03.007.

Ares, G., Aschemann-Witzel, J., Curutchet, M. R., Antúnez, L., Machín, L., Vidal, L., Martínez, J. and Giménez, A. (2018). Nutritional warnings and product substitution or abandonment: Policy implications derived from a repeated purchase simulation, *Food Quality and Preference* 65: 40–48. https://doi.org/10.1016/j.foodqual.2017.12.001.

Arrúa, A., Machín, L., Curutchet, M. R., Martínez, J., Antúnez, L., Alcaire, F., Giménez, A. and Ares, G. (2017). Warnings as a directive front-of-pack nutrition labelling scheme: Comparison with the Guideline Daily Amount and traffic-light systems, *Public Health Nutrition* 20(13): 2308–2317. https://doi.org/10.1017/S1368980017000866.

Asbridge, S. C. M., Pechey, E., Marteau, T. M. and Hollands, G. J. (2021). Effects of pairing health warning labels with energy-dense snack foods on food choice and attitudes: online experimental study, *Appetite* 160: 105090. https://doi.org/10.1016/j.appet.2020.105090.

Bablani, L., Ni Mhurchu, C., Neal, B., Skeels, C. L., Staub, K. E. and Blakely, T. (2020). The impact of voluntary front-of-pack nutrition labelling on packaged food reformulation: A difference-in-differences analysis of the Australasian Health Star Rating scheme, *PLoS Medicine* 17(11): e1003427. https://doi.org/10.1371/journal.pmed.1003427.

Ballco, P., de-Magistris, T. and Caputo, V. (2019). Consumer preferences for nutritional claims: an exploration of attention and choice based on an eye-tracking choice experiment, *Food Research International* 116: 37–48. https://doi.org/10.1016/j.foodres.2018.12.031.

Ballco, P., Caputo, V. and de-Magistris, T. (2020a). Consumer valuation of European nutritional and health claims: do taste and attention matter?, *Food Quality and Preference* 79: 103793. https://doi.org/10.1016/j.foodqual.2019.103793.

Ballco, P., Jurado, F. and Gracia, A. (2020b). Do health claims add value to nutritional claims? Evidence from a close-to-real experiment on breakfast biscuits, *Food Quality and Preference* 85: 103968. https://doi.org/10.1016/j.foodqual.2020.103968.

Bartels, M., Tillack, K. and Lin, C. T. J. (2018). Communicating nutrition information at the point of purchase: an eye-tracking study of shoppers at two grocery stores in the United States, *International Journal of Consumer Studies* 42(5): 557–565. https://doi.org/10.1111/ijcs.12474.

Baum, J. I., Gaines, B. L., Kubas, G. C., Mitchell, C. F. and Russell, S. L. (2017). Educational nutrition messaging at breakfast reduces snack intake and influences snack preferences in adult men and women, *Appetite* 117: 67–73. https://doi.org/10.1016/j.appet.2017.06.009.

Becker, M. W., Bello, N. M., Sundar, R. P., Peltier, C. and Bix, L. (2015). Front of pack labels enhance attention to nutrition information in novel and commercial brands, *Food Policy* 56: 76–86. https://doi.org/10.1016/j.foodpol.2015.08.001.

Benson, T., Lavelle, F., Bucher, T., McCloat, A., Mooney, E., Egan, B., Collins, C. E. and Dean, M. (2018). The impact of nutrition and health claims on consumer perceptions and portion size selection: results from a nationally representative survey, *Nutrients* 10(5): 656. https://doi.org/10.3390/nu10050656.

Benson, T., Lavelle, F., McCloat, A., Mooney, E., Bucher, T., Egan, B. and Dean, M. (2019). Are the claims to blame? A qualitative study to understand the effects of nutrition

© Burleigh Dodds Science Publishing Limited, 2024. All rights reserved.

and health claims on perceptions and consumption of food, *Nutrients* 11(9): 2058. https://doi.org/10.3390/nu11092058.

Benson, T., Bucher, T., Oughton, R., McCloat, A., Mooney, E., Farrell, S. and Dean, M. (2022). The effects of nutrition and health claims on the nutrient composition of single and subsequent meal servings, *Appetite* 176: 106105. https://doi.org/10.1016/j.appet.2022.106105.

Bernard, J. C., Duke, J. M. and Albrecht, S. E. (2019). Do labels that convey minimal, redundant, or no information affect consumer perceptions and willingness to pay?, *Food Quality and Preference* 71: 149–157. https://doi.org/10.1016/j.foodqual.2018.06.012.

Betini, M., Volpato, E. S., Anastácio, G. D., de Faria, R. T. and El Dib, R. (2014). Choosing the right journal for your systematic review, *Journal of Evaluation in Clinical Practice* 20(6): 834–836. https://doi.org/10.1111/jep.12196.

Bialkova, S., Sasse, L. and Fenko, A. (2016). The role of nutrition labels and advertising claims in altering consumers' evaluation and choice, *Appetite* 96: 38–46. https://doi.org/10.1016/j.appet.2015.08.030.

Bignell, J. (2002). *Media Semiotics: an Introduction*, Manchester University Press, Manchester, UK.

Billich, N., Blake, M. R., Backholer, K., Cobcroft, M., Li, V. and Peeters, A. (2018). The effect of sugar-sweetened beverage front-of-pack labels on drink selection, health knowledge and awareness: an online randomised controlled trial, *Appetite* 128: 233–241. https://doi.org/10.1016/j.appet.2018.05.149.

Bossuyt, S., Custers, K., Tummers, J., Verbeyst, L. and Oben, B. (2021). Nutri-Score and Nutrition Facts Panel through the eyes of the consumer: correct healthfulness estimations depend on transparent labels, fixation duration, and product equivocality, *Nutrients* 13(9): 2915. https://doi.org/10.3390/nu13092915.

Boztuğ, Y., Juhl, H. J., Elshiewy, O. and Jensen, M. B. (2015). Consumer response to monochrome Guideline Daily Amount nutrition labels, *Food Policy* 53: 1–8. https://doi.org/10.1016/j.foodpol.2015.03.002.

Bromberg, M., Sinai, T., Keinan-Boker, L., Endevelt, R. and Frankenthal, D. (2022). Current use of nutrition facts tables and attitudes towards new red and green front-of-package labels among Israeli consumers, *International Journal of Food Sciences and Nutrition* 73(2): 230–237. https://doi.org/10.1080/09637486.2021.1955841.

Bryła, P. (2020). Who reads food labels? Selected predictors of consumer interest in front-of-package and back-of-package labels during and after the purchase, *Nutrients* 12(9): 2605. https://doi.org/10.3390/nu12092605.

Cabrera, M., Machín, L., Arrúa, A., Antúnez, L., Curutchet, M. R., Giménez, A. and Ares, G. (2017). Nutrition warnings as front-of-pack labels: Influence of design features on healthfulness perception and attentional capture, *Public Health Nutrition* 20(18): 3360–3371. https://doi.org/10.1017/S136898001700249X.

Carbonneau, E., Perron, J., Drapeau, V., Lamarche, B., Doucet, É., Pomerleau, S. and Provencher, V. (2015). Impact of nutritional labelling on 10-d energy intake, appetite perceptions and attitudes towards food, *British Journal of Nutrition* 114(12): 2138–2147. https://doi.org/10.1017/S0007114515003918.

Carter, K. A. and Gonzalez-Vallejo, C. (2018). Nutrient-specific system versus full fact panel: testing the benefits of nutrient-specific front-of-package labels in a student sample, *Appetite* 125: 512–526. https://doi.org/10.1016/j.appet.2018.03.001.

© Burleigh Dodds Science Publishing Limited, 2024. All rights reserved.

Cavaliere, A., Ricci, E. C. and Banterle, A. (2015). Nutrition and health claims: who is interested? An empirical analysis of consumer preferences in Italy, *Food Quality and Preference* 41: 44–51. https://doi.org/10.1016/j.foodqual.2014.11.002.

Cavaliere, A., De Marchi, E. and Banterle, A. (2016). Does consumer health-orientation affect use of nutrition facts panel and claims? An empirical analysis in Italy, *Food Quality and Preference* 54: 110–116. https://doi.org/10.1016/j.foodqual.2016.07.008.

Cawley, J., Sweeney, M. J., Sobal, J., Just, D. R., Kaiser, H. M., Schulze, W. D., Wethington, E. and Wansink, B. (2015). The impact of a supermarket nutrition rating system on purchases of nutritious and less nutritious foods, *Public Health Nutrition* 18(1): 8–14. https://doi.org/10.1017/S1368980014001529.

Centurión, M., Machín, L. and Ares, G. (2019). Relative impact of nutritional warnings and other label features on cereal bar healthfulness evaluations, *Journal of Nutrition Education and Behavior* 51(7): 850–856. https://doi.org/10.1016/j.jneb.2019.01.021.

Cheah, Y. K., Moy, F. M. and Loh, D. A. (2015). Socio-demographic and lifestyle factors associated with nutrition label use among Malaysian adults, *British Food Journal* 117(11): 2777–2787. https://doi.org/10.1108/BFJ-01-2015-0001.

Choi, H., Northup, T. and Reid, L. N. (2021). How health consciousness and health literacy influence evaluative responses to nutrient-content claimed messaging for an unhealthy food, *Journal of Health Communication* 26(5): 350–359. https://doi.org/10.1080/10810730.2021.1946217.

Clarke, N., Pechey, E., Mantzari, E., Blackwell, A. K. M., De-Loyde, K., Morris, R. W., Munafò, M. R., Marteau, T. M. and Hollands, G. J. (2020), Impact of health warning labels on snack selection: An online experimental study, *Appetite*, 154: 104744. https://doi.org/10.1016/j.appet.2020.104744.

Cooper, S. L., Butcher, L. M., Scagnelli, S. D., Lo, J., Ryan, M. M., Devine, A. and O'Sullivan, T. A. (2020). Australian consumers are willing to pay for the Health Star Rating front-of-pack nutrition label, *Nutrients* 12(12): 3876. https://doi.org/10.3390/nu12123876.

Dallas, S. K., Liu, P. J. and Ubel, P. A. (2015). Potential problems with increasing serving sizes on the nutrition facts label, *Appetite* 95: 577–584. https://doi.org/10.1016/j.appet.2015.08.012.

Dang, A. and Nichols, B. S. (2023). Consumer response to positive nutrients on the facts up front (FUF) label: A comparison between healthy and unhealthy foods and the role of nutrition motivation, *Journal of Marketing Theory and Practice* 31(2): 223–242. https://doi.org/10.1080/10696679.2021.2020662.

De Gennaro, B. C., Roselli, L., Bimbo, F., Carlucci, D., Cavallo, C., Cicia, G., Del Giudice, T., Lombardi, A., Paparella, A. and Vecchio, R. (2021). Do Italian consumers value health claims on extra-virgin olive oil?, *Journal of Functional Foods* 81: 104461. https://doi.org/10.1016/j.jff.2021.104461.

De Temmerman, J., Heeremans, E., Slabbinck, H. and Vermeir, I. (2021). The impact of the Nutri-Score nutrition label on perceived healthiness and purchase intentions, *Appetite* 157: 104995. https://doi.org/10.1016/j.appet.2020.104995.

Defago, D., Geng, J. F., Molina, O. and Santa Maria, D. (2020). Can traffic light nutritional labels induce healthier consumer choices? Experimental evidence from a developing country, *International Journal of Consumer Studies* 44(2): 151–161. https://doi.org/10.1111/ijcs.12554.

Delivett, C. P., Klepacz, N. A., Farrow, C. V., Thomas, J. M., Raats, M. M. and Nash, R. A. (2020). Front-of-pack images can boost the perceived health benefits of dietary products, *Appetite* 155: 104831. https://doi.org/10.1016/j.appet.2020.104831.

© Burleigh Dodds Science Publishing Limited, 2024. All rights reserved.

Deliza, R., de Alcantara, M., Pereira, R. and Ares, G. (2020). How do different warning signs compare with the guideline daily amount and traffic-light system?, *Food Quality and Preference* 80: 103821. https://doi.org/10.1016/j.foodqual.2019.103821.

Devia, G., Forli, S., Vidal, L., Curutchet, M. R. and Ares, G. (2021). References to home-made and natural foods on the labels of ultra-processed products increase healthfulness perception and purchase intention: insights for policy making, *Food Quality and Preference* 88: 104110. https://doi.org/10.1016/j.foodqual.2020.104110.

Donovan, C. M., O'Donovan, C. B., O'Mahony, S., Lyons, O. C., Flynn, M. A. T. and Collins, N. A. (2020). True or false – The Irish consumer's attitude to nutrition and health claims, *Proceedings of the Nutrition Society* 79(OCE3). Nutrition Society Live 2020, 14–15th July, E769. https://doi.org/10.1017/S0029665120007557.

Egnell, M., Ducrot, P., Touvier, M., Allès, B., Hercberg, S., Kesse-Guyot, E. and Julia, C. (2018a). Objective understanding of Nutri-Score front-of-package nutrition label according to individual characteristics of subjects: comparisons with other format labels, *PLoS ONE* 13(8): e0202095. https://doi.org/10.1371/journal.pone.0202095.

Egnell, M., Kesse-Guyot, E., Galan, P., Touvier, M., Rayner, M., Jewell, J., Breda, J., Hercberg, S. and Julia, C. (2018b). Impact of front-of-pack nutrition labels on portion size selection: an experimental study in a French cohort, *Nutrients* 10(9): 1268. https://doi.org/10.3390/nu10091268.

Egnell, M., Talati, Z., Hercberg, S., Pettigrew, S. and Julia, C. (2018c). Objective Understanding of Front-of-Package Nutrition Labels: an international comparative experimental study across 12 countries, *Nutrients* 10(10): 1542. https://doi.org/10.3390/nu10101542.

Egnell, M., Talati, Z., Gombaud, M., Galan, P., Hercberg, S., Pettigrew, S. and Julia, C. (2019). Consumers' responses to front-of-pack nutrition labelling: results from a sample from the Netherlands, *Nutrients* 11(8): 1817. https://doi.org/10.3390/nu11081817.

Embling, R., Price, M., Lee, M. and Wilkinson, L. (2019). Food-variety-focused labelling does not increase ideal portion size, expected fullness or snack intake, *Food Quality and Preference* 73: 46–55. https://doi.org/10.1016/j.foodqual.2018.12.005.

Fenko, A., Nicolaas, I. and Galetzka, M. (2018). Does attention to health labels predict a healthy food choice? An eye-tracking study, *Food Quality and Preference* 69: 57–65. https://doi.org/10.1016/j.foodqual.2018.05.012.

Fialon, M., Egnell, M., Talati, Z., Galan, P., Dreano-Trecant, L., Touvier, M., Pettigrew, S., Hercberg, S. and Julia, C. (2020). Effectiveness of different front-of-pack nutrition labels among Italian consumers: results from an online randomized controlled trial, *Nutrients* 12(8): 2307. https://doi.org/10.3390/nu12082307.

Fichera, E. and von Hinke, S. (2020). The response to nutritional labels: evidence from a quasi-experiment, *Journal of Health Economics* 72: 102326. https://doi.org/10.1016/j.jhealeco.2020.102326.

Fink, A. (2014). *Conducting Research Literature Reviews: From the Internet to Paper* (4th edn.), Sage, Beverly Hills.

Finkelstein, E. A., Ang, F. J. L., Doble, B., Wong, W. H. M. and van Dam, R. M. (2019). A randomized controlled trial evaluating the relative effectiveness of the Multiple Traffic Light and Nutri-Score front of package nutrition labels, *Nutrients* 11(9): 2236. https://doi.org/10.3390/nu11092236.

Finkelstein, E. A., Doble, B., Ang, F. J. L., Wong, W. H. M. and van Dam, R. M. (2021). A randomized controlled trial testing the effects of a positive front-of-pack label with

© Burleigh Dodds Science Publishing Limited, 2024. All rights reserved.

or without a physical activity equivalent label on food purchases, *Appetite* 158: 104997. https://doi.org/10.1016/j.appet.2020.104997.

Fisher, G. (2018). Nutrition labelling reduces valuation of food through multiple health and taste channels, *Appetite* 120: 500–504. https://doi.org/10.1016/j.appet.2017.09.013.

Folkvord, F., Bergmans, N. and Pabian, S. (2021). The effect of the Nutri-Score label on consumer's attitudes, taste perception and purchase intention: an experimental pilot study, *Food Quality and Preference* 94: 104303. https://doi.org/10.1016/j.foodqual.2021.104303.

Follett, L., Bryant, A., Beck, D. and Rice, S. (2022). Investigating the short and long-term effects of "nutritional-score" pricing on food pantry selections, *Appetite* 168: 105692. https://doi.org/10.1016/j.appet.2021.105692.

Food Safety Authority of Ireland and FSAI (2021). Information on nutrition and health claims Available at: https://www.fsai.ie/publications_nutrition_healthclaims/ (Last accessed 15/7/2022).

Franco-Arellano, B., Vanderlee, L., Ahmed, M., Oh, A. and L'Abbe, M. (2020). Influence of front-of-pack labelling and regulated nutrition claims on consumers' perceptions of product healthfulness and purchase intentions: A randomized controlled trial, *Appetite* 149: 104629. https://doi.org/10.1016/j.appet.2020.104629.

Gabor, A. M., Stojnic, B. and Ostic, D. B. (2020). Effects of different nutrition labels on visual attention and accuracy of nutritional quality perception - Results of an experimental eye-tracking study, *Food Quality and Preference* 84: 103948. https://doi.org/10.1016/j.foodqual.2020.103948.

Goiana-da-Silva, F., Cruz-e-Silva, D., Nobre-da-Costa, C., Nunes, A. M., Fialon, M., Egnell, M., Galan, P., Julia, C., Talati, Z., Pettigrew, S., Darzi, A., Araujo, F. and Hercberg, S. (2021). Nutri-Score: the most efficient front-of-pack nutrition label to inform Portuguese consumers on the nutritional quality of foods and help them identify healthier options in purchasing situations, *Nutrients* 13(12): 4335. https://doi.org/10.3390/nu13124335.

González-Vallejo, C., Lavins, B. D. and Carter, K. A. (2016). Analysis of nutrition judgments using the Nutrition Facts Panel, *Appetite* 105: 71–84. https://doi.org/10.1016/j.appet.2016.05.014.

Goodman, S., Vanderlee, L., Acton, R., Mahamad, S. and Hammond, D. (2018). The impact of front-of-package label design on consumer understanding of nutrient amounts, *Nutrients* 10(11): 1624. https://doi.org/10.3390/nu10111624.

Gracia, A. and Barreiro-Hurle, J. (2019). Making sense of information overload: consumer ranking of nutritional claims in cereal based products, *Nutrients* 11(12): 2858. https://doi.org/10.3390/nu11122858.

Gröppel-Klein, A., Franke, C. and Spilski, A. (2020). Do sustainability labels reinforce health claims - and vice versa?, *Marketing ZFP* 42: 3–18. https://doi.org/10.15358/0344-1369-2020-2.

Hackl, C., Lueth, D. and Di Bartolo, T. (2022). *Navigating the Metaverse – A Guide to Limitless Possibilities in a Web 3.0 World*, Wiley, Hoboken, NJ. ISBN: 978-1-119-89899-3.

Hagmann, D. and Siegrist, M. (2020). Nutri-Score, multiple traffic light and incomplete nutrition labelling on food packages: effects on consumers' accuracy in identifying healthier snack options, *Food Quality and Preference* 83: 103894. https://doi.org/10.1016/j.foodqual.2020.103894.

© Burleigh Dodds Science Publishing Limited, 2024. All rights reserved.

Hamlin, R. P., McNeill, L. S. and Moore, V. (2015). The impact of front-of-pack nutrition labels on consumer product evaluation and choice: An experimental study, *Public Health Nutrition* 18(12): 2126–2134. https://doi.org/10.1017/S1368980014002997.

Hamlin, R. and McNeill, L. (2016). Does the Australasian "Health Star Rating" front of pack nutritional label system work?, *Nutrients* 8(6): 327. https://doi.org/10.3390/nu8060327.

Hamlin, R. and McNeill, L. (2018). The impact of the Australasian 'Health Star Rating', front-of-pack nutritional label, on consumer choice: A longitudinal study, *Nutrients* 10(7): 906. https://doi.org/10.3390/nu10070906.

Hassan, H. F. and Dimassi, H. (2017). Usage and understanding of food labels among Lebanese shoppers, *International Journal of Consumer Studies* 41(5): 570–575. https://doi.org/10.1111/ijcs.12368.

Hélias, A., van der Werf, H. M. G., Soler, L. G., Aggeri, F., Dourmad, J.-Y., Julia, C., Nabec, L., Pellerin, S., Ruffieux, B. and Trystram, G. (2022). Implementing environmental labelling of food products in France, *The International Journal of Life Cycle Assessment* 27(7): 926–931. https://doi.org/10.1007/s11367-022-02071-8.

Huang, J. P., Peng, Y. B. and Wan, X. A. (2021). The color-flavor incongruency effect in visual search for food labels: an eye-tracking study, *Food Quality and Preference* 88: 104078. https://doi.org/10.1016/j.foodqual.2020.104078.

Hung, Y. and Verbeke, W. (2019). Consumer evaluation, use and health relevance of health claims in the European Union, *Food Quality and Preference* 74: 88–99. https://doi.org/10.1016/j.foodqual.2019.01.002.

Hutton, T. R. and Gresse, A. (2022). Objective understanding of five front-of-pack labels among consumers in Nelson Mandela Bay, South Africa, *South African Journal of Clinical Nutrition* 35(3): 108–114. https://doi.org/10.1080/16070658.2021.1991674.

Jáuregui, A., White, C. M., Vanderlee, L., Hall, M. G., Contreras-Manzano, A., Nieto, C., Sacks, G., Thrasher, J. F., Hammond, D. and Barquera, S. (2022). Impact of front-of-pack labels on the perceived healthfulness of a sweetened fruit drink: A randomised experiment in five countries, *Public Health Nutrition* 25(4): 1094–1104. https://doi.org/10.1017/S1368980021004535.

Jo, J. and Lusk, J. L. (2018). If it's healthy, it's tasty and expensive: effects of nutritional labels on price and taste expectations, *Food Quality and Preference* 68: 332–341. https://doi.org/10.1016/j.foodqual.2018.04.002.

Kergoat, M., Meyer, T. and Legal, J. B. (2019). Influence of "health" versus "commercial" physical activity message on snacking behavior, *Journal of Consumer Marketing* 37(2): 170–179. https://doi.org/10.1108/JCM-07-2018-2765.

Khandpur, N., de Morais Sato, P., Mais, L. A., Bortoletto Martins, A. P., Spinillo, C. G., Garcia, M. T., Urquizar Rojas, C. F. and Jaime, P. C. (2018). Are front-of-package warning labels more effective at communicating nutrition information than traffic-light labels? A randomized controlled experiment in a Brazilian sample, *Nutrients* 10(6): 688. https://doi.org/10.3390/nu10060688.

Klopčič, M., Slokan, P. and Erjavec, K. (2020). Consumer preference for nutrition and health claims: A multi-methodological approach, *Food Quality and Preference* 82: 103863. https://doi.org/10.1016/j.foodqual.2019.103863.

Koen, N., Wentzel-Viljoen, E., Nel, D. and Blaauw, R. (2018a). Consumer knowledge and use of food and nutrition labelling in South Africa: A cross-sectional descriptive

© Burleigh Dodds Science Publishing Limited, 2024. All rights reserved.

study, *International Journal of Consumer Studies* 42(3): 335–346. https://doi.org/10.1111/ijcs.12422.

Koen, N., Wentzel-Viljoen, E. and Blaauw, R. (2018b). Price rather than nutrition information the main influencer of consumer food purchasing behaviour in South Africa: A qualitative study, *International Journal of Consumer Studies* 42(4): 409–418. https://doi.org/10.1111/ijcs.12434.

Kontopoulou, L., Karpetas, G., Fradelos, E. C., Papathanasiou, I. V., Malli, F., Papagiannis, D., Mantzaris, D., Fialon, M., Julia, C. and Gourgoulianis, K. I. (2021). Online consumer survey comparing different front-of-pack labels in Greece, *Nutrients* 14(1): 46. https://doi.org/10.3390/nu14010046.

Kraus, S., Breier, M. and Dasí-Rodríguez, S. (2020). The art of crafting a systematic literature review in entrepreneurship research, *International Entrepreneurship and Management Journal* 16(3): 1023–1042. https://doi.org/10.1007/s11365-020-00635-4.

Kumar, N. and Kapoor, S. (2017). Do labels influence purchase decisions of food products? Study of young consumers of an emerging market, *British Food Journal* 119(2): 218–229. https://doi.org/10.1108/BFJ-06-2016-0249.

Leathwood, P. D., Richardson, D. P., Sträter, P., Todd, P. M. and van Trijp, H. C. (2007). Consumer understanding of nutrition and health claims: Sources of evidence, *British Journal of Nutrition* 98(3): 474–484. https://doi.org/10.1017/S000711450778697X.

Liu, R. D., Hoefkens, C. and Verbeke, W. (2015). Chinese consumers' understanding and use of a food nutrition label and their determinants, *Food Quality and Preference* 41: 103–111. https://doi.org/10.1016/j.foodqual.2014.11.007.

Lombardi, A., Carlucci, D., Cavallo, C., De Gennaro, B., Del Giudice, T., Giannoccaro, G., Paparella, A., Roselli, L., Vecchio, R. and Cicia, G. (2021). Do consumers understand health claims on extra-virgin olive oil?, *Food Research International* 143: 110267. https://doi.org/10.1016/j.foodres.2021.110267.

Lundeberg, P. J., Graham, D. J. and Mohr, G. S. (2018). Comparison of two front-of-package nutrition labeling schemes, and their explanation, on consumers' perception of product healthfulness and food choice, *Appetite* 125: 548–556. https://doi.org/10.1016/j.appet.2018.02.027.

Ma, G. and Zhuang, X. (2021). Nutrition label processing in the past 10 years: contributions from eye tracking approach, *Appetite* 156: 104859. https://doi.org/10.1016/j.appet.2020.104859.

Machín, L., Cabrera, M., Curutchet, M. R., Martínez, J., Giménez, A. and Ares, G. (2017). Consumer perception of the healthfulness of ultra-processed products featuring different front-of-pack nutrition labeling schemes, *Journal of Nutrition Education and Behavior* 49(4): 330–338.e1. https://doi.org/10.1016/j.jneb.2016.12.003.

Machín, L., Aschemann-Witzel, J., Curutchet, M. R., Gimenez, A. and Ares, G. (2018a). Does front-of-pack nutrition information improve consumer ability to make healthful choices? Performance of warnings and the traffic light system in a simulated shopping experiment, *Appetite* 121: 55–62. https://doi.org/10.1016/j.appet.2017.10.037.

Machín, L., Aschemann-Witzel, J., Curutchet, M. R., Giménez, A. and Ares, G. (2018b). Traffic light system can increase healthfulness perception: implications for policy making, *Journal of Nutrition Education and Behavior* 50(7): 668–674. https://doi.org/10.1016/j.jneb.2018.03.005.

Machín, L., Curutchet, M. R., Gimenez, A., Aschemann-Witzel, J. and Ares, G. (2019). Do nutritional warnings do their work? Results from a choice experiment involving snack

© Burleigh Dodds Science Publishing Limited, 2024. All rights reserved.

products, *Food Quality and Preference* 77: 159–165. https://doi.org/10.1016/j.foodqual.2019.05.012.

Mandle, J., Tugendhaft, A., Michalow, J. and Hofman, K. (2015). Nutrition labelling: A review of research on consumer and industry response in the global South, *Global Health Action* 8: 25912. https://doi.org/10.3402/gha.v8.25912.

Mantzari, E., Pechey, R., Codling, S., Sexton, O., Hollands, G. J. and Marteau, T. M. (2020). The impact of 'on-pack' pictorial health warning labels and calorie information labels on drink choice: A laboratory experiment, *Appetite* 145: 104484. https://doi.org/10.1016/j.appet.2019.104484.

Marr, B. (2022). *How to Create Your Digital Avatar for the Metaverse, Forbes*. Available at: https://www.forbes.com/sites/bernardmarr/2022/05/27/how-to-create-your-digital-avatar-for-the-metaverse/ (Last accessed 19/07/2022).

Mazzu, M. F., Romani, S., Baccelloni, A. and Gambicorti, A. (2021a). A cross-country experimental study on consumers' subjective understanding and liking on front-of-pack nutrition labels, *International Journal of Food Sciences and Nutrition* 72(6): 833–847. https://doi.org/10.1080/09637486.2021.1873918.

Mazzu, M. F., Romani, S. and Gambicorti, A. (2021b). Effects on consumers' subjective understanding of a new front-of-pack nutritional label: A study on Italian consumers, *International Journal of Food Sciences and Nutrition* 72(3): 357–366. https://doi.org/10.1080/09637486.2020.1796932.

Mazzu, M. F., Romani, S., Baccelloni, A. and Lavini, L. (2022). Introducing the Front-Of-Pack Acceptance Model: The role of usefulness and ease of use in European consumers' acceptance of front-of-pack labels, *International Journal of Food Sciences and Nutrition* 73(3): 378–395. https://doi.org/10.1080/09637486.2021.1980866.

McCrickerd, K., Tang, C. S. and Forde, C. G. (2020). The independent and combined impact of front-of-pack labelling and sensory quality on calorie estimations and portion selection of commercial food products, *Food Quality and Preference* 79: 103766. https://doi.org/10.1016/j.foodqual.2019.103766.

Mediano Stoltze, F., Busey, E., Taillie, L. S. and Dillman Carpentier, F. R. (2021). Impact of warning labels on reducing health halo effects of nutrient content claims on breakfast cereal packages: A mixed-measures experiment, *Appetite* 163: 105229. https://doi.org/10.1016/j.appet.2021.105229.

Medina-Molina, C. and Pérez-González, B. (2020). Nutritional labelling and purchase intention interaction of interpretative food labels with consumers' beliefs and decisions, *British Food Journal* 123(2): 754–770. https://doi.org/10.1108/BFJ-04-2020-0353.

Menger-Ogle, A. D. and Graham, D. J. (2018). The influence of front-of-package nutrition claims on food perceptions and purchase intentions among Nepali consumers, *Food Quality and Preference* 66: 160–170. https://doi.org/10.1016/j.foodqual.2017.12.017.

Miklavec, K., Pravst, I., Grunert, K. G., Klopčič, M. and Pohar, J. (2015). The influence of health claims and nutritional composition on consumers' yoghurt preferences, *Food Quality and Preference* 43: 26–33. https://doi.org/10.1016/j.foodqual.2015.02.006.

Miklavec, K., Pravst, I., Raats, M. M. and Pohar, J. (2016). Front of package symbols as a tool to promote healthier food choices in Slovenia: accompanying explanatory claim can considerably influence the consumer's preferences, *Food Research International* 90: 235–243. https://doi.org/10.1016/j.foodres.2016.10.052.

© Burleigh Dodds Science Publishing Limited, 2024. All rights reserved.

Miller, C., Ettridge, K., Pettigrew, S., Wittert, G., Wakefield, M., Coveney, J., Roder, D., Martin, J., Brownbill, A. and Dono, J. (2022). Warning labels and interpretive nutrition labels: impact on substitution between sugar and artificially sweetened beverages, juice and water in a real-world selection task, *Appetite* 169: 105818. https://doi.org/10.1016/j.appet.2021.105818.

Moreira, M. J., Garcia-Diez, J., de Almeida, J. M. M. M. and Saraiva, C. (2019). Evaluation of food labelling usefulness for consumers, *International Journal of Consumer Studies* 43(4): 327–334. https://doi.org/10.1111/ijcs.12511.

Narver, J. C. and Slater, S. T. (1990). The effect of a market orientation on business profitability, *Journal of Marketing* 54(4): 20–35. https://doi.org/10.2307/1251757.

Neal, B., Crino, M., Dunford, E., Gao, A., Greenland, R., Li, N., Ngai, J., Ni Mhurchu, C., Pettigrew, S., Sacks, G., Webster, J. and Wu, J. H. Y. (2017). Effects of different types of front-of-pack labelling information on the healthiness of food purchases: A randomised controlled trial, *Nutrients* 9(12): 1284. https://doi.org/10.3390/nu9121284.

Neuhofer, Z., McFadden, B. R., Rihn, A., Wei, X., Khachatryan, H. and House, L. (2020). Can the updated nutrition facts label decrease sugar-sweetened beverage consumption?, *Economics and Human Biology* 37: 100867. http://doi.org/10.1016/j.ehb.2020.100867.

Ni Mhurchu, C., Eyles, H., Jiang, Y. N. and Blakely, T. (2018). Do nutrition labels influence healthier food choices? Analysis of label viewing behaviour and subsequent food purchases in a labelling intervention trial, *Appetite* 121: 360–365. https://doi.org/10.1016/j.appet.2017.11.105.

Nikolova, H. D. and Inman, J. J. (2015). Healthy choice: the effect of simplified point-of-sale nutritional information on consumer food choice behavior, *Journal of Marketing Research* 52(6): 817–835. https://doi.org/10.1509/jmr.13.0270.

Nobrega, L., Ares, G. and Deliza, R. (2020). Are nutritional warnings more efficient than claims in shaping consumers' healthfulness perception?, *Food Quality and Preference* 79: 103749. https://doi.org/10.1016/j.foodqual.2019.103749.

Nyilasy, G., Lei, J., Nagpal, A. and Tan, J. (2016). Colour correct: The interactive effects of food label nutrition colouring schemes and food category healthiness on health perceptions, *Public Health Nutrition* 19(12): 2122–2127. https://doi.org/10.1017/S1368980016000483.

Oh, G. G., Huh, Y. E. and Mukhopadhyay, A. (2021). Informed indulgence: The effects of nutrition information provision and dietary restraint on consecutive food consumption decisions, *Psychology and Health* 36(11): 1314–1335. https://doi.org/10.1080/08870446.2020.1841764.

Orquin, J. L. and Scholderer, J. (2015). Consumer judgments of explicit and implied health claims on foods: misguided but not misled, *Food Policy* 51: 144–157. https://doi.org/10.1016/j.foodpol.2015.01.001.

Pabst, E., Corsi, A. M., Vecchio, R., Annunziata, A. and Loose, S. M. (2021). Consumers' reactions to nutrition and ingredient labelling for wine - A cross-country discrete choice experiment, *Appetite* 156: 104843. https://doi.org/10.1016/j.appet.2020.104843.

Packer, J., Russell, S. J., Ridout, D., Hope, S., Conolly, A., Jessop, C., Robinson, O. J., Stoffel, S. T., Viner, R. M. and Croker, H. (2021). Assessing the effectiveness of front of pack labels: findings from an online randomised-controlled experiment

© Burleigh Dodds Science Publishing Limited, 2024. All rights reserved.

in a representative British sample, *Nutrients* 13(3): 900. https://doi.org/10.3390/nu13030900.

Packer, J., Russell, S. J., Ridout, D., Conolly, A., Jessop, C., Viner, R. M. and Croker, H. (2022). Secondary outcomes of a front-of-pack-labelling randomised controlled experiment in a representative British sample: understanding, ranking speed and perceptions, *Nutrients* 14(11): 2188. https://doi.org/10.3390/nu14112188.

Pelly, F. E., Swanepoel, L., Rinella, J. and Cooper, S. (2020). Consumers' perceptions of the Australian Health Star Rating labelling scheme, *Nutrients* 12(3): 704. https://doi.org/10.3390/nu12030704.

Pettigrew, S., Dana, L. M., Talati, Z., Tian, M. Y. and Praveen, D. (2021). The role of colour and summary indicators in influencing front-of-pack food label effectiveness across seven countries, *Public Health Nutrition* 24(11): 3566–3570. https://doi.org/10.1017/S1368980020004966.

Pichierri, M., Peluso, A. M., Pino, G. and Guido, G. (2020). Communicating the health value of extra-virgin olive oil: An investigation of consumers' responses to health claims, *British Food Journal* 123(2): 492–508. https://doi.org/10.1108/BFJ-03-2020-0198.

Rønnow, H. N. (2020). The effect of front-of-pack nutritional labels and back-of-pack tables on dietary quality, *Nutrients* 12(6): 1704. https://doi.org/10.3390/nu12061704.

Rosenblatt, D. H., Bode, S., Dixon, H., Murawski, C., Summerell, P., Ng, A. and Wakefield, M. (2018). Health warnings promote healthier dietary decision making: effects of positive versus negative message framing and graphic versus text-based warnings, *Appetite* 127: 280–288. https://doi.org/10.1016/j.appet.2018.05.006.

Russell, C. G., Burke, P. F., Waller, D. S. and Wei, E. (2017). The impact of front-of-pack marketing attributes versus nutrition and health information on parents' food choices, *Appetite* 116: 323–338. https://doi.org/10.1016/j.appet.2017.05.001.

Rybak, G., Burton, S., Johnson, A. M. and Berry, C. (2021). Promoted claims on food product packaging: comparing direct and indirect effects of processing and nutrient content claims, *Journal of Business Research* 135: 464–479. https://doi.org/10.1016/j.jbusres.2021.06.036.

Saab, N. (2021). New research supports informative food labels for improved nut allergy care, *Food Safety Magazine* (Last accessed 21 September) Available at: https://www.food-safety.com/articles/7354-new-research-supports-informative-food-labels-for-improved-nut-allergy-care (Last accessed 2/6/2022).

Sarda, B., Julia, C., Serry, A. J. and Ducrot, P. (2020). Appropriation of the front-of-pack nutrition label Nutri-Score across the French population: evolution of awareness, support, and purchasing behaviors between 2018 and 2019, *Nutrients* 12(9): 2887. https://doi.org/10.3390/nu12092887.

Schubert, E., Smith, E., Brydevall, M., Lynch, C., Ringin, E., Dixon, H., Kashima, Y., Wakefield, M. and Bode, S. (2021). General and specific graphic health warning labels reduce willingness to consume sugar-sweetened beverages, *Appetite* 161: 105141. https://doi.org/10.1016/j.appet.2021.105141.

Siegrist, M., Leins-Hess, R. and Keller, C. (2015). Which front-of-pack nutrition label is the most efficient one? The results of an eye-tracker study, *Food Quality and Preference* 39: 183–190. https://doi.org/10.1016/j.foodqual.2014.07.010.

Siegrist, M., Hartmann, C. and Lazzarini, G. A. (2019). Healthy choice label does not substantially improve consumers' ability to select healthier cereals: results of an

© Burleigh Dodds Science Publishing Limited, 2024. All rights reserved.

online experiment, *British Journal of Nutrition* 121(11): 1313–1320. https://doi.org/10.1017/S0007114519000448.

Skubisz, C. (2017). Naturally good: front-of-package claims as message cues, *Appetite* 108: 506–511. https://doi.org/10.1016/j.appet.2016.10.030.

Smed, S., Edenbrandt, A. K., Koch-Hansen, P. and Jansen, L. (2017). Who is the purchaser of nutrition-labelled products?, *British Food Journal* 119(9): 1934–1952. https://doi.org/10.1108/BFJ-11-2016-0552.

Smed, S., Edenbrandt, A. K. and Jansen, L. (2019). The effects of voluntary front-of-pack nutrition labels on volume shares of products: The case of the Dutch Choices, *Public Health Nutrition* 22(15): 2879–2890. https://doi.org/10.1017/S1368980019001423.

Steinhauser, J., Janssen, M. and Hamm, U. (2019a). Who buys products with nutrition and health claims? A purchase simulation with eye tracking on the influence of consumers' nutrition knowledge and health motivation, *Nutrients* 11(9): 2199. https://doi.org/10.3390/nu11092199.

Steinhauser, J., Janssen, M. and Hamm, U. (2019b). Consumers' purchase decisions for products with nutrition and health claims: what role do product category and gaze duration on claims play?, *Appetite* 141: 104337. https://doi.org/10.1016/j.appet.2019.104337.

Snyder, H. (2019). Literature review as a research methodology: an overview and guidelines, *Journal of Business Research* 104: 333–339. https://doi.org/10.1016/j.jbusres.2019.07.039.

Soederberg Miller, L. M. and Cassady, D. L. (2015). Review of the literature on the effects of nutrition knowledge on food label use: a review of the literature, *Appetite* 92: 207–216. https://doi.org/10.1016/j.appet.2015.05.029.

Soederberg Miller, L. M., Sutter, C., Wilson, M. D., Bergman, J. J., Beckett, L. A. and Gibson, T. N. (2019). Assessment of an e-training tool for college students to improve accuracy and reduce effort associated with reading nutrition labels, *Journal of American College Health: J of ACH* 67(5): 441–448. https://doi.org/10.1080/07448481.2018.1484369.

Song, J., Brown, M. K., Tan, M., MacGregor, G. A., Webster, J., Campbell, N. R. C., Trieu, K., Ni Mhurchu, C., Cobb, L. K. and He, F. J. (2021). Impact of color-coded and warning nutrition labelling schemes: A systematic review and network meta-analysis, *PLoS Medicine* 18(10): e1003765. https://doi.org/10.1371/journal.pmed.1003765.

Taillie, L. S., Higgins, I. C. A., Lazard, A. J., Miles, D. R., Blitstein, J. L. and Hall, M. G. (2022). Do sugar warning labels influence parents' selection of a labeled snack for their children? A randomized trial in a virtual convenience store, *Appetite* 175: 106059. https://doi.org/10.1016/j.appet.2022.106059.

Talati, Z., Pettigrew, S., Kelly, B., Ball, K., Neal, B., Dixon, H., Shilton, T. and Miller, C. (2018). Can front-of-pack labels influence portion size judgements for unhealthy foods?, *Public Health Nutrition* 21(15): 2776–2781. https://doi.org/10.1017/S1368980018001702.

Talati, Z., Egnell, M., Hercberg, S., Julia, C. and Pettigrew, S. (2019). Consumers' perceptions of five front-of-package nutrition labels: an experimental study across 12 countries, *Nutrients* 11(8): 1934. https://doi.org/10.3390/nu11081934.

Temple, J. L., Ziegler, A. M. and Epstein, L. H. (2016). Influence of price and labeling on energy drink purchasing in an experimental convenience store, *Journal of Nutrition Education and Behavior* 48(1): 54–59.e1. https://doi.org/10.1016/j.jneb.2015.08.007.

© Burleigh Dodds Science Publishing Limited, 2024. All rights reserved.

Terry, Q. and Keeney, S. D. S. (2022). *The Metaverse Handbook – Innovating for the Internet's Next Tectonic Shift*, Wiley, Hoboken, NJ. ISBN: 978-1-119-89254-0.

Thiene, M., Scarpa, R., Longo, A. and Hutchinson, W. G. (2018). Types of front pack food labels: do obese consumers care? Evidence from Northern Ireland, *Food Policy* 80: 84–102. https://doi.org/10.1016/j.foodpol.2018.09.004.

Tonnesen, M. T., Hansen, S., Laasholdt, A. V. and Lahteenmaki, L. (2022). The impact of positive and reduction health claims on consumers' food choices, *Food Quality and Preference* 98: 104526. https://doi.org/10.1016/j.foodqual.2022.104526.

Tortora, G., Machín, L. and Ares, G. (2019). Influence of nutritional warnings and other label features on consumers' choice: results from an eye-tracking study, *Food Research International* 119: 605–611. https://doi.org/10.1016/j.foodres.2018.10.038.

United Nations (2020). World population ageing 2020: highlights. Available at: https://www.un.org/development/desa/pd/sites/www.un.org.development.desa.pd/files/undesa_pd-2020_world_population_ageing_highlights.pdf (Last accessed 14/6/2022).

US Food and Drug Administration - FDA (2022). Label claims for conventional foods and dietary supplements. Available at: https://www.fda.gov/food/food-labeling-nutrition/label-claims-conventional-foods-and-dietary-supplements (Last accessed 14/7/2022).

van Buul, V. J. and Brouns, F. J. P. H. (2015). Nutrition and health claims as marketing tools, *Critical Reviews in Food Science and Nutrition* 55(11): 1552–1560. https://doi.org/10.1080/10408398.2012.754738.

van den Akker, K., Bartelet, D., Brouwer, L., Luijpers, S., Nap, T. and Havermans, R. (2022). The impact of the Nutri-Score on food choice: A choice experiment in a Dutch supermarket, *Appetite* 168: 105664. https://doi.org/10.1016/j.appet.2021.105664.

Vanderlee, L., Franco-Arellano, B., Ahmed, M., Oh, A., Lou, W. and L'Abbe, M. R. (2021). The efficacy of 'high in' warning labels, health star and traffic light front-of-package labelling: An online randomised control trial, *Public Health Nutrition* 24(1): 62–74. https://doi.org/10.1017/S1368980020003213.

Van Rijmenam, M. (2022). *Step into the Metaverse – How the Immersive Internet Will Unlock a Trillion-Dollar Social Economy*, Wiley, Hoboken, NJ. ISBN: 978-1-119-88757-7.

Vasiljevic, M., Pechey, R. and Marteau, T. M. (2015). Making food labels social: the impact of colour of nutritional labels and injunctive norms on perceptions and choice of snack foods, *Appetite* 91: 56–63. https://doi.org/10.1016/j.appet.2015.03.034.

Velasco Vizcaíno, F. and Velasco, A. (2019). The battle between brands and nutritional labels: how brand familiarity decreases consumers' alertness toward traffic light nutritional labels, *Journal of Business Research* 101: 637–650. https://doi.org/10.1016/j.jbusres.2019.02.054.

Velázquez, A. L., Alcaire, F., Vidal, L., Varela, P., Næs, T. and Ares, G. (2021). The influence of label information on the snacks parents choose for their children: individual differences in a choice based conjoint test, *Food Quality and Preference* 94: 104296. https://doi.org/10.1016/j.foodqual.2021.104296.

Ventsel, M., Pechey, E., De-Loyde, K., Pilling, M. A., Morris, R. W., Maistrello, G., Ziauddeen, H., Marteau, T. M., Hollands, G. J. and Fletcher, P. C. (2022). Effect of health warning labels on motivation towards energy-dense snack foods: Two experimental studies, *Appetite*, 175: 106084. https://doi.org/10.1016/j.appet.2022.106084.

© Burleigh Dodds Science Publishing Limited, 2024. All rights reserved.

Vidal, G., Machín, L., Aschemann-Witzel, J. and Ares, G. (2019). Does message framing matter for promoting the use of nutritional warnings in decision making?, *Public Health Nutrition* 22(16): 3025–3034. https://doi.org/10.1017/S1368980019002507.

Wennström, P. and Mellentin, J. (2003). *The Food and Health Marketing Handbook*, New Nutrition Business, Hammersmith, London, UK.

Wilson, L. and Lusk, J. L. (2020). Consumer willingness to pay for redundant food labels, *Food Policy* 97: 101938. https://doi.org/10.1016/j.foodpol.2020.101938.

Zhang, X. M., Liu, Y., Gu, Y., Wang, S. R. and Chen, H. (2020). Red for "Stop": "Traffic-Light" nutrition labels decrease unhealthy food choices by increasing activity and connectivity in the frontal lobe, *Nutrients* 12(1): 128. https://doi.org/10.3390/nu12010128.

© Burleigh Dodds Science Publishing Limited, 2024. All rights reserved.

Chapter 7

Trends in consumer preference for locally sourced food products

Matthew Gorton and Barbara Tocco, Newcastle University, UK; Péter Csillag, Eco-Sensus Research and Communication Non-profit Ltd., Hungary; Jelena Filipović, University of Belgrade, Serbia; and John White, University of Plymouth, UK

1 Introduction

Over the last two decades, academic and consumer interest in local foods has grown substantially (Kneafsey et al., 2013; Bianchi and Mortimer, 2015; Feldmann and Hamm, 2015; Enthoven and Van den Broeck, 2021; Banerjee and Quinn, 2022). Local food systems can generate higher margins for producers, better-paying jobs, and increased local economic multipliers compared to 'long' supermarket-led supply chains (Malak-Rawlikowska et al., 2019) while also delivering benefits for consumers (Kneafsey et al., 2013; Denver et al., 2019; O'Neill et al., 2022). The latter relate to improved access to unique high-quality food products, with superior intrinsic characteristics (e.g. freshness, seasonality, and good farming methods). European citizens overwhelmingly value local foods - and report placing increasing importance on 'locational' factors, such as traditional knowledge, know-how, and geographical area in their food choices (Kneafsey et al., 2013; European Commission, 2020), especially since the coronavirus pandemic (EIT Food, 2021). Yet, within the

http://dx.doi.org/10.19103/AS.2023.0129.10
© Burleigh Dodds Science Publishing Limited, 2024. All rights reserved.

academic literature, there is a lack of agreement regarding how to define local foods, and consumers typically struggle to define precisely what makes a food local (Eriksen, 2013; Enthoven and Van den Broeck, 2021; Banerjee and Quinn, 2022). Moreover, consumers' positive attitudes to local foods often do not translate into sales (Brečić et al., 2021) – by most definitions, local foods' market share in Europe remains small, albeit with important variations across states (European Parliament, 2016). In the US, local food channels account for less than 4% of total farm sales despite overwhelmingly favourable consumer attitudes (Martinez and Park, 2021). Consequently, this chapter reviews the literature regarding the conceptualisation and consumer understandings of local foods, motivations, and barriers to purchase, as well as customer profiles, drawing largely on evidence from Europe and North America. A case study of cooperative labelling for local and regional food markets in southern Hungary illustrates the market opportunities and frequently encountered barriers faced by local food initiatives. The chapter concludes by discussing a future agenda for research.

2 Defining local food

There are no universally accepted definitions of local foods or local food systems (Martinez et al., 2010; Kneafsey et al., 2013; Augère-Granier, 2016). However, it is possible to distinguish three broad approaches to defining 'local' within the literature. The most common approach defines local food geographically. This can be in terms of the geographical distance between the point of production and consumption (e.g. within a pre-defined kilometre radius) or that which is produced and consumed within a specific geographical entity, such as a local authority district, region, or country (Bingen et al., 2011; Racine et al., 2013). In some cases, the specified geographical area can be particularly large – the 2008 US Farm Act defines local food as that which is sold within 400 miles of its origin or within the state in which it is produced (Martinez et al., 2010). While a formal definition of 'local' is also missing in European legislation, geographical based rules for the direct selling of products via farmers' markets typically exist at the national level to define the eligibility of local markets for support as part of rural development programmes (Kneafsey et al., 2013). For instance, in France, the radius is set at 80 km around the farm, with variations up to 200 km. In Hungary, regulations specify a 40 km radius. In Italy, where 'zero kilometre' often denotes local food, the typical radius is set at 70 km, although regions are autonomous in establishing their own rules, sometimes setting a ceiling for a carbon footprint in transport, other times considering the regional administrative level where the farm is based (the latter, for instance, is the criterion used by Coldiretti's *Campagna Amica*, the largest network of farmers' markets in

© Burleigh Dodds Science Publishing Limited, 2024. All rights reserved.

the country). This often results in contrasting rules for selling via farmers' markets (geographical distance vs. regional boundary). Such a geographical approach, however, does not distinguish across types and methods of food production systems, so that, all that is produced and consumed within the spatial boundary counts as local, even industrial-scale food manufacturing. When defined in this manner, the claims or benefits of local food, for instance, regarding sustainability and food quality, may be limited and problematic (Enthoven and Van den Broeck, 2021; Young, 2022).

A second approach defines local foods in terms of supply chain characteristics and relational proximity (Eriksen, 2013). According to this approach, either producers and consumers directly exchange or only a single intermediary is present (Mount, 2012). According to this approach, short food supply chains, in contrast to long/globalised, conventional, or supermarket-led supply chains, produce local foods (Kneafsey et al., 2013). Short food supply chains are typically of marginal importance in terms of their share of total food sales in Western Economies, and encompass, amongst other forms, farmers' markets, organic box schemes, pick your own, and community-supported agriculture (CSA) (Plakias et al., 2020). While most short food supply chains typically operate within a tightly defined geographical area, this may not always be the case. For instance, a food producer could sell directly to consumers worldwide via digital platforms and social media channels.

The third definitional approach focuses on the qualities and values embodied by the food (Eriksen, 2013). According to Duram and Oberholtzer (2010, p. 100), the values associated with local food 'typically include environmental sustainability, social justice, organic production, support of local and regional farmers, as well as eating seasonally'. Consequently, local foods are those that deliver wider socio-economic and environmental benefits to communities as well as consumers (Eriksen, 2013). The latter may be in the form of increased consumer awareness about food production and better access to quality food, in terms of freshness and nutritional/health benefits. According to this approach, the proportion of food produced/sold within European geographical territories that can be classed as local is uncertain, as defining local food in terms of qualities and values presents problems regarding the criteria for exclusion and trade-offs between different sustainability dimensions (Young, 2022) – the value of different food systems for communities is rarely appraised and may not be easily quantified (Bellassen et al., 2022). Generally, there is a lack of cross-country comparable data regarding the outcomes of local food systems (Malak-Rawlikowska et al., 2019), with the empirical evidence showing considerable heterogeneity across types of supply chains, production systems, and regional contexts (Vittersø et al., 2019; Enthoven and Van den Broeck, 2021).

© Burleigh Dodds Science Publishing Limited, 2024. All rights reserved.

3 Consumer understanding of local food and motivations for purchase

Consumers' understanding of local foods typically combines elements of all three definitional approaches previously discussed (Megicks et al., 2012; Kumar and Smith, 2017; Banerjee and Quinn, 2022). For instance, Banerjee and Quinn (2022) found that consumers regarded local food as embodying more than merely the geographical/social distance between production and consumption. Rather, consumers understand local foods as a multi-layered concept, with several meanings and associations around authenticity and provenance, including supporting the local economy, self and group identity, and quality attributes such as freshness, taste, and sense of enjoyment (Tregear and Ness, 2005; Chambers et al., 2007; Cerjak et al., 2014; Memery et al., 2015; Banerjee and Quinn, 2022; O'Neill et al., 2022). Local foods can also engender emotional connections, linked to social pride and nostalgia (Cerjak et al., 2014; Banerjee and Quinn, 2022). For many consumers, local foods are morally superior (Memery et al., 2015; Banerjee and Quinn, 2022). However, while consumers' associations of local foods are overwhelmingly positively weighted, the associations of local foods are rarely exclusive. For instance, while consumers may regard local foods as tasty and nostalgia inducing, many non-local foods may also possess such associations.

Regarding local foods, both intrinsic and extrinsic motivations are salient. Intrinsic motivations relate to superior sensory properties, as perceived by consumers, particularly regarding freshness and taste (Weatherell et al., 2003; Chambers et al., 2007; Megicks et al., 2012). Consumers may also regard local foods as safer, familiar, and more trustworthy (Banerjee and Quinn, 2022) and offer superior traceability and transparency in production and marketing practices (Birch et al., 2018). Extrinsic motivations, which relate to separable outcomes beyond the act of eating (Bingen et al., 2011), encompass supporting local producers (Toler et al., 2009), environmental protection, and community cohesion. Fulfilling extrinsic motivations may lead to consumers feeling a 'warm glow' of personal satisfaction (Giebelhausen et al., 2016). Consequently, Banerjee and Quinn (2022) argue that a sense of responsibility (moral obligation) towards supporting the local food economy generates amongst purchasers positive emotional feelings such as pride in the local food products they consume (moral satisfaction). However, the salience of specific intrinsic and extrinsic motivations varies across consumers, and generally, intrinsic motivations are of greater importance (Weatherell et al., 2003; Megicks et al., 2012; Birch et al., 2018).

4 Consumer profiles and predicting behaviour

Several studies profile local food purchasers, considering attitudinal, socio-economic, and demographic characteristics. Evidence regarding the latter

© Burleigh Dodds Science Publishing Limited, 2024. All rights reserved.

two factors is inconclusive and often conflicting (Feldmann and Hamm, 2015; Birch et al., 2018), although there is some evidence that wealthier (Brown et al., 2009), rurally located (Megicks et al., 2012; Racine et al., 2013), and female (Weatherell et al., 2003; Megicks et al., 2012) consumers are more likely to purchase local foods. *Ceteris paribus*, those with relatively low levels of educational attainment are less likely to purchase local foods (Cranfield et al., 2012; Evola et al., 2022). However, results regarding the effects of gender, age, and income lack consistency and strong explanatory power when seeking to understand the differences between purchasers and non-purchasers of local foods (Zepeda and Li, 2006; Cranfield et al., 2012; Feldmann and Hamm, 2015; Birch et al., 2018).

A drawback of the literature is a reliance on single-country studies, which limits the ability to understand the roles of culture and structural factors in consumer behaviour. Within Europe, generally, local food sales are greater in countries characterised by a predominance of small farms (European Parliament, 2016; Evola et al., 2022). Yet, the relationships between consumers' cultural identities and purchases of local food remain underexplored (Miocevic et al., 2022).

Attitudes and other food-related behaviours appear to be more reliable predictors of local food purchases. Specifically, consumers' degree of food involvement, whether they grow foods, and prepare meals from scratch are significant in explaining local food purchases (Tregear and Ness, 2005; Zepeda and Li, 2006; Cranfield et al., 2012). Attitudes towards the benefits of buying locally, protecting the environment, supporting local food producers, and consumer ethnocentrism are also significant predictors (Cranfield et al., 2012; Bianchi and Mortimer, 2015; Hanus, 2020; Cicia et al., 2021; Evola et al., 2022). However, importantly, positive attitudes toward local foods may not always translate into purchases (Brečić et al., 2021) and/or willingness to pay a premium (Zepeda and Li, 2006), drawing attention to major barriers to purchase.

5 Barriers to purchase

Generally, local foods sold via farmers' markets, farm shops, and other independent small-scale outlets are more expensive compared to supermarket alternatives, and in some cases, the differential can be substantial (Thomson et al., 2021). While direct comparisons of prices are difficult due to differences in quality consumers typically regard price as the most formidable barrier to purchasing local foods (Chambers et al., 2007; Birch and Memery, 2020). Improving the price competitiveness of local foods without sacrificing the intrinsic and extrinsic qualities which motivate purchase remains a key challenge. If unaddressed, there is a danger that local foods merely provide 'exclusive' products for 'exclusive' customers. However, even when local food prices are

© Burleigh Dodds Science Publishing Limited, 2024. All rights reserved.

comparable to non-local supermarket alternatives, sales may be disappointing (Campbell and Fairhurst, 2016), prompting a consideration of other barriers.

A lack of convenience and availability are important barriers that may lead to a discrepancy between consumers' food choices and preferences (Dörrich et al., 2021). Across Europe, supermarkets and other 'convenience'/grocery retail outlets are ubiquitous, with far fewer local food outlets. While consumers generally regard visiting farmers' markets and other (small) local food retailers as enjoyable and worthy, most feel it is inconvenient to shop there on a regular basis, compared to supermarkets (Chambers et al., 2007; Megicks et al., 2012; Ritter et al., 2019). Consequently, it is not something they can do routinely due to a lack of time and opportunity (Chambers et al., 2007). Compounding this, farmers' markets and local food outlets often stock a smaller range of products, further weakening their attractiveness (Thomson et al., 2021). Consequently, much local food shopping falls into an occasional leisure activity, supplementing everyday supermarket purchases (Amilien et al., 2018). In this context, food hubs are emerging as an innovative business model to consolidate (aggregate and distribute) products from a wide range of small- and mid-sized producers (Barham et al., 2012; Guzman and Reynolds, 2019), offering a 'one-stop' option either for in-person visits or through online ordering and home delivery. While these market arrangements can improve the availability of supplies and consumer convenience via increased economical and logistical efficiency, they currently remain, however, marginal (Cleary et al., 2019).

While regarding local foods favourably, consumers possess multiple requirements and goals, such as frugality, familiarity, and convenience, which affect their purchases (Chambers et al., 2007; Brečić et al., 2021). These goals often conflict, with choices involving trade-offs between competing objectives. McEachern et al. (2010) argue that most local food purchasers are thus flexible consumers, balancing the desire to buy locally against the achievement of other goals relating to time, price, and convenience. Conflicts between goals may not be resolved consistently, so that behaviour can be unpredictable and heavily context dependent (Bargh, 2017; Brečić et al., 2021). Similarly, Amilien et al. (2018) note how these differences, or apparent inconsistencies, between what people say and do relate to contextual factors, constraints, and competing objectives.

The contexts in which decisions occur affect consumer behaviour by altering the salience of competing goals at the point of decision-making (Kay et al., 2004; Bargh, 2017). For instance, supermarkets, with their predominance of sale and discount signs, prime the salience of frugality in consumers' minds (Chartrand et al., 2008). In such a context, local foods competing against cheaper, imported alternatives are less likely to be chosen (Brečić et al., 2021). However, in-store point-of-sale materials reminding consumers of the benefits of local foods can counteract this and significantly increase sales of local foods even in the presence of cheaper alternatives (Brečić et al., 2021).

© Burleigh Dodds Science Publishing Limited, 2024. All rights reserved.

Consumer motives for purchasing local foods also may be justified according to contrasting ethical concerns and, as such, relate to different justifications or sustainability dimensions (Amilien et al., 2018). Balancing the trade-offs between different priorities and conflicting interests may lead to unintended effects (Oglethorpe and Heron, 2009). A notable example is the misconception about food miles and the environmental impacts of local foods. Specifically, a major driver behind consumer demand for local foods is environmental benefits (Megicks et al., 2012; Memery et al., 2015; Birch et al., 2018; O'Neill et al., 2022), with supposedly shorter food miles and associated carbon footprints. However, local food systems typically generate many bespoke vehicle journeys (e.g. customers' dedicated trips to farmers' markets and producers' delivery via vans) so that their carbon emissions exceed those of more concentrated and efficient supermarket systems (Malak-Rawlikowska et al., 2019). Consequently, food miles, measured in terms of distance travelled, can be a poor guide to carbon emissions (Oglethorpe and Heron, 2009). Questionable environmental benefits undermine the appeal of local food systems to consumers and policymakers (Oglethorpe and Heron, 2009; Young, 2022). Addressing how to reduce carbon emissions is a major challenge for local food systems, particularly as their customer base skews to 'concerned consumers' for whom environmental issues are highly salient (Weatherell et al., 2003).

To illustrate the market opportunities and constraints faced by local food marketing initiatives, the next section looks at a specific case study.

6 Cooperative labelling for local food markets in Szekszárd, Hungary

Szekszárd is a small town (population of 34 000) in Tolna County, southern Hungary. Tolna County is sparsely populated, and agri-food production dominates its economy. The county possesses fertile soils and long traditions in arable and viticultural production. Some of the county's local products are well known within Hungary, especially the wines Kadarka, Kékfrankos, and Bikavér, Szekszárdi márc (a honey-based syrup) and 'őrölt fűszerpaprika' (ground pepper). However, most agri-food enterprises are small-scale, micro-enterprises, which typically are short of capital, support relatively low-paying jobs, and have poor market access. Tolna County's population is declining with hyper- and supermarkets as well as hard discounters increasingly dominating food sales. These problems are common to much of rural Hungary and the wider Central and Eastern European region (Benedek and Balázs, 2015).

Confronted with these problems, a pilot action was conducted to stimulate producer cooperation and regional food labelling in the district of Szekszárd, led by Eco-Sensus, a local not-for-profit enterprise. The intervention involved in 2011 the creation of a trademark 'Quality local food – Szekszárd and its region'

© Burleigh Dodds Science Publishing Limited, 2024. All rights reserved.

which can only be applied to the packaging of food originating from registered users in Tolna County. Eco-Sensus encouraged food producers to register with the scheme, keeping producers' registration fees to around €30 annually. Today, there are 105 registered producers in the system (Eco-Sensus, 2022), but the activity of the initiative shrunk to providing a free online database of producers because there was a general unwillingness to pay registration and membership fees.

In the first 2 to 3 years of the initiative when supporting regular marketing activity occurred, several producers used the logo on their products, but later the scheme proved not to be self-financing, and producers abandoned the logo. The reason behind such producer behaviour was the established distrust of numerous, simultaneously existing, and perceived 'useless' regional labels and the diverse fees associated with them. Moreover, there was a lack of belief that a local label would generate direct financial benefits for producers, amongst which some free-riding behaviour occurred, with an attitude of 'I do not want to pay but I will participate for free and later if I see it works, I will pay for the service'. Supporting the initiative, Eco-Sensus undertook consumer research, which provides further insights into the opportunities and constraints affecting local foods.

Eco-Sensus conducted a cross-sectional survey in 2019 with a random sample of 250 adults living in the Szekszárd district to evaluate consumer awareness, preferences, and attitudes for local food, as well as engagement in local/regional food initiatives. Particular focus was around perceptions of local food certification trademarks, and specifically the Eco-Sensus 'quality local food – Szekszárd and region' logo. The data, which was weighed to be representative of the adult population on key demographics (age, gender, educational level, household size, and income), allowed for comparison with a previous survey conducted in 2011, prior to the introduction of the local food branding initiative. To complement the consumer research, semi-structured interviews with 30 local food producers in Szekszárd occurred also in 2019, to explore constraints to local production and identify market opportunities.

Key findings suggest that consumers value, in order of importance, food that is seasonal, fresh, high-quality, and healthy. Albeit on the rise, consumers' preference and knowledge of local food, however, remain moderate. In terms of local food trademarks, consumers find them puzzling and often confuse them with local product brands – this was specifically the case for the Eco-Sensus local certification, whereby most respondents (over 70%) demonstrated low awareness and understanding. Despite such low recognition and thus questionable effectiveness of the trademark, when asked about the perceptions towards such a trademark in certifying local products, most respondents (over 75%) declared that the design was appropriate.

© Burleigh Dodds Science Publishing Limited, 2024. All rights reserved.

Producers' perceptions and attitudes towards local food trademarks also revealed general scepticism and distrust, given the multitude of labels on the market (e.g. geographical designations, and production standards), as well as poor compliance/low inspection from certification bodies and thus, credibility, across operators. This contributed to a negative spill-over effect on the Eco-Sensus local certification, which witnessed weakened effectiveness in the last 5–6 years. A critical issue relates to the 'passive' way in which producers (do not) use the label – in other words, despite having the certification, they often do not display it on packaging or do not actively communicate it via bespoke marketing and other promotional activities.

The pilot action case study highlights three key lessons for local food initiatives. First, 'chicken and egg' limitations often occur on the supply and demand sides – producers are unwilling to join an initiative without clear financial incentives stemming from increased consumer demand, but without producers' brand messaging and promotion, a local food initiative, consumer knowledge and use remains minimal. Secondly, positive consumer attitudes towards local foods may not be reflected in consumer behaviour, and local food initiatives require a clearer focus on how to overcome attitude–behaviour gaps. Thirdly, while attitudes to local food are overwhelmingly positive, this may not be the case for all local food initiatives. Many producers are sceptical of the benefits and some poorly designed and regulated schemes have further negatively reinforced attitudes. Moreover, consumers often feel confused and overwhelmed by the array of food labels, which impedes understanding and consumer uptake.

7 Conclusion and future trends in research

Recent work on local foods opens up three main avenues for future research. First, existing studies overwhelmingly focus on 'bricks and mortar' marketing channels such as farmers' markets and retail stores (Milford et al., 2021). In contrast, direct sales via the internet or social media and other digital media platforms attract much less attention (Delicato et al., 2019). While the attitudinal, socio-economic, and demographic characteristics of those who buy local foods via farmers' markets and CSA have been investigated extensively (Evola et al., 2022), it is unclear whether online sales serve different and potentially larger market segments, particularly as they may offer more convenient shopping (Hobbs, 2020). Moreover, the economic, social, and environmental outcomes of digital marketing channels remain largely undocumented, so that it remains unclear whether they offer superior outcomes to producers and communities than conventional supermarket-led supply chains.

Secondly, most consumer research on local foods draws on survey, focus group, and interview evidence (e.g. Tregear and Ness, 2005; Chambers

© Burleigh Dodds Science Publishing Limited, 2024. All rights reserved.

et al., 2007; Megicks et al., 2012; Banerjee and Quinn, 2022). While providing important insights, such methods may not fully capture actual consumer behaviour. In this context, action-research merits particular attention, in its potential to test research-informed interventions and demonstrate the potential of marketing strategies. Specifically, there is a scarcity of communication research on local foods, with few attempts to manipulate labelling and (digital) promotional materials (Abrams and Soukup, 2017). This limits the ability of academic research to generate actionable insights for local food managers and businesses. Consequently, Abrams and Soukup (2017, p. 14) call for further research using field experiments and intervention studies as well as studies to gain 'a better understanding of local food messaging' and its effectiveness.

Finally, food hubs, by consolidating the local food offerings from a wide array of producers, may offer improved convenience and availability to consumers (Guzman and Reynolds, 2019). However, logistics for food hubs is often problematic, with substantial costs in bringing together the goods of geographically dispersed producers and serving consumers with home delivery, especially when demand is limited and spatially fragmented. Attempts to improve consumer demand for food hubs, via app-based delivery services and digital technology, and encourage low-carbon emissions warrant further research.

8 Acknowledgements

This chapter draws on research funded by the European Union's Horizon 2020 research and innovation programme 'Strength2Food' under grant agreement No 678024, and the National Innovation Centre for Rural Enterprise (NICRE) funded by Research England.

9 Where to look for further information

- Bianchi, C. and Mortimer, G. (2015). Drivers of local food consumption: A comparative study, *British Food Journal* 117(9), 2282–2299. https://eprints.qut.edu.au/85659/25/85659.pdf.
- Chambers, S., Lobb, A., Butler, L., Harvey, K. and Traill, W. B. (2007). Local, national and imported foods: A qualitative study, *Appetite* 49(1), 208–213. https://www.sciencedirect.com/science/article/pii/S0195666307000190.
- Enthoven, L. and Van den Broeck, G. (2021). Local food systems: Reviewing two decades of research, *Agricultural Systems* 193, 103226. https://www.sciencedirect.com/science/article/pii/S0308521X21001797.
- Kneafsey, M., Venn, L., Schmutz, U., Balázs, B., Trenchard, L., Eyden-Wood, T., Bos, E., Sutton, G. and Blackett, M. (2013). *Short Food Supply Chains*

© Burleigh Dodds Science Publishing Limited, 2024. All rights reserved.

and Local Food Systems in the EU: A State of Play of Their Socio-Economic Characteristics. Luxembourg: Publications Office of the European Union. https://ideas.repec.org/p/ipt/iptwpa/jrc80420.html.

- Young, C. (2022). Should you buy local? *Journal of Business Ethics* 176(2), 265–281. https://link.springer.com/article/10.1007/s10551-020-04701-3.

10 References

Abrams, K. M. and Soukup, C. (2017). Matching local food messages to consumer motivators: an experiment comparing the effects of differently framed messages, *Journal of Applied Communications* 101(4), 1–19.

Amilien, V., Roos, G., Arfini, F., Biasini, B., Csillag, P., Duboys de Labarre, M., Filipović, J., Haugrønning, V., Kuč, V., La Spina, M., Menozzi, D., Meyer, K., Ognjanov, G., Simons, J., Tocco, B., Török, Á., Veljković, S. and Wavresky, P. (2018). Ethnographic study: Qualitative research findings on European consumers' food practices linked to sustainable food chains and food quality schemes, *Strength2Food Deliverable D8.2.* [Online]. Available at: https://www.strength2food.eu/wp-content/uploads/2018/09/D8.2-Qualitative-Research-Findings-on-European-Consumers%E2%80%99-Food-Practices-and-FQS-compressed-protected.pdf.

Augère-Granier, M.-L. (2016). *Short Food Supply Chains and Local Food Systems in the EU: Briefing September 2016*. Available at: http://www.europarl.europa.eu/RegData/etudes/BRIE/2016/586650/EPRS_BRI(2016)586650_EN.pdf (Accessed: 3 February 2019). Brussels: European Parliamentary Research Service. [Online].

Banerjee, R. and Quinn, B. (2022). Exploring consumer constructions of local food: Meanings and influences, *European Journal of Marketing*, ahead-of-print 56(5), 1269–1300.

Bargh, J. A. (2017). *Before You Know It: the Unconscious Reasons We Do What We Do*. London: William Heinemann.

Barham, J., Tropp, D., Enterline, K., Farbman, J., Fisk, J. and Kiraly, S. (2012). *Regional Food Hub Resource Guide*. Washington, DC: United States Department of Agriculture, Agricultural Marketing Service. [Online]. Available at: https://www.ams.usda.gov/sites/default/files/media/Regional%20Food%20Hub%20Resource%20Guide.pdf.

Bellassen, V., Drut, M., Hilal, M., Bodini, A., Donati, M., de Labarre, M. D., Filipović, J., Gauvrit, L., Gil, J. M., Hoang, V., Malak-Rawlikowska, A., Mattas, K., Monier-Dilhan, S., Muller, P., Napasintuwong, O., Peerlings, J., Poméon, T., Tomić Maksan, M., Török, Á., Veneziani, M., Vittersø, G. and Arfini, F. (2022). The economic, environmental and social performance of European certified food, *Ecological Economics* 191, 107244.

Benedek, Z. and Balázs, B. (2015). Current status and future prospect of local food production in Hungary: A spatial analysis, *European Planning Studies* 24(3), 1–18.

Bianchi, C. and Mortimer, G. (2015). Drivers of local food consumption: A comparative study, *British Food Journal* 117(9), 2282–2299.

Bingen, J., Sage, J. and Sirieix, L. (2011). Consumer coping strategies: A study of consumers committed to eating local, *International Journal of Consumer Studies* 35(4), 410–419.

Birch, D. and Memery, J. (2020). Tourists, local food and the intention-behaviour gap, *Journal of Hospitality and Tourism Management* 43, 53–61.

© Burleigh Dodds Science Publishing Limited, 2024. All rights reserved.

Birch, D., Memery, J. and De Silva Kanakaratne, M. (2018). The mindful consumer: balancing egoistic and altruistic motivations to purchase local food, *Journal of Retailing and Consumer Services* 40, 221–228.

Brečić, R., Sinčić Ćorić, D., Lučić, A., Gorton, M. and Filipović, J. (2021). Local food sales and point of sale priming: Evidence from a supermarket field experiment, *European Journal of Marketing* 55(13), 41–62.

Brown, E., Dury, S. and Holdsworth, M. (2009). Motivations of consumers that use local, organic fruit and vegetable box schemes in Central England and southern France, *Appetite* 53(2), 183–188.

Campbell, J. M. and Fairhurst, A. E. (2016). Reducing the intention-to-behaviour gap for locally produced foods purchasing: the role of store, trust, and price, *International Journal of Retail and Distribution Management* 44(5), 508–523.

Cerjak, M., Haas, R., Brunner, F. and Tomić, M. (2014). What motivates consumers to buy traditional food products? Evidence from Croatia and Austria using word association and laddering interviews, *British Food Journal* 116(11), 1726–1747.

Chambers, S., Lobb, A., Butler, L., Harvey, K. and Traill, W. B. (2007). Local, national and imported foods: A qualitative study, *Appetite* 49(1), 208–213.

Chartrand, T. L., Huber, J., Shiv, B. and Tanner, R. J. (2008). Nonconscious goals and consumer choice, *Journal of Consumer Research* 35(2), 189–201.

Cicia, G., Furno, M. and Del Giudice, T. (2021). Do consumers' values and attitudes affect food retailer choice? Evidence from a national survey on farmers' market in Germany, *Agricultural and Food Economics* 9(1), 3.

Cleary, R., Goetz, S., Thilmany, D. and Ge, H. (2019). Excess competition among food hubs, *Journal of Agricultural and Resource Economics* 44, 141–163.

Cranfield, J., Henson, S. and Blandon, J. (2012). The effect of attitudinal and sociodemographic factors on the likelihood of buying locally produced food, *Agribusiness* 28(2), 205–221.

Delicato, C., Collison, M., Myronyuk, I., Symochko, T. and Boyko, N. (2019). Is local better? Consumer value in food purchasing and the role of Short Food Supply Chains, *Studies in Agricultural Economics* 121(2), 75–83.

Denver, S., Jensen, J. D., Olsen, S. B. and Christensen, T. (2019). 'Consumer Preferences for 'Localness' and organic food production, *Journal of Food Products Marketing* 25(6), 668–689.

Dörrich, J., Kastenhofer, K., Gorton, M., Lemke, F., Nicolau, M. and Bethge, J. (2021). Behaviour change interventions for more sustainable food consumption. A review of three REWE International cases studies. Wuppertal: VALUMICS Project Deliverable 8.3.

Duram, L. and Oberholtzer, L. (2010). A geographic approach to place and natural resource use in local food systems, *Renewable Agriculture and Food Systems* 25(2), 99–108.

Eco-Sensus (2022), *Producer Database*. Available at: https://www.szekszarditermek.hu/termeloi-adatbazis-projekt?field_termek_kategoria_tid=All&field_telepules_tid=All&combine=&page=4 (Accessed: 12th June).

EIT Food (2021). *COVID-19 Impact on Consumer Food Behaviours in Europe*. Brussels: European Institute of Innovation and Technology Food.

Enthoven, L. and Van den Broeck, G. (2021). Local food systems: reviewing two decades of research, *Agricultural Systems* 193, 103226.

Eriksen, S. N. (2013). Defining local food: Constructing a new taxonomy – three domains of proximity, *Acta Agriculturae Scandinavica, Section B – Soil and Plant Science* 63 (sup1), 47–55, *Section B*.

© Burleigh Dodds Science Publishing Limited, 2024. All rights reserved.

European Commission (2020). *Special Barometer 504: Europeans, Agriculture and the CAP*. Brussels: European Commission. [Online]. Available at: https://ec.europa.eu/commfrontoffice/publicopinion/index.cfm/survey/getsurveydetail/instruments/special/surveyky/2229.

European Parliament (2016). *Short Food Supply Chains and Local Food Systems in the EU*. Available at: https://www.europarl.europa.eu/RegData/etudes/BRIE/2016/586650/EPRS_BRI(2016)586650_EN.pdf. Brussels: European Parliamentary Research Service. [Online].

Evola, R. S., Peira, G., Varese, E., Bonadonna, A. and Vesce, E. (2022). Short food supply chains in Europe: scientific research directions, *Sustainability* 14(6), 3602.

Feldmann, C. and Hamm, U. (2015). 'Consumers' perceptions and preferences for local food: a review, *Food Quality and Preference* 40, 152–164.

Giebelhausen, M., Chun, H. H., Cronin, J. J. and Hult, G. T. M. (2016). Adjusting the warm-glow thermostat: how incentivizing participation in voluntary green programs moderates their impact on service satisfaction, *Journal of Marketing* 80(4), 56–71.

Guzman, P. and Reynolds, C. (2019). *Food Hubs in the UK: Where Are We and What Next?* London: Food Research Collaboration. Food Policy Discussion Paper. [Online]. Available at: https://foodresearch.org.uk/publications/food-hubs/.

Hanus, G. (2020). Ethnocentrism in polish consumer food behaviour as a determinant of short supply chain development, *European Journal of Sustainable Development* 9(4), 169.

Hobbs, J. E. (2020). Food supply chains during the COVID-19 pandemic, *Canadian Journal of Agricultural Economics/Revue Canadienne d'Agroeconomie* 68(2), 171–176.

Kay, A. C., Wheeler, S. C., Bargh, J. A. and Ross, L. (2004). Material priming: the influence of mundane physical objects on situational construal and competitive behavioral choice, *Organizational Behavior and Human Decision Processes* 95(1), 83–96.

Kneafsey, M., Venn, L., Schmutz, U., Balázs, B., Trenchard, L., Eyden-Wood, T., Bos, E., Sutton, G. and Blackett, M. (2013). *Short Food Supply Chains and Local Food Systems in the EU. A State of play of Their Socio-Economic Characteristics*. Available at: https://publications.jrc.ec.europa.eu/repository/handle/JRC80420.

Kumar, A. and Smith, S. (2017). Understanding local food consumers: theory of planned behavior and segmentation approach, *Journal of Food Products Marketing*, 1–20.

Malak-Rawlikowska, A., Majewski, E., Wąs, A., Borgen, S. O., Csillag, P., Donati, M., Freeman, R., Hoàng, V., Lecoeur, J.-L., Mancini, M. C., Nguyen, A., Saïdi, M., Tocco, B., Török, Á., Veneziani, M., Vittersø, G. and Wavresky, P. (2019). Measuring the economic, environmental, and social sustainability of short food supply chains, *Sustainability* 11(15), 4004.

Martinez, S., Da Pra, M., Pollack, S., Ralston, K. and Smith, T. (2010). *Local Food Systems: Concepts, Impacts, and Issues*. Washington, DC: United States Department of Agriculture.

Martinez, S. W. and Park, T. (2021). *Marketing Practices and Financial Performance of Local Food Producers: A Comparison of Beginning and Experienced Farmers*. Washington, DC: United States Department of Agriculture, Economic Research Service. [Online]. Available at: https://www.ers.usda.gov/webdocs/publications/101786/eib-225.pdf?v=9439.1.

McEachern, M. G., Warnaby, G., Carrigan, M. and Szmigin, I. (2010). Thinking locally, acting locally? Conscious consumers and farmers' markets, *Journal of Marketing Management* 26(5–6), 395–412.

© Burleigh Dodds Science Publishing Limited, 2024. All rights reserved.

Megicks, P., Memery, J. and Angell, R. J. (2012). Understanding local food shopping: unpacking the ethical dimension, *Journal of Marketing Management* 28(3-4), 264-289.

Memery, J., Angell, R., Megicks, P. and Lindgreen, A. (2015). Unpicking motives to purchase locally-produced food: Analysis of direct and moderation effects, *European Journal of Marketing* 49(7/8), 1207-1233.

Milford, A. B., Lien, G. and Reed, M. (2021). Different sales channels for different farmers: local and mainstream marketing of organic fruits and vegetables in Norway, *Journal of Rural Studies* 88, 279-288.

Miocevic, D., Brecic, R. and Zdravkovic, S. (2022). Exploring the chain of effects between local identity and expatriate consumers' preference for local food brands, *Journal of Product and Brand Management* 31(5), 718-730.

Mount, P. (2012). Growing local food: Scale and local food systems governance, *Agriculture and Human Values* 29(1), 107-121.

O'Neill, C., Hashem, S., Moran, C. and McCarthy, M. (2022). Thou shalt not waste: unpacking consumption of local food, *Sustainable Production and Consumption* 29, 851-861.

Oglethorpe, D. and Heron, G. (2009). Investigating the practicalities and benefits of local food production, and identifying any unintended effects and trade-offs (Defra. Project FO 0104 Final Report). London: DEFRA.

Plakias, Z. T., Demko, I. and Katchova, A. L. (2020). Direct marketing channel choices among US farmers: Evidence from the Local Food Marketing Practices Survey, *Renewable Agriculture and Food Systems* 35(5), 475-489.

Racine, E. F., Mumford, E. A., Laditka, S. B. and Lowe, A. E. (2013). Understanding characteristics of families who buy local produce, *Journal of Nutrition Education and Behavior* 45(1), 30-38.

Ritter, G., Walkinshaw, L. P., Quinn, E. L., Ickes, S. and Johnson, D. B. (2019). An assessment of perceived barriers to farmers' market access, *Journal of Nutrition Education and Behavior* 51(1), 48-56.

Thomson, J. L., Goodman, M. H., Landry, A. S. and Walls, T. I. (2021). 'Farmers' Market versus Grocery Store Produce: results of the Delta Produce Sources Study, *Journal of Hunger and Environmental Nutrition*, 1-14.

Toler, S., Briggeman, B. C., Lusk, J. L. and Adams, D. C. (2009). Fairness, farmers markets, and local production, *American Journal of Agricultural Economics* 91(5), 1272-1278.

Tregear, A. and Ness, M. (2005). Discriminant analysis of consumer interest in buying locally produced foods, *Journal of Marketing Management* 21(1-2), 19-35.

Vittersø, G., Torjusen, H., Laitala, K., Tocco, B., Biasini, B., Csillag, P., de Labarre, M. D., Lecoeur, J.-L., Maj, A., Majewski, E., Malak-Rawlikowska, A., Menozzi, D., Török, Á. and Wavresky, P. (2019). Short food supply chains and their contributions to sustainability: participants' views and perceptions from 12 European cases, *Sustainability* 11(17), 4800.

Weatherell, C., Tregear, A. and Allinson, J. (2003). In search of the concerned consumer: UK public perceptions of food, farming and buying local, *Journal of Rural Studies* 19(2), 233-244.

Young, C. (2022). Should you buy local?, *Journal of Business Ethics* 176(2), 265-281.

Zepeda, L. and Li, J. (2006). Who buys local food?, *Journal of Food Distribution Research* 37, 1-11.

© Burleigh Dodds Science Publishing Limited, 2024. All rights reserved.

Chapter 8

The effects of consumer perception of food safety and quality in food purchase decisions

Anne Wilcock, University of Guelph, Canada; Kathryn A. Boys, North Carolina State University, USA; and Brita Ball, University of Guelph, Canada

1 Introduction

Consumers expect their food to be safe and of high quality. However, food safety issues can and do occur. In extreme situations, unsafe food may lead to outbreaks that result in foodborne illnesses, hospitalizations, and even death. Such outbreaks have increased consumer awareness of food safety and quality which has, in turn, affected the decisions consumers make when they purchase food.

This chapter presents a broad overview of topics that offer insight into consumer perceptions of food safety and quality, and how they relate to subsequent purchase decisions. It begins with a discussion of the numerous factors that consumers consider as they make decisions about what food to purchase. How do food standards, certifications, and labeling influence their decision-making? Are consumers strongly inclined to purchase organic foods?

http://dx.doi.org/10.19103/AS.2023.0129.17
© Burleigh Dodds Science Publishing Limited, 2024. All rights reserved.

Are consumer perceptions of genetically modified foods positive, negative, or neutral when it comes to purchasing behavior? Food safety risk perceptions have a direct influence on consumers' willingness to purchase food. Are consumers willing to pay a premium for foods that are perceived to be safer?

We examine consumers' intent to purchase specific types of food and how intent is correlated with purchase behavior *per se*. Do food recalls affect consumers' short and longer-term purchase decisions about the food that has been recalled? Consumers purchase food from different types of vendors (e.g. farm gate versus retail); do consumer perceptions of food safety and quality differ among these vendors?

Finally, the recent COVID-19 pandemic is known to have affected consumer perceptions of safety in general. This chapter ends with a discussion of how it has affected consumer perceptions of food safety and, ultimately, their purchase behavior. Recognizing the broad scope of this topic, the discussion in this chapter focuses on recent studies of food demand by North American and European consumers.

2 Factors affecting demand for food products

Demand for food products is derived from consumers' willingness and, importantly, their ability to pay for food products. A long and deep literature has examined factors that affect consumer demand for food products. This issue is usually examined from the perspectives of either food consumers or the primary purchaser of food for a household. Considerations that affect this demand are extensive and include many factors including: (1) a wide array of socio-cultural and economic factors which shape consumer opinions and preferences; (2) aspects of the food consumption setting; (3) the type and level of information that a consumer or buyer has about a particular food product; and (4) the attributes of the product itself. For a given consumer or buyer, the relative importance of these influences in the decision-making process varies and can interact in complex ways.

2.1 Socio-cultural and economic factors

At its foundation, consumer food preferences are based on predispositions for particular flavors, textures, and other sensory attributes of a particular food. Beyond these biologically-shaped preferences, however, a wide and diverse variety of factors shape consumer perceptions for food. Contento (2011) has developed a well-used framework that organizes these determinants into those that are sensory affective; those related to a consumer's beliefs, norms,

© Burleigh Dodds Science Publishing Limited, 2024. All rights reserved.

attitudes, and skills; and those that are influenced by social and environmental considerations.[1] These multifaceted determinants, and examples of specific factors embodied within them, are presented in Fig. 1.

Sensory-affective motivations relate to the manner in which previous experiences condition an individual's preferences for specific food items. Food practices are learned primarily through diffusion from family members (often parents) to children. In addition to previous direct experiences with food, here the roles of food safety knowledge and consumer conditioning concerning satiety and other food consumption behaviors are recognized. While not explicitly stated, food requirements due to food sensitivities and allergies, and diet-related health conditions would presumably also be embodied within this category.

The influence of inter- and intra-personal drivers on food choice is captured by the next sphere. The role and relative importance of these drivers change over one's lifetime. As a child, family and social networks directly affect food choice by directly providing food to a child and a child's natural own observation of food consumption behaviors. Family and social networks also shape food choices through support (or lack thereof) from others when making food choices and, as an adult, through negotiation with others when food is shared (Monterrosa et al., 2020). Finally, social and environmental forces include characteristics of the physical and built environment, social and cultural environment, economic environment, and informational environment.

Importantly, this framework does not offer insights or assumptions into the relative importance of either the broad categories of drivers or their component parts. In considering consumer demand, however, economic considerations do require particular consideration. Demand is generated by consumer willingness to pay for a product and, crucially, consumers' ability to do so. Consumers, of course, also have options to make trade-offs between the cost and extent to which a food embodies particular attributes such as quality and convenience of form.

2.2 Food consumption setting

Consumers have numerous options as to where they obtain their food. Aside from preparing food at home, prepared foods are widely available through restaurants (quick service or 'fast-food', limited-service, full-service), commercial cafeterias, convenience and grocery stores, specialty food shops, vending machines, and institutional food service settings such as schools,

[1] Other frameworks which examine factors influencing food choice and their interrelationships include Sobal et al., (2009) and Chen and Antonelli (2020).

© Burleigh Dodds Science Publishing Limited, 2024. All rights reserved.

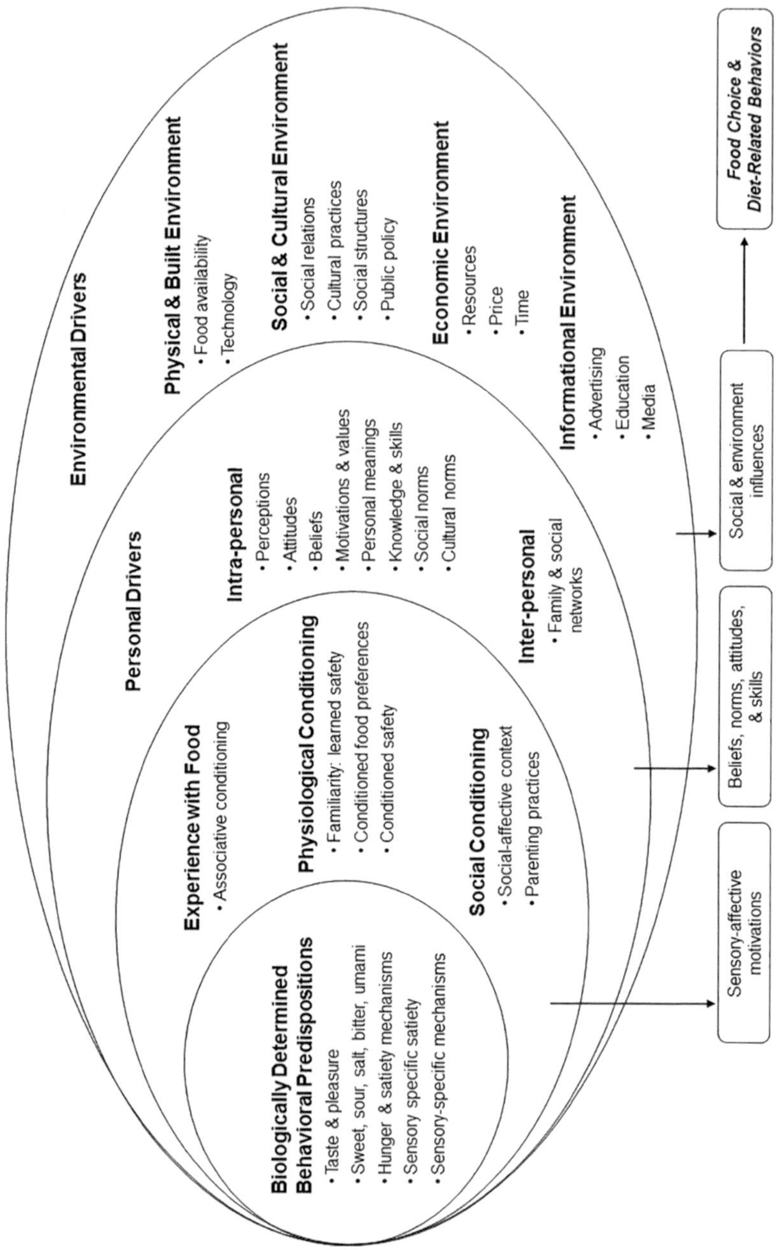

Figure 1 Determinants of consumer perception of food. Source: Adapted from: Contento (2011).

© Burleigh Dodds Science Publishing Limited, 2024. All rights reserved.

hospitals, and prisons.[2] Eating food away from home (FAFH) is a large and growing component of consumers' diets. Indeed, 2010 US consumer spending on FAFH has exceeded spending on food prepared at home (FAH; Saksena et al., 2018). Not surprisingly, the frequency of FAFH consumption has been found to vary across several population characteristics, including age, gender, race, income, and employment status; overall, there has been found to be a positive correlation between socioeconomic status and the number of FAFH acquisitions per week (Saksena et al., 2018).

There is extensive evidence that consumers make different food choices in FAH and FAFH settings. In a careful review of international studies that have examined this issue, Lachat et al. (2012) reported that eating outside of one's home was associated with higher total energy intake and energy contribution of fat to one's daily diet. Some of these reviewed studies have also found consumption of FAFH was associated with lower intake of vitamins and other micronutrients. Similarly, in a unique and broad study of US consumers, Saksena et al. (2018) found that FAFH was of consistently lower nutritional quality than food at home. These results held for all households across income levels, and for those participating in a US federal program designed to increase the food purchasing power of low-income participants [Supplemental Nutrition Assistance Program (SNAP)]. Other studies have found that consumers with healthier home diets tend to pay less attention to nutrition when away from home (Binkley, 2019).

Multifaceted reasons likely account for these differences in food choices. In restaurant settings, several aspects of the restaurant's ambiance have been found to affect food choice. It is thought that these external factors can indirectly manipulate people's cognitive, emotional, or physiologic reactions and thus unconsciously affect their mood, appetite, the food's attractiveness, and ultimately their food choice (Stroebele and De Castro, 2004). Changes in food choice have been detected with differing amounts of food accessibility (seating availability and proximity to food and beverages), ambient temperatures, music, sounds, and lighting (Stroebele and De Castro, 2004; Biswas et al., 2017; Zellner et al., 2017). Other aspects of the setting related to characteristics of the food itself, such as the smell of food, its temperature, food presentation such as plate size and material, and interactions between food colors and other objects (e.g. table settings, décor) have also been found to be significant (Stroebele and De Castro, 2004; Boesveldt and de Graaf, 2017; Cho et al., 2019). A variety of other factors, including menu design (Filimonau and Krivcova 2017; Bacon and Krpan, 2018), social considerations such as the number of people present

[2] Clients of institutional food services have, at best, a limited array of food options. Given this, the nature of the demand for food in these settings is quite different than in usual market conditions and is considered beyond the scope of this discussion.

© Burleigh Dodds Science Publishing Limited, 2024. All rights reserved.

while food is purchased or consumed, and a consumer's relationship to these people (Higgs and Ruddock, 2020), also affect food demand.

2.3 Product information

The information available to food consumers (or buyers) at the time they are making decisions about products, such as the regularity of the shape of a piece of fruit or the extent of its bruising, may also influence consumer choice. Search characteristics are those that consumers can directly examine and evaluate prior to purchase or consumption of a product; the regularity of the shape of a piece of fruit or the extent of its bruising are examples of this. Many of the attributes that consumers may prefer in their foods are not possible to verify prior to consuming them - and sometimes not even after they are consumed. Experience characteristics are those that the buyer can evaluate only after a good has been purchased, with flavors, spiciness, and product mouthfeel being common examples. Finally, credence characteristics are attributes of a product that cannot be determined even after its use or consumption. A product's organic status offers a useful example of this. For a variety of personal health, environmental, and other reasons, consumers may prefer to purchase organic foods. It is not possible though, by taste or other objective tests,[3] to verify that a food has been organically produced. Geographical indications (GIs), which identify a good as originating from a region (e.g. Florida oranges or Irish whiskey),[4] and social or ethical production characteristics such as fair-trade coffee and cage-free eggs are other credence attributes.

Many of these attributes, particularly credence characteristics, reflect higher quality products and command a price premium. Consumers may opt not to buy products that meet their preferences because they are unaware of, cannot be assured of, or are deceived about the quality of a product (Caswell and Anders, 2011). In these instances, consumers rely on information from food product labels - which frequently require a third party to verify adherence to a prescribed set of production or processing practices - to provide assurance that the product contains (or doesn't contain) features of interest.

2.4 Product attributes

An extensive array of product characteristics may be important to consumers. In addition to the characteristics of the food itself such as its nutritional attributes and flavors, the product's quality, packaging, branding, and other

[3] It is possible, of course, to test for the presence of chemicals (e.g. fertilizers and fungicides) commonly used in plant and animal production. The absence of them, however, does not guarantee that the product is organic.

[4] More specifically, GIs are 'indications which identify a good as originating in the territory of a country or from a region or locality within that territory, where a given quality, reputation, or other characteristic of the good is essentially attributable to its geographic region' (USPTO, 2021).

© Burleigh Dodds Science Publishing Limited, 2024. All rights reserved.

product information and features also matter to some consumers. The quality perception framework, drawn from applied psychology, consumer behavior, and marketing literatures, offers a useful lens through which to identify and consider these food attributes.[5] This framework explores which cues consumers use to judge a product's quality prior to and after consuming (or using) it. With this knowledge, firms have a better sense of which product characteristics consumers would benefit from having more information about, and which should be enhanced or otherwise adjusted (Caswell et al., 2002). These factors may be broadly categorized as intrinsic attributes, which are part of the physical product, or extrinsic quality indicators and cues which are marketing tools used to signal the characteristics about the product (Steenkamp and van Trijp, 1996).

A useful classification for intrinsic and extrinsic quality attributes of food was proposed and later updated by Caswell (2006) and Caswell and Anders (2011), respectively. This classification has been further updated in Table 1. As described here, the major categories of intrinsic food characteristics are those related to food safety attributes, nutrition attributes, sensory and organoleptic attributes, value and function attributes, and process attributes. Extrinsic indicators include objective test and measurement indicators and a variety of information cues. In considering food products, for which many characteristics valued by consumers are credence attributes, these extrinsic indicators are quite important in informing consumer perceptions about a product and helping them to choose between product alternatives. Well-known extrinsic indicators include price, brand name, country of origin, and product labels and certifications. In establishing their expectations about a product's quality, consumers implicitly weigh these intrinsic attributes, and extrinsic indicators and cues. For most consumers, it has been found that intrinsic rather than extrinsic attributes carry more weight in their formation of product quality perceptions (Steenkamp, 1989).

3 Distinct categories of food products

3.1 Local foods

The origin of the local food movement in the US originates from several overlapping environmental and social justice movements. Food production systems that consumers view as having 'local' characteristics date back several decades. Sustainable farming approaches in the form of biodynamic farming (Steiner, 2011) and the progressive conservation movement can be traced back to at least the 1920s (Harwood, 1990). Formal federal government action to

[5] As described by Caswell et al. (2002), the consumer interface framework is another useful alternative.

© Burleigh Dodds Science Publishing Limited, 2024. All rights reserved.

Table 1 Intrinsic and extrinsic attributes of food products

Intrinsic quality attributes				
1. Food safety attributes	2. Nutrition attributes	3. Sensory, organoleptic	4. Value, function	5. Process
Foodborne pathogens	Calories	Flavor	Compositional integrity	Animal welfare
Heavy metals and toxins	Protein	Texture	Size	Authenticity of place (including geographic indications)
Pesticide or drug residue	Fat and cholesterol content	Mouthfeel	Style	Authenticity of process
Soil and water contaminants	Vitamins and minerals	Color	Preparation/convenience	Legality of production practices
Food preservatives	Carbohydrate and fiber content	Appearance/blemishes	Package materials	Traceability
Physical hazards	Phytonutrients including antioxidants	Odor (smell, aroma)	Perishability	Biotechnology
Spoilage and botulism	Food additives and compounds	Freshness	Form (e.g. fresh and frozen)	Organic
Irradiation and fumigation	'Free-of' particular ingredients or nutrients (e.g. sugar and gluten)	Other	Other	Worker Safety
Other	Other			Other
Extrinsic quality indicators and cues				
1. Test and measurement indicators		2. Cues		
Quality management systems		Price		
Certification		Brand name, Certification Symbols		
Records		Manufacturer name		
Traceability		Store name		
Quality signaling/labeling		Packaging		
Minimum quality standards		Advertising		
Occupational licensing		Country of origin		
Other		Distribution outlet		
		Warranty, Reputation, Past Purchase Experience, Other Information.		

© Burleigh Dodds Science Publishing Limited, 2024. All rights reserved.

promote conservation and to protect agricultural resources was embedded in early agricultural legislation including the Soil Conservation and Domestic Allotment Act (1936), the Soil Bank Program (1956; Helms 1985), and the Food Security Act (1985; Dimitri et al., 2005).

From a consumer perspective, the US local food movement is thought to primarily derive from a long history of political (or ethical) consumerism through which people worked to change the marketplace (Micheletti, 2013). In the US, there is a long tradition of boycott and 'buycott' campaigns which respectively seek to direct consumer purchasing power either away from, or intentionally toward, specific companies or products. Buy Local Food campaigns have their roots in the 'Buy Union', 'Buy Black', and, most directly, 'Buy American' campaigns (Allen and Hinrichs, 2007). Against this backdrop, the interests of those with globalization concerns and those who wished to support alternative food networks, intersected and forged support for local food systems (Allen and Hinrichs, 2007). By the 1990s, interest in local food had gained notable market traction.

Given these origins, it is not surprising that local food consumers, and their motivations for purchasing these products, vary both by place and within place. In a review of global literature examining consumers' perceptions and preferences for local food, Feldmann and Hamm (2015) found that preferences for these products were shaped by consumer attitudes and beliefs. More specifically, these authors found that options concerning the quality of local food products (freshness and taste), consumers' personal health, food safety, care for the environment, and support of the local economy most frequently resulted in positive attitudes toward local foods.

Many studies have also explored the relationship between demographic characteristics, attitudes, and purchasing behavior toward local foods. While demographic characteristics have not always been found to be significantly related to local foods, there have been reports in which demographic groups had supportive attitudes toward local food products. Generally, consumers who are older, wealthier, and live in rural areas are more likely to support these products (Feldmann and Hamm, 2015). Other studies have reported that female consumers, those with children, and those with at least a college degree (who would also likely have relatively good incomes) are also more inclined to purchase these products. Not surprisingly, consumer research has found complementarity between consumer preferences for local foods and other food attributes such as being 'natural' or 'organic'. Studies have been largely consistent in reporting that consumers have a stronger preference, as measured by a higher willingness-to-pay, for local than for other food attributes (Loureiro and Hine, 2002; Thilmany et al., 2008; Onken et al., 2011; Meas et al., 2013, among others).

© Burleigh Dodds Science Publishing Limited, 2024. All rights reserved.

It is important to distinguish between those who consume local foods by choice and those who do so out of necessity (Boys and Blank, 2016). These latter consumers are seen in at least two settings. First, in many locations, particularly in developing countries, consuming local foods is a necessity that is rooted in underdeveloped supply chain networks or limited availability of food. Importantly, in some such settings, local food is not trusted and is seen as an inferior food source due to pollution of local soil and water (e.g. Kenya - Freidberg and Goldstein, 2011). Even in these settings, however, some consumers and tourists may seek out these foods for their cultural heritage or novelty (e.g. Turkey - Okumus et al., 2007; Dominica - Boys et al., 2014).

Less studied, however, is a second group who live in developed countries but who are recipients of locally sourced foods through social support programs. These individuals are considered 'non-traditional consumers of local food products' (Boys and Blank, 2016). A number of social service programs are explicitly designed to incentivize the consumption of local food products among 'economically disadvantaged' consumers. In the US, for example, these programs include the federal government's Women, Infants, and Children Farmers Market Nutrition Program (WIC-FMNP) and the Senior Farmers Markets Nutrition Program (SFMNP). Participants in these programs receive coupons which can be used to purchase a variety of products at farmers' markets including raw produce, honey, and herbs. Several US state governments have also implemented incentive programs. For example, in South Carolina, a portion of local food purchases by participants in the needs-based SNAP is matched to allow additional food purchases (Draper and Jones, 2013). While information concerning participation and redemption rates in these programs is tracked, very little is known about consumer decision-making among subsidized food options and trade-offs across subsidized local and non-local foods.

3.2 Organic foods

The intent to purchase organic food and beverages is influenced by different factors. These include trust and values such as food safety risk perception (FSRP) and environmental or animal welfare concerns, as well as other attributes such as the perception that organic foods are more nutritious than non-organically produced foods (Acee-Eke and Mac-Kingsley, 2020; Bourn and Prescott, 2002; Curvelo et al., 2019; Eyinade et al., 2021). Sociodemographic factors, lifestyle, social consciousness, and price also influence purchase intent and/or behavior (Chauke and Duh, 2019; Dettmann and Dimitri, 2010; Liang, 2016; Lim et al., 2014; Paul and Rana, 2012; Rana and Paul, 2017).

Research to gain a clear understanding of the diverse reasons for purchasing organic food is complex. The various approaches all have

© Burleigh Dodds Science Publishing Limited, 2024. All rights reserved.

limitations, and comparisons among countries are difficult. Willer et al. (2022) confirmed that many people purchase organic food for reasons of health and wellness. In the US, one of the main reasons for purchasing organic foods is to avoid foods produced using genetic engineering (i.e. GMOs). Organic consumers in Germany prioritize animal welfare, while organic consumers in France and Denmark do so because of environmental concerns. Demographics also confound results. For example, Dettmann and Dimitri (2010) used the 2004 Nielsen Homescan data to show significant differences with respect to race, education, and income among American consumers who purchased organic vegetables. These results were not consistent from one study to the next. In jurisdictions where the organic designation is regulated and standards are enforced, consumers may trust that products bearing an organic logo or labeling are verified as being organically produced. Yet, Bellows et al. (2008, p. 5) indicated that the evolution of organic standards, labeling, and definitions is 'confusing or frustrating to consumers'. Lack of trust reduces the willingness to pay a premium price.

Compared to consumers who are resistant or neutral to organic foods, those who perceive organic foods to be more healthful than conventionally grown foods are more likely to purchase and to pay a premium for organic foods (Wilcock and Ball, 2014). Rana and Paul (2017) found that health consciousness is the strongest predictor of purchase intention, while environmental concerns are a primary motivator for a positive attitude toward organic foods. Katt and Meixner (2020) found that both health consciousness and hedonic shopping value were stronger predictors of purchase intention than environmental concern among 394 US consumers shopping at discount food stores, while price had a negative effect. On the other hand, among 210 'middle-class' US consumers, Shahriari et al. (2019) identified that the positive relationships between health and purchase intention and also between authenticity/naturalness and purchase intention were fully mediated by attitude towards organic food, while attitude partially mediated the negative relationship between price and purchase intention. Attitude toward organic food and social norms positively affected purchase intention, while the direct effect of price had a negative effect. The authors did not consider environmental concerns or perceived behavioral control (Shahriari et al., 2019), the latter a key construct in the theory of planned behavior.

In a study of 206 Portuguese consumers using the theory of planned behavior (Teixeira et al., 2022), attitude toward organic food was the only predictor of intent to purchase. Perceived quality and health consciousness contributed to a positive attitude about organic food, whereas environmental concern was not significant. In another European study of 158 Italian and 49 French consumers, Guido et al. (2010) found the Italian

© Burleigh Dodds Science Publishing Limited, 2024. All rights reserved.

consumers' attitude, social norms, perceived behavioral control, and moral norms predicted purchase intention, the latter representing an extension of the theory of planned behavior. Furthermore, social norms were partially mediated by moral norms. Among the French consumers, however, only moral norms predicted purchase intention and mediated the relationship between social norms and purchase intention. The strength of the causal link between moral norms and intention to purchase organic foods suggests that these consumers are primarily influenced by social pressures and their personal ethical concern for health, the environment, and animal welfare (Guido et al., 2010).

3.3 Genetically modified ('bioengineered') food

The natural process of plant and animal reproduction leads to genetic changes that help species adapt to their environments. Bioengineering techniques can be used to accelerate conventional breeding processes and have made it possible to transfer genes from one variety or species to another (Sikora and Rzymski, 2021; Wilcock and Ball, 2014). Genetically modified organisms (GMOs) produced through bioengineering have protected crops from pests, disease, and/or drought, added nutrients, increased yield, and/or provided other benefits. Since 2020, more than 90% of soybeans, corn, canola, and sugar beets grown worldwide have originated from bioengineered seeds (Food and Drug Administration, 2022). Bioengineered animal species (i.e. salmon and pigs) have been approved for food use in some countries (Food and Drug Administration, 2022).

The technology underlying genetic modification in the food industry is poorly understood by consumers. Many people do not have enough information to establish their own opinions and instead accept the opinions of people they trust (Fernbach and Light, 2020; Sloman and Fernbach, 2018). According to Lusk et al. (2018), US consumers have a preconceived notion that there exist food safety risks associated with genetically modified (GM) food. A study by Caputo et al. (2020) confirmed that US consumers had very little understanding of genetic engineering. However, when the genetic engineering of foods was explained to them, their acceptance of the technology increased. While this finding could not be confirmed by their study, the authors suggested that consumers who understand innovations underlying novel foods were more accepting of them. The most extreme opponents of GM food know the least about GMOs but think they know the most (Fernbach et al., 2020).

The use of bioengineering to make genetic changes in crops and food animals has caused controversy and concern among some consumers (Sikora

© Burleigh Dodds Science Publishing Limited, 2024. All rights reserved.

and Rzymski, 2021; Wilcock and Ball, 2014). Although various GM foods have been shown to be safe for consumption, some consumers may still reject GMOs because they do not trust information sources or remain opposed to bioengineered foods for environmental, ethical, or philosophical reasons (Sharratt and Chopra, 2019; Sikora and Rzymski, 2021; Wilcock and Ball, 2014). Concern about the escape of transgenes from bioengineered organisms has been borne out by the discovery of multiple unapproved and/or deregulated transgenic plants, some of which have reached the food supply or become persistent in the environment (Aono et al., 2011; Marvier and Van Acker, 2005; Sharratt and Chopra, 2019).

When GMOs were relatively new to the market, Brewer and Rojas (2008) found that among 405 US consumers, 80.9% (n = 361) believed that GM foods that were found safe by the Food and Drug Administration (FDA) were safe to eat. However, 31.7% (n = 366) would not knowingly purchase GM foods, and 39.3% (n = 356) indicated that they would pay more for food that was not GM. Importantly, 22.6% (n = 354) indicated they believed GM foods are not safe to eat under any circumstances.[6] Food safety is just one factor consumers consider in the acceptance of bioengineered foods (Wilcock and Ball, 2014). In a more detailed national study of 509 US consumers, Han and Harrison (2007) found that consumers who believed GM foods were safe to eat were 8.98 times more willing to purchase GM foods than to not purchase, and those who believed bioengineering provided benefit to society due to production efficiencies were 3.01 times more likely to purchase than to not purchase. Consumers who had ethical concerns about GM plants and animals were 0.41 times less willing to purchase them.

In 2001, the Eurobarometer survey of 16029 people in the European Union found that 56.4% believed GMOs were dangerous, 17.1% believed they were not dangerous and 26.5% did not know (Bonny, 2004). Among European nations, France was the most suspicious about scientific and technological progress, with 49.2% of French respondents believing that science and technology improve farming and food production, 31.6% believing they do not, and 19.2% indicating they did not know. In contrast, respondents in Sweden, Denmark, and Finland were the least suspicious with 75.7%, 78.0%, and 78.3% of respondents, respectively, believing in improvements and 12.4%, 12.5%, and 9.7% not believing in improvements (Bonny, 2004).

In 2002, an Ipsos-Reid poll of 6500 people in 12 countries reported that consumers in European and Asia-Pacific nations were more concerned about the safety of GMOs than American consumers. Fifty percent of respondents

[6] Given these values, some consumers who indicated that GM foods were not safe to eat under any circumstances also indicated that GM foods found to be safe by the FDA are safe to eat.

© Burleigh Dodds Science Publishing Limited, 2024. All rights reserved.

from European and Asia-Pacific nations were strongly concerned about GM food safety while 44% of North American consumers were strongly concerned (Bonny, 2004).

In 2007, experimental willingness-to-pay studies of 209 French consumers, Noussair et al. (2007) categorized consumers as Unwilling (34.9%), Reluctant (42.2%), Indifferent (18.1%), and Favorable (4.9%). The unwilling consumers refused to buy GMO-containing products. Favorable consumers were willing to pay the same price for GMOs as for non-GMO products. The experiments showed that 89% of participants would purchase products containing up to 1% GMOs which, in the European Union, would be exempt from mandatory GMO labeling.

In 2022, Kim et al. (2022) reported that GMO labels increased consumer willingness to pay for non-GMO products, while reducing the importance that consumers attributed to price. The color and style of labels influenced consumer responses to GMO labeling with consumers with neutral attitudes toward GMOs being influenced most by the label design.

The US implemented a mandatory GMO labeling system in 2022 for food that contains detectable genetic material that cannot be created through conventional breeding or found in nature. Even a neutral GMO label may lead consumers to focus on the negative aspects of GMOs, pay less attention to price, and become more reluctant to purchase in the product category.

Gene editing allows precise changes and reduces the risk of inserting undesirable genes (Nguyen et al., 2022). When gene editing is used, the resulting novel modified organism may not be required to go through the same extensive testing to demonstrate the safety and nutritiousness of GMOs. Governments will need to make decisions about whether to label gene-edited foods as GMOs even if there is no detectable foreign DNA (Cyranoski, 2015).

Consumer acceptance of gene editing technology has not been extensively studied. The technology may change how consumers and regulatory authorities perceive GM foods. It has been suggested that 'new breeding techniques' (i.e. GE and GMO) may be controversial with consumers and environmental groups (Caputo et al., 2020). Consumers may have ethical concerns or may have precautionary attitudes because of their lack of familiarity with GE and GMO. These concerns cross borders and are internationally present (Qaim, 2020).

Pruitt et al. (2021) conducted a study to determine whether or not health-conscious consumers (who exhibited 'healthy behaviors' as defined by physical activity over a two-day period) were willing to pay for bioengineered foods. The authors were unable to establish a relationship between physical activity and consumer preferences for these foods.

In terms of purchase behavior, will consumers' responses to GE foods be similar to what they were to foods containing GMOs? How will information be

© Burleigh Dodds Science Publishing Limited, 2024. All rights reserved.

shared, and will it influence consumers' preferences? These questions will be most important in regions such as the European Union which have traditionally reacted negatively to foods containing GMOs (Colson and Rousu, 2013; Marette et al., 2021).

To determine whether consumers would treat GE foods like foods containing GMOs, Marette et al. (2021) conducted a study of the willingness of consumers in the US and France to pay for apples that do not brown after they had been cut into. Consumers in both countries considered the apples negatively, but more negatively for apples described as containing GMOs than those identified as genetically engineered. This suggests that the negative reaction to the GMO products was stronger than for the GE products.

What can be concluded about the impact of genetic engineering on the willingness of consumers to purchase these foods? That remains to be seen. However, it seems clear that consumers are more likely to accept such products if the underlying technology is clearly explained to them, and the personal benefits, as well as the benefits to society and the environment, are made clear.

3.4 Geographical indications

Specific quality labels that designate foods as unique and exclusive to a region or country can influence consumers in their food purchase decisions. Geographic certification marks in the US (e.g. Idaho potatoes) and the European Union's Protected Designation of Origin (PDO) (e.g. Prosciutto di Parma) and Protected Geographical Indication (PGI) (e.g. Gouda Holland) are three examples of geographic indicators (Wang et al., 2021). These indicators are meant to guarantee authentic products that protect producers from imitations and support local economies (Grunert and Aachmann, 2016; Likoudis et al., 2016; Martinelli and De Canio, 2018; van Bussel et al., 2022).

Van der Lans et al. (2001) found a geographic indication cue and PDO labeling of olive oil indirectly influenced 152 Italian consumers in Rome and two smaller towns in olive-producing areas. Female respondents, especially those living in the producing regions, preferred region-of-origin oil (ROO) (extra virgin olive oil = −3.32, ROO =0.88, $P < 0.01$), but may not have placed value on PDO protection labeling introduced in the European Union.

Likoudis et al. (2016) found that 70% and 40% of Greek consumers ($n = 615$) in Athens were aware of PDOs and PGIs, respectively. Although consumers did not fully understand the labeling, those who were aware of it and considered it important were more likely to buy PDO/PGI. In addition, consumers who expressed sustainable behavior intended to buy PDO/PGI.

© Burleigh Dodds Science Publishing Limited, 2024. All rights reserved.

Using four Eurobarometer surveys, including over 110 000 consumers, Goudis and Skuras (2021) found a continued lack of awareness about protected designations across Europe. Regional differences were reported to be significant, with consumers in Italy, France, Spain, Greece, Czech Republic, and Slovakia having high logo awareness. Logo awareness was also highest with highly educated consumers and those who tend to buy local food to be 'climate friendly'.

4 Food safety risks and consumer perceptions of risk

True food safety risks ('technical risks') and risks as perceived by consumers are different. A technical risk is a combination of the likelihood of occurrence of a specific hazard and the severity of the consequences. Its calculation is based on possible outcomes and the probability of each. Its shortcomings are twofold: (1) the identity and probability of outcomes may not be known so the calculation is based on uncertainty rather than risk, and (2) individuals exposed to a hazard tend to focus on the potential severity rather than the likelihood of occurrence when they assess the significance of exposure to risk or uncertainty. For these reasons, calculated risk has been deemed insufficient for the management of food safety issues (Yeung and Morris, 2001).

In contrast, perceived risk influences how consumers think about food safety. It cannot be calculated mathematically. While it differs widely among consumers, it is the more practical approach because it influences consumer purchase decisions.

Sociodemographic factors, knowledge of food safety, and trust have all been correlated with consumer concerns about food safety and perceived food safety risk. Early studies on this topic relied on regional samples. While dated, a nationally representative study by Lin (1995) analyzed the results of interviews with primary meal planners or purchasers in households across the US in order to determine if demographic and socioeconomic factors influenced perceived food safety. Respondents' food safety knowledge and attitudes differed. All things being equal, females were statistically more likely to believe that food safety was important in food shopping than males. Older meal planners considered food safety to be more important than did their younger counterparts. Education was also shown to play a role; it was hypothesized that more educated meal planners were exposed to more information on food safety or were more able to understand the relationship between food safety and health. However, meal planners with the most education considered food safety less seriously even though they were aware of the risks. Not surprisingly, food safety was significantly more important to meal planners or purchasers in households with vulnerable consumers (the young, the elderly, or someone who was pregnant). There was no significant difference between white and

© Burleigh Dodds Science Publishing Limited, 2024. All rights reserved.

non-white meal planners, but there were regional differences between meal planners across the US. Those in the South and Northeastern US held stronger beliefs about food safety than did those in the West and Midwest.

The number of studies involving food safety risk perception has steadily increased, possibly because of the increased visibility of foodborne outbreaks or because of the recognition of the importance of risk perception by food manufacturers. A meta-analysis by Nardi et al. (2020) reviewed 128 empirical studies, with approximately 35% of those studies published in 2015 or later. This attests to the increased attention devoted to this topic. It is on this meta-analysis that the remainder of this section is based.

Consistent with the early study by Lin (1995), one of the key drivers of food safety risk perception that was identified in the recent meta-analysis by Nardi et al. (2020) was sociodemographic characteristics of consumers. Others were knowledge, trust, and subjective characteristics.

4.1 Sociodemographic characteristics

Sociodemographic characteristics related to increased food safety risk perception include the size of the household, the presence of children in the household, age, gender, and income (Nardi et al., 2020). This is consistent with statements in the literature about consumers being more concerned about risk when they were responsible for other members of their household. The gender difference was explained on the basis of women commonly being more responsible for food preparation in the home than men.

4.2 Knowledge

The meta-analysis (Nardi et al., 2020) identified a negative association between consumers' perception of knowledge about food safety and food safety risk perception. In other words, 'what individuals think they know about food safety make(s) them safer about their food consumption' (Nardi et al., 2020, p. 9). It is important to recognize, however, that what consumers think they know about food safety is not necessarily the same as what they actually do know.

4.3 Trust

Trust in the food supply is a result of knowledge of food safety, suppliers, and the food supply chain. Not surprisingly, the perception of food safety risk is lower among consumers who trust the government and members of the food supply chain who are involved in ensuring the safety of the food supply (Nardi et al., 2020).

© Burleigh Dodds Science Publishing Limited, 2024. All rights reserved.

4.4 Subjective characteristics

Food safety risk perception is related to consumers' concerns about the likelihood of – and severity of – adverse effects associated with food consumption (Nardi et al., 2020). Perceived control, perceived benefits, and preference for natural food all reduce consumers' food safety risk perception.

5 Consumer perceptions of risk and willingness to purchase

It is clear that the topic of food safety risk perception has been of great interest to researchers over the past 20 years. Two opposing factors may explain this interest. On the supply side, new technologies have raised consumer concern about its long-term effects on human health and on the environment. On the demand side, food producers are facing challenges due to consumers' loss of confidence in the food supply chain. How do these factors influence consumers' willingness to buy food products?

A survey of German consumers (Röhr et al., 2005) showed that there were two distinct segments of consumers in terms of their willingness to pay for food safety: price-sensitive consumers and safety-sensitive consumers. It is reasonable to assume that the food safety risk perception of these two groups of consumers differs as does their willingness to pay for food.

The beef sector has received considerable attention, possibly because of BSE (bovine spongiform encephalopathy) or 'mad cow disease' which was particularly prevalent in the UK in the early 1990s. Since food safety is a credence attribute, the meat industry developed systems to increase beef consumption using approaches such as certification schemes and traceability. Have these approaches been successful?

In a nationwide survey of 650 Spanish shoppers (reported in Angulo et al., 2005), 63% were more concerned about food safety than they were 5 years earlier and, of these, 52% said that their shopping habits had been substantially influenced by social media. Respondents reported that they had lost the greatest confidence in meat, canned food, preserved food, and ready-to-eat foods (Angulo et al., 2005). Despite the high percentage of consumers who expressed a loss of confidence in the safety of beef, 72.5% said they were not willing to pay a premium to buy beef with a traceability certificate. The authors explained this result on the basis of food safety (1) being a requirement that food must meet, (2) being something consumers may be concerned about but not enough to pay a premium for, and (3) not being guaranteed by a traceability certificate.

© Burleigh Dodds Science Publishing Limited, 2024. All rights reserved.

Latvala and Kola (2004) conducted a survey in Finland in which consumers were asked if they would be willing to pay for additional information about the quality and safety of beef. If they indicated they would, then they were asked how much of a premium they would be willing to pay. Of the 1640 respondents, 22% were the primary household shoppers and 54% shared food shopping responsibilities. Fifty-nine percent of respondents said that they were willing to pay a premium for additional information. Those unwilling to do so indicated that they were satisfied with the information that was already present, that they could not afford to pay more, or that they either did not trust the system or did not have enough information about it.

With so many foods on the market, it is simply not possible to investigate the correlation between risk perception and willingness to purchase each food individually. That relationship has, however, been examined for a variety of categories (Nardi et al., 2020). For example, consumers perceive risks resulting from technological modification of food to be greater than microbiological risks, hence the former has a greater effect on willingness to buy than the latter. There is less of a negative effect of food safety risk perception on willingness to buy foods that consumers consider to be healthy. The effect of risk perception on willingness to buy is less for food that has a shorter than a longer shelf life. It may be that the lower perception of risk is because products with a shorter shelf life are believed to contain fewer preservatives.

A few studies have gone beyond investigating the relationship between food safety risk perception and willingness to purchase and have attempted to specifically determine how much of a premium consumers would be willing to pay. Sckokai et al. (2014) surveyed 973 Italian consumers about their perceptions of the risk of mycotoxins in a simulated bottle of milk as well as their willingness to pay a premium price for milk with 'reduced mycotoxins'. Participants were told that the milk was from 'cows fed with . . . maize certified for farmers and processors applying GAPs (good agricultural practices) and GMPs (good manufacturing practices) such that a lower risk of mycotoxin contamination could be expected' (Sckokai et al., 2014, p. 6). Respondents indicated a willingness to pay an average of 29% more for the milk that reportedly contained reduced mycotoxins. Three demographic groups stood out as different based on age, education, and income. Respondents who were between the ages of 44 and 54 years indicated a greater than average willingness to pay a premium while consumers over 65 years expressed a lesser than average willingness to pay. Consumers seeking their first job also indicated a less than average willingness to pay a premium. Consumers who had completed tertiary education were more willing to pay a premium than were those who had less than an elementary school education.

© Burleigh Dodds Science Publishing Limited, 2024. All rights reserved.

Not surprisingly, consumers in the two highest income classes said they were more willing to pay a premium than those reporting the lowest incomes or not reporting their incomes at all. The larger premium may be explained by the respondents having more disposable income than other respondents in the study.

6 Consumer response to recalls

Food recalls involve the removal of foods from the marketplace. Recalled items are those which have left the possession of a food manufacturer but are still in downstream points of the supply chain (e.g. in food retailers) or which have been purchased and are in the homes of consumers. Products may be recalled for a variety of reasons. Most commonly, allergens that are not included on product labels are the cause of recalls. The presence of disease-causing microorganisms, foreign material (e.g. metal), processing or handling defects, mislabeling, and spoilage are other common causes of food recalls (Page, 2018).

While recalling food from the marketplace may be one of the most effective ways to avoid (reduce) human health impacts, doing so comes at a significant cost in terms of expenses to implement the recall and related corrective actions, income from lost sales, and brand and reputation damage.[7]

To date, most of the research examining the impact of recalls has been focused on firms that issue a recall and on the industries experiencing them. Most commonly these studies considered the effectiveness of a recall (Hooker et al., 2005; Seys et al., 2015; Yu and Hooker, 2020), the economic losses to the company issuing the recall due to price changes (Marsh et al., 2004), or effects on corporate valuations (i.e. stock returns; Pozo and Schroeder, 2016; Boys and Lee, 2019).

Implicitly market price or corporate value changes are derived, at least in part, from current and future expectations of demand (sales, willingness to pay) for the recalled product. Many other factors including the reason for and size of the recall, the potential severity of the health threat, whether the recall was voluntary or mandatory, if the firm had a recall in recent years, and the reputation of the company have also been examined and found to affect market response to recall events (Boys and Lee, 2019; Pozo and Schroeder, 2016; Chen et al., 2009; Rupp, 2004).

[7] Recall costs can include costs to investigate the source of the problem; recovery and disposal of the recalled product (which may be covered by recall insurance); facility cleaning, training operations and facility improvements, and increases in insurance premiums. Recalls can impact sales due to declines in demand and possible discontinued products, brand and firm reputation damage, consumer claims (refunds), government fines, and potential product liability, criminal, and civil lawsuits (Boys and Lee, 2019).

© Burleigh Dodds Science Publishing Limited, 2024. All rights reserved.

This information does not offer insight into the consumer perceptions of recalled products and how they might affect their willingness to purchase and willingness to pay for recalled and implicated-but-not-recalled[8] products. Intuitively, consumers could respond in one of two ways. Some may view a food recall as evidence that a particular company (or brand, product line, or product) is careless in its production or handing processes and as such may opt to not purchase the products in the future. Alternatively, some may believe that, because a company that has issued a recall is under increased scrutiny, their products may be produced with extra care in the future. As such, demand for these products may not decrease, but may increase, following a recall.

Empirical evidence directly examining consumer response to recalls is quite limited. Using grocery scanner data, Shang and Tonsor (2017) examined the impact of Food Safety Inspection Service (FSIS) recalls due to bacterial contamination on US consumer beef demand. These authors found that recalls did, contemporaneously, significantly reduce the demand for recalled beef. Results of this study also revealed statistically significant differences in the one type of meat on the demand for other meat products (e.g. pork recall on chicken demand). Also, while a majority of food safety recalls were found to have similar effects on consumer meat demand across regions, this was not universally the case; in particular, consumer response to chicken products was found to vary across geographic regions. The effect of the recall on consumer demand was also found to vary across time. In this study, for most products, demand for the recalled product was only affected for a relatively short time (a few months). This result is generally consistent with other firm- and market-level impacts of recalls which, unless the recalled product could have serious human health effects, were not usually long term (Boys and Lee, 2019).

Given the nature of the data used by Shang and Tonsor (2017), it was not possible to discern what motivated these differences in consumer demand. Other studies, however, do offer some insight. In particular, the source and nature of the information about food safety events have been found to be important in shaping consumer response to them. For example, in the USA, the majority of recalls are initiated by the food manufacturing firm – with a small number being required by the FSIS for USDA recalls, or the FDA for recalls of food products regulated by that agency.[9] In conducting a social listening analysis of social and online media, Jung et al. (2021) found that online

[8] Products which were considered as a potential cause of a food safety outbreak but which were ultimately not the source of it.

[9] In the USA, several federal agencies have regulatory oversight over the food industry. The US Department of Agriculture (USDA)'s Food Safety Inspection Service (FSIS) has authority concerning domestically produced and imported meat, poultry, and processed egg products. Processed food products, pet food, and animal feed are regulated by the Department of Health and Human Services (DHHS)' Food and Drug Administration (FDA). This oversight is mainly exercised through inspections and audits. While companies will typically 'voluntarily' issue a recall, when human health is at risk, through the Food Safety Modernization Act (2011), these agencies now have the authority to mandate a recall.

© Burleigh Dodds Science Publishing Limited, 2024. All rights reserved.

public discourse was much more responsive to reports of foodborne illness issued by the Centers for Disease Control and Prevention (CDC) than to recall announcements from either the FDA or the FSIS. In instances where there was significant potential for human health impacts, consumer studies have found that food safety concerns reduced people's consumption of the affected products, resulting in a structural change in the demand for these products.[10]

7 Effect of the COVID-19 pandemic on food safety behavior and purchasing

The COVID-19 pandemic has had widespread economic and social impacts on consumers. To date, there has been very little information published on how these impacts have affected consumer behavior regarding food and food purchases.

It is logical to begin this discussion by delving into some of the underlying reasons that impacted how consumers perceived risks during the pandemic. Social distancing changed the way people interacted with one another and the way business was conducted. Policies and regulations were based on a true lack of understanding of risks. Information that had not been validated was shared on social media, causing a great deal of confusion among consumers. In addition, daily reports about COVID-19-related illnesses, hospitalizations, and deaths appeared in the media, which likely influenced how consumers perceived risks, including risks associated with food (Charlebois and Music, 2021).

While it is clear that the pandemic had a severe impact on the Canadian food supply, what is less clear is how consumers connected COVID-19 with food safety. Charlebois and Music (2021) proposed a model linking COVID-19 to risk perception with the epidemiology of the virus and food safety surveillance as mediating variables (Fig. 2). This relationship between COVID-19 and risk perception will have influenced food purchasing behavior since it is well known that how consumers perceive risk impacts their purchase decisions (Barrena and Sánchez, 2010).

As has been discussed previously in this chapter, the purchase of food by consumers is a result, at least in part, of the values of the main shopper in the household. In 2020, the International Food Information Council (IFIC, 2020) reported taste, price, healthfulness, convenience, and environmental sustainability to be the most important determinants of food choices. Lusk and Briggeman (2009) had proposed a broader set of food values that influenced food choice but stated that the main values were food safety, nutrition, taste, and price. While the top values have been relatively unchanged over time (IFIC,

[10] For example, structural changes in demand were found to have occurred following the 2003 discovery of bovine spongiform encephalopathy (BSE) in Canada (Yang and Goddard, 2011) and the US (Taylor et al., 2016).

© Burleigh Dodds Science Publishing Limited, 2024. All rights reserved.

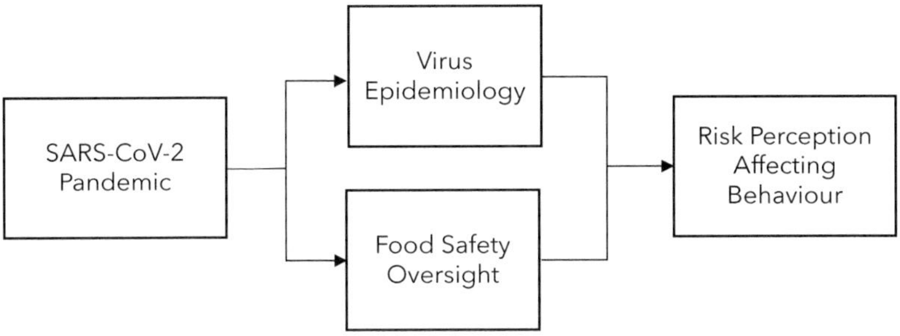

Figure 2 Relationship between COVID-19 pandemic and consumer risk perception. Source: Adapted from Charlebois and Music (2021).

2020) (Lusk and Briggeman, 2009), the relative importance of these values changed during the COVID-19 pandemic.

Ellison et al. (2021) surveyed 1370 US consumers at four different times, with each survey spaced approximately two weeks apart. The four surveys were administered from mid-March to late April 2020 and corresponded to what they called 'stocking up', 'shelter(ing) in place', 'Coronavirus Aid, Relief, and Economic Security (CARES) payment distribution', and, finally, 'reopening plans'. As the authors pointed out, the rapidity of the onset of the pandemic led to rapid changes in consumer behavior unlike what had ever been seen in previous times of uncertainty. Taste remained a strong determinant of food purchase but there was a decrease in the importance of nutrition and price.

During the pandemic, risk perceptions and trusted sources of information about food safety were investigated in a large US study of a total of 3584 consumers (a minimum of 700 per month for five months) (Thomas and Feng, 2021). Participants in the study were most trusting of the FDA, the CDC, and healthcare professionals and least trusting of social media for COVID-19 and food safety information. The authors concluded that increased risk perception during the pandemic may have led consumers to increase good handwashing practices, but was this related to food safety or just greater attention to personal hygiene?

Although there is no evidence that food was a likely source of transmission of COVID-19, not all consumers were convinced. For some consumers, there was a perception that it was possible to contract the virus through food. For some consumers, this confusion may have been due to early reporting about food being contaminated with SARS-CoV-2, which was then passed on to those who handled the food (Thomas and Feng, 2021). Alternatively, Crew (2020) suggested that this confusion was due to the terminology used in technical and scientific publications being poorly understood by consumers.

Whatever the source of this misunderstanding, this change in the perception of sources of food safety risk may have prompted consumers to change their food handling practices - in both beneficial and unhelpful ways

© Burleigh Dodds Science Publishing Limited, 2024. All rights reserved.

(Thomas and Feng, 2021). In addition, COVID-19 was also found to induce changes in consumer food procurement practices. Among studies examining this issue is one by Byrd et al. (2021), who conducted a nationwide survey of US consumers during the height of the pandemic. These authors also found that consumers were concerned about contracting COVID-19 from food - especially food that was served in restaurants. Interestingly, almost 50% of survey respondents indicated that they were 'very concerned' about getting the virus from food served in restaurants compared to restaurant food in general. Food packaging was also of 'moderate or high concern' to most survey respondents, perhaps because of reports that the virus could be acquired from touching their faces after contacting contaminated surfaces. Household income was also important in that the more consumers were concerned about their ability to afford food, the more they reported concern about getting the virus from food and food packaging. The authors also investigated whether there was a relationship between consumer demographics and concern about acquiring COVID-19 from restaurant food and packaging. Women and consumers in high health risk categories were more concerned about contracting the virus from restaurant food and packaging than were men and consumers in low health risk categories. Survey respondents residing in jurisdictions that restricted dining in restaurants were less concerned about contracting COVID-19 from food (in general) and food delivered by restaurants than were those who lived in jurisdictions that had no such restrictions.

To date, by necessity, studies have examined the short-term impacts of the COVID-19 pandemic on consumer risk perception and food procurement and consumption behaviors. It remains to be seen what the longer-term consequences of the pandemic are on these behaviors, and consumer trust in public-health (which may extend to food-safety) authorities.

8 Conclusion

A complex array of factors affects consumer perception of risk and quality of foods. Economic considerations and socio-cultural factors, such as an individual's values and ethics, shape consumer preferences and opinions concerning specific food products. The setting in which the food is consumed, and with whom one consumes food, will also shape demand for food. In particular, there are considerable differences in the choices an individual makes when consuming food at home compared to in away-from-home settings (e.g. restaurants). The information available to consumers during the food purchase process, and the way this information is communicated, also shapes consumer decisions. Consumers evaluate products they are interested in purchasing by specific product attributes, which can be categorized as search, experience, or credence (SEC) attributes, that are important to them or to those for whom

© Burleigh Dodds Science Publishing Limited, 2024. All rights reserved.

the food is being purchased. Credence attributes, i.e. those that cannot be verified even after purchase and consumption, are particularly important for food products. As such, food labels - the amount and nature of information communicated through product labels, the manner in which it is communicated, and the extent one trusts that information – serve an important role in consumer assessment of a food's quality and the risk associated with its consumption.

For many consumers, characteristics of the setting in which food is produced, and the specific nature of inputs used in the process of food production and manufacturing, are central to their assessment of a food's quality and risk. Foods produced in an area near the consumer (local foods), organically grown and produced foods, genetically modified foods, and geographical indications of the place a food is grown or produced are among the many production considerations that may affect consumer demand for food products. Instances where a company has either voluntarily or involuntarily withdrawn products from the marketplace will, at least in the short-term, affect consumer perceptions of the risk of specific food products and the company that produced them. For industries and products that experience repeated food safety failures, such as the case of raw bean sprouts, consumer perception of the risk associated with these products may be adversely affected in the medium - or long-term.

Moving forward, scientific advances, technological innovation, changing environmental circumstances, government policy, and food markets will continue to shape consumer understanding and perceptions of the quality and safety of foods. Several of these considerations will become increasingly important in the coming years. Continued innovation will lead to new food sources and production systems. For example, there is considerable global interest in developing foods using edible insects and jellyfish, plant-based alternatives, seaweeds, and cell-based products (e.g. animal-free, or lab-grown meats); each of these items, of course, brings its own quality, microbial and chemical food safety considerations (FAO, 2022). Technological advances are also offering expanded opportunities for food production in non-traditional settings. With continued urbanization and increasing food insecurity in many urban areas, there is ongoing interest in growing foods in urban spaces. While urban agriculture can improve food quality by shortening supply chains (getting perishable products to market sooner), there is concern regarding the soils used in urban agriculture which may have real or perceived contamination issues.

Other trends are decidedly more problematic. Climate change is already inducing farmers to make changes to both the types of crops and the specific varieties that they opt to grow (Alkharabsheh et al., 2021; Tully et al., 2019). Changes in air and soil temperature and water availability change the mix and prevalence of insect species and soil microbiome in local environments. Both can have adverse impacts on the quality and safety of food. For example, new

© Burleigh Dodds Science Publishing Limited, 2024. All rights reserved.

or more insects in a production environment may lead to additional insect damage to fruits and vegetables. Even if this damage is only cosmetic, it can still lead to changes in consumers' perceptions about the quality of the food. To the extent that the soil microbiome includes an increased prevalence of human pathogens, these may transfer to the surface of foods such as cucurbits (e.g. melons, cucumbers) and may pose an increased or additional food safety risk. Also concerning is the globally increasing trend in food fraud, the economically motivated adulteration of food in which an ingredient is intentionally removed, left out, or substituted for another ingredient. At minimum, food fraud can lead to a loss of consumer trust in the food industry and supply chain (Ehmke et al., 2019; Esteki et al., 2019).

Companies experiencing fraud also suffer from economic and social losses due to recalls and reputation damage, thus impacting the overall industry (van Ruth et al., 2018). In instances where the missing or substituted ingredient was needed to fortify the nutritional composition of the product (i.e. baby formula) or is an allergen (i.e. the addition of peanut shells as a bulking agent in spices), food fraud can pose serious public health risks.

Continued innovation – in food production, processing, manufacturing, distribution and storage systems – will continue. Complimenting this, technological and data innovations, such as food-sensing technologies (handheld devices, Internet of Things), Artificial Intelligence, and Blockchain will be needed to fully capture and leverage the efficiencies and food safety benefits which may be offered by these food system changes (FAO, 2022; Krzyzanowski Guerra and Boys, 2021). Simultaneous to these changes, policy and regulatory innovation will be needed to govern and safeguard these changes. Despite the many changes and challenges the agri-food industry will undoubtedly face in the future, food quality and safety will remain issues of high importance to consumers in all walks of life.

9 Where to look for further information

- European Food Safety Authority, '2022 Eurobarometer on Food Safety in the EU', Sept. 28, 2022, https://www.efsa.europa.eu/en/corporate/pub/eurobarometer22, Accessed October 15, 2022.
- International Food Information Council, '2021 Food Trends: From our Eating and Purchasing Habits to Food Safety, Covid-19 Still looms Large on Food Decisions and Health Goals', Jan. 6, 2021, https://ific.org/media-information/press-releases/2021-food-trends-from-our-eating-and-purchasing-habits-to-food-safety-covid-19-still-looms-large-on-food-decisions-and-health-goals/, Accessed Oct. 14, 2022.
- U.S. Food & Drug Administration, 'Shopping for Food during the Covid-19 Pandemic – Information for Consumers', May 1, 2020, https://www

© Burleigh Dodds Science Publishing Limited, 2024. All rights reserved.

.fda.gov/food/food-safety-during-emergencies/shopping-food-during-covid-19-pandemic-information-consumers, Accessed October 14, 2022.

- U.S. Food & Drug Administration, 'Getting Smarter about Food Safety: The Pandemic and Lessons Learned', June 23, 2020, https://www.fda.gov/consumers/consumer-updates/getting-smarter-about-food-safety-pandemic-and-lessons-learned, Accessed October 14, 2022.
- Special Issue of *Applied Economic Perspectives and Policy*: Special Issue on Covid-19. Available online at: https://onlinelibrary.wiley.com/toc/20405804/2021/43/1.
- U.S. Food & Drug Administration, 'New Era of Smarter Food Safety Summit on E-Commerce: Ensuring the Safety of Foods Ordered Online and Delivered Directly to Customers', Oct. 19-21, 2021, https://www.fda.gov/food/workshops-meetings-webinars-food-and-dietary-supplements/new-era-smarter-food-safety-summit-e-commerce-ensuring-safety-foods-ordered-online-and-delivered, Accessed October 14, 2022.

10 References

Acee-Eke, B. C. and Mac-Kingsley, I. (2020). Consumers' demand for organic food products and purchase intentions: Empirical evidence from a consumer survey in Nigeria. *International Journal of Management and Marketing Systems* 13(7), 1-18.

Alkharabsheh, H. M., Seleiman, M. F., Hewedy, O. A., Battaglia, M. L., Jalal, R. S., Alhammad, B. A., Schillaci, C., Ali, N. and Al-Doss, A. (2021). Field crop responses and management strategies to mitigate soil salinity in modern agriculture: A review. *Agronomy* 11(11), 2299. https://doi.org/10.3390/agronomy11112299.

Allen, P. and Hinrichs, C. (2007). *Buying into "Buying Local": Engagements of United States Local Food Initiatives*. Maye, D., Holloway, L. and Kneafsey, M. (Eds). Emerald Group Publishing Limited.

Angulo, A. M., Gil, J. M. and Tamburo, L. (2005). Food safety and consumers' willingness to pay for labelled beef in Spain. *Journal of Food Products Marketing* 11(3), 89-105. https://doi.org/10.1300/J038v11n03_06.

Aono, M., Wakiyama, S., Nagatsu, M., Kaneko, Y., Nishizawa, T., Nakajima, N., Tamaoki, M., Kubo, A. and Saji, H. (2011). Seeds of a possible natural hybrid between herbicide-resistant Brassica napus and Brassica rapa detected on a riverbank in Japan. *GM Crops* 2(3), 201-210. https://doi.org/10.4161/gmcr.2.3.18931.

Attwood, S. and Hajat, C. (2020). How will the COVID-19 pandemic shape the future of meat consumption? *Public Health Nutrition* 23(17), 3116-3120. https://doi.org/10.1017/S136898002000316X.

Bacon, L. and Krpan, D. (2018). (Not) Eating for the environment: The impact of restaurant menu design on vegetarian food choice. *Appetite* 125, 190-200. https://doi.org/10.1016/j.appet.2018.02.006.

Barrena, R. and Sánchez, M. (2010). Frequency of consumption and changing determinants of purchase decision: From attributes to values in the organic food

© Burleigh Dodds Science Publishing Limited, 2024. All rights reserved.

market. *Spanish Journal of Agricultural Research* 8(2), 251. https://doi.org/10.5424/sjar/2010082-1178.

Bellows, A. C., Onyango, B., Diamond, A. and Hallman, W. K. (2008). Understanding consumer interest in organics: Production values vs. purchasing behavior. *Journal of Agricultural and Food Industrial Organization* 6(1). https://doi.org/10.2202/1542-0485.1169.

Binkley, J. K. (2019). Nutrition and Food Choice: Home vs. restaurants. *Journal of Consumer Affairs* 53(3), 1146–1166. https://doi.org/10.1111/joca.12221.

Biswas, D., Szocs, C., Chacko, R. and Wansink, B. (2017). Shining light on atmospherics: How ambient light influences food choices. *Journal of Marketing Research* 54(1), 111–123. https://doi.org/10.1509/jmr.14.0115.

Boesveldt, S. and de Graaf, K. (2017). The differential role of smell and taste for eating behavior. *Perception* 46(3–4), 307–319. https://doi.org/10.1177/0301006616685576.

Bonny, S. (2004). Factors explaining opposition to GMOs in France and the rest of Europe. In: Evenson, R. E. and Santaniello, V. (Eds). *Consumer Acceptance of Genetically Modified Foods*. CAB International.

Bourn, D. and Prescott, J. (2002). A comparison of the nutritional value, sensory qualities, and food safety of organically and conventionally produced foods. *Critical Reviews in Food Science and Nutrition* 42(1), 1–34. https://doi.org/10.1080/10408690290825439.

Boys, K. A. and Lee, D. (2019). *Market Incentives for Safe Foods: The Effect of Food Recalls on Food Industry Stock Prices*.

Boys, K. A., Willis, D. B. and Carpio, C. E. (2014). Consumer willingness to pay for organic and locally grown produce on Dominica: Insights into the potential for an "Organic Island." *Environment, Development and Sustainability* 16(3), 595–617. https://doi.org/10.1007/s10668-013-9496-3.

Boys, K. and Blank, S. (2016). *The Evolution of Local Foods: A Retrospective and Prospective Consideration* Lang, M. and Stanton, J. (Eds). The meaning of local foods: A food marketing management perspective, The Institute of Food Products Marketing.

Brewer, M. S. and Rojas, M. (2008). Consumer attitudes toward issues in food safety. *Journal of Food Safety* 28(1), 1–22. https://doi.org/10.1111/j.1745-4565.2007.00091.x.

Byrd, K., Her, E., Fan, A., Almanza, B., Liu, Y. and Leitch, S. (2021). Restaurants and COVID-19: What are consumers' risk perceptions about restaurant food and its packaging during the pandemic? *International Journal of Hospitality Management* 94. https://doi.org/10.1016/j.ijhm.2020.102821.

Caputo, V., Lusk, J. and Kilders, V. (2020). *Consumer Acceptance of Gene Edited Foods: A Nationwide Survey on US Consumer Beliefs, Knowledge, Understanding, and Willingness to Pay for Gene-Edited Foods under Different Information Treatments*.

Caswell, J. A. (2006). Quality assurance, information tracking, and consumer labeling. *Marine Pollution Bulletin* 53(10–12), 650–656. https://doi.org/10.1016/j.marpolbul.2006.08.007.

Caswell, J. A. and Anders, S. M. (2011). Private versus third party versus government labeling. In: Lusk, J. L., Roosen, J. and Shogren, J. F. (Eds). *The Oxford Handbook of the Economics of Food Consumption and Policy*. Oxford University Press, pp. 471–498.

Caswell, J. A., Noelke, C. M. and Mojduszka, E. M. (2002). Unifying two frameworks for analyzing quality and quality assurance for food products. In: Krissoff, B., Bohman,

© Burleigh Dodds Science Publishing Limited, 2024. All rights reserved.

M. and Caswell, J. A. (Eds). *Global Food Trade and Consumer Demand for Quality*. Springer, Boston, MA. pp. 43–61. https://doi.org/10.1007/978-1-4757-5329-5_3.

Charlebois, S. and Music, J. (2021). SARS-CoV-2 pandemic and food safety oversight: Implications in Canada and coping strategies. *Foods* 10, 22–41. https://doi.org/10.3390/foods10102241.

Chauke, D. X. and Duh, H. I. (2019). Marketing and socio-psychological factors influencing organic food purchase and post-purchase outcomes. *Journal of Food Products Marketing* 25(9), 896–920. https://doi.org/10.1080/10454446.2019.1697980.

Chen, Y., Ganesan, S. and Liu, Y. (2009). Does a firm's product-recall strategy affect its financial value? An examination of strategic alternatives during product-harm crises. *Journal of Marketing* 73(6), 214–226. https://doi.org/10.1509/jmkg.73.6.214.

Chen, P. J. and Antonelli, M. (2020). Conceptual models of food choice: Influential factors related to foods, individual differences, and society. *Foods* 9(12), 1898. https://doi.org/10.3390/foods9121898.

Cho, C. H., Mattila, A., Bordi, P. and Kwon, E. (2019). It tastes better when Bach meets red: The role of music and plate color on food evaluation. *British Food Journal* 122(1), 14–25. https://doi.org/10.1108/BFJ-02-2018-0100.

Colson, G. and Rousu, M. C. (2013). What do consumer surveys and experiments reveal and conceal about consumer preferences for genetically modified foods? *GM Crops and Food* 4(3), 158–165. https://doi.org/10.4161/gmcr.26322.

Contento, I. R. (2011). *Nutrition Education: Linking Research, Theory, and Practice* (2nd edn.). Jones and Bartlett Publishers.

Crew, S. (2020). Food safety risk during the pandemic. *Food Science and Technology* 34(2), 14–17. https://doi.org/10.1002/fsat.3402_4.x.

Curvelo, I. C. G., Watanabe, E. A. de M. and Alfinito, S. (2019). Purchase intention of organic food under the influence of attributes, consumer trust and perceived value. *Revista de Gestao* 26(3), 198–211. https://doi.org/10.1108/REGE-01-2018-0010.

Cyranoski, D. (2015). CRISPR tweak may help gene-edited crops bypass biosafety regulation. *Nature*. https://doi.org/10.1038/nature.2015.18590.

Dettmann, R. L. and Dimitri, C. (2010). Who's buying organic vegetables? Demographic characteristics of U.S. consumers. *Journal of Food Products Marketing* 16(1), 79–91. https://doi.org/10.1080/10454440903415709.

Dimitri, C., Effland, A. and Conklin, N. (2005). *The 20th Century Transformation of U.S. Agriculture and Farm Policy*. USDA Economic Information Bulletin (vol. 3).

Draper, C. and Jones, S. (2013). SNAP double bucks programs: Increasing access to produce & supporting farmers & the local economy. *University of South Carolina: Center for Research in Nutrition and Health Disparities*.

Ehmke, M. D., Bonanno, A., Boys, K. A. and Smith, T. G. (2019). Food fraud: Economic insights into the dark side of incentives. *Australian Journal of Agricultural and Resource Economics* 63(4), 685–700. https://doi.org/10.1111/1467-8489.12346.

Ellison, B., McFadden, B., Rickard, B. J. and Wilson, N. L. W. (2021). Examining food purchase behavior and food values during the COVID-19 pandemic. *Applied Economic Perspectives and Policy* 43(1), 58–72. https://doi.org/10.1002/aepp.13118.

Esteki, M., Regueiro, J. and Simal-Gándara, J. (2019). Tackling fraudsters with global strategies to expose fraud in the food chain. *Comprehensive Reviews in Food Science and Food Safety* 18(2), 425–440. https://doi.org/10.1111/1541-4337.12419.

© Burleigh Dodds Science Publishing Limited, 2024. All rights reserved.

Eyinade, G. A., Mushunje, A. and Yusuf, S. F. G. (2021). The willingness to consume organic food: A review. *Food and Agricultural Immunology* 32(1), 78–104. https://doi.org/10.1080/09540105.2021.1874885.

FAO (2022). *Thinking about the Future of Food Safety – A Foresight Report*. Rome. https://doi.org/10.4060/cb8667en.

Feldmann, C. and Hamm, U. (2015). Consumers' perceptions and preferences for local food: A review. *Food Quality and Preference* 40, 152–164. https://doi.org/10.1016/j.foodqual.2014.09.014.

Fernbach, P. M. and Light, N. (2020). Knowledge is shared. *Psychological Inquiry* 31(1), 26–28. https://doi-org.subzero.lib.uoguelph.ca/10.1080/1047840X.2020.1722601.

Fernbach, P. M., Light, N., Scott, S. E., Inbar, Y. and Rozin, P. (2020). Extreme opponents of genetically modified foods know the least but think they know the most. *Nature Human Behaviour* 3, 251–256. https://doi.org/10.1038/s41562-018-0520-3.

Filimonau, V. and Krivcova, M. (2017). Restaurant menu design and more responsible consumer food choice: An exploratory study of managerial perceptions. *Journal of Cleaner Production* 143, 516–527. https://doi.org/10.1016/j.jclepro.2016.12.080.

Food and Drug Administration (2022). GMO crops, animal food, and beyond. Available at: https://www.fda.gov/food/agricultural-biotechnology/gmo-crops-animal-food-and-beyond.

Freidberg, S. and Goldstein, L. (2011). Alternative food in the global south: Reflections on a direct marketing initiative in Kenya. *Journal of Rural Studies* 27(1), 24–34. https://doi.org/10.1016/j.jrurstud.2010.07.003.

Goudis, A. and Skuras, D. (2021). Consumers' awareness of the EU's protected designations of origin logo. *British Food Journal* 123(13), 1–18. https://doi.org/10.1108/BFJ-02-2020-0156.

Grunert, K. G. and Aachmann, K. (2016). Consumer reactions to the use of EU quality labels on food products: A review of the literature. *Food Control* 59, 178–187. https://doi.org/10.1016/j.foodcont.2015.05.021.

Guerra, K. and Boys, K. A. (2021). A new food chain: Adoption and policy implications for blockchain use in agri-food industries. *Applied Economic Perspectives and Policy* 44(1), 324–349.

Guido, G., Prete, M. I., Peluso, A. M., Maloumby-Baka, R. and Buffa, C. (2010). The role of ethics and product personality in the intention to purchase organic food products: A structural equation modeling approach: rivista internazionale di scienze. *International Review of Economics* 57(1), 79–102. https://doi.org/10.1007/s12232-009-0086-5.

Han, J. H. and Harrison, W. R. (2007). Factors influencing urban consumers' acceptance of genetically modified foods. *Review of Agricultural Economics* 29(4), 700–719. https://doi.org/10.1111/j.1467-9353.2007.00382.x.

Harwood, R. R. (1990). A history of sustainable agriculture. *Sustainable Agricultural Systems*.

Helms, J. D. (1985). *Brief History of the USDA Soil Bank Program*. United States Department of Agriculture Natural Resources Conservation Service.

Higgs, S. and Ruddock, H. (2020). Social influences on eating. In: *Handbook of Eating and Drinking*. Springer International Publishing, (277–291). https://doi.org/10.1007/978-3-030-14504-0_27.

© Burleigh Dodds Science Publishing Limited, 2024. All rights reserved.

Hooker, N. H., Teratanavat, R. P. and Salin, V. (2005). Crisis management effectiveness indicators for US meat and poultry recalls. *Food Policy* 30(1), 63–80. https://doi.org/10.1016/j.foodpol.2004.11.003.

IFIC (2020). *IFIC-Food-and-Health-Survey-2020*.

Jung, J., Bir, C., Widmar, N. O. and Sayal, P. (2021). Initial reports of foodborne illness drive more public attention than do food recall announcements. *Journal of Food Protection* 84(7), 1150–1159.

Kim, Y., Kim, S. and Arora, N. (2022). GMO labeling policy and consumer choice. *Journal of Marketing* 86(3), 21–39. https://doi.org/10.1177/00222429211064901.

Lachat, C., Nago, E., Verstraeten, R., Roderbroid, D., Van Camp, J. and Kolsteren, P. (2012). Eating out of home and its association with dietary intake: A systematic review of the evidence. *Obesity Reviews* 13(4), 329–346. https://doi.org/10.1111/j.1467-789X.2011.00953.x.

Latvala, T. and Kola, J. (2004). Consumer willingness to pay for additional information on food quality and safety. 84th EAAE Seminar. Available at: https://www.researchgate.net/publication/23510048. Food Safety in a Dynamic World.

Lee, D. and Boys, K. A. (2018). Market incentives for safe foods: An examination of the effect of food recalls on firms' stock prices. Poster presented to the 2018 AAEA Annual Meeting, Washington, DC, August 5–7, 2018. https://doi.org/10.22004/ag.econ.273894.

Liang, R.-D. (2016). Predicting intentions to purchase organic food: The moderating effects of organic food prices. *British Food Journal* 118(1), 183–199. https://doi.org/10.1108/BFJ-06-2015-0215.

Likoudis, Z., Sdrali, D., Costarelli, V. and Apostolopoulos, C. (2016). Consumers' intention to buy protected designation of origin and protected geographical indication foodstuffs: The case of Greece. *International Journal of Consumer Studies* 40(3), 283–289. https://doi.org/10.1111/ijcs.12253.

Lim, K. H., Hu, W., Maynard, L. J. and Goddard, E. (2014). A taste for safer beef? How much does consumers' perceived risk influence willingness to pay for country-of-origin labeled beef. *Agribusiness* 30(1), 17–30. https://doi.org/10.1002/agr.21365.

Lim, W. M., Yong, J. L. S. and Suryadi, K. (2014). Consumers' perceived value and willingness to purchase organic food. *Journal of Global Marketing* 27(5), 298–307. https://doi.org/10.1080/08911762.2014.931501.

Lin, C.-T. J. (1995). Demographic and socioeconomic influences on the importance of food safety in food shopping. *Agricultural and Resource Economics Review* 24(2), 190–198. https://doi.org/10.1017/S1068280500008832.

Loureiro, M. L. and Hine, S. (2002). Discovering niche markets: A comparison of consumer willingness to pay for local (Colorado grown), organic, and GMO-free products. *Journal of Agricultural and Applied Economics* 34(3), 477–487. https://doi.org/10.1017/S1074070800009251.

Lusk, J. L. and Briggeman, B. C. (2009). Food values. *American Journal of Agricultural Economics* 91(1), 184–196. https://doi.org/10.1111/j.1467-8276.2008.01175.x.

Lusk, J. L., McFadden, B. R. and Wilson, N. (2018). Do consumers care how a genetically engineered food was created or who created it? *Food Policy* 78, 81–90. https://doi.org/10.1016/j.foodpol.2018.02.007.

© Burleigh Dodds Science Publishing Limited, 2024. All rights reserved.

Marette, S., Disdier, A. C. and Beghin, J. C. (2021). A comparison of EU and US consumers' willingness to pay for gene-edited food: Evidence from apples. *Appetite* 159, 105064. https://doi.org/10.1016/j.appet.2020.105064.

Marsh, T. L., Schroeder, T. C. and Mintert, J. (2004). Impacts of meat product recalls on consumer demand in the USA. *Applied Economics* 36(9), 897–909. https://doi.org/10.1080/0003684042000233113.

Martinelli, E. and de Canio, F. (2018). Does PDO/PGI labels contribute to consumers' intention to buy premium private labels products? An empirical survey. In: *Springer Proceedings in Business and Economics*, pp. 37–45. https://doi.org/10.1007/978-3-319-92084-9_6.

Marvier, M. and van Acker, R. C. V. (2005). Can crop transgenes be kept on a leash? *Frontiers in Ecology and the Environment* 3(2), 99–106. https://doi.org/10.2307/3868516.

Meas, T., Hu, W., Batte, M. T., Woods, T. and Ernst, S. (2013). *Local Is the New Organic: Do Consumers Agree?* Agricultural and Applied Economics Association.

Micheletti, M. (2013). *Political Virtue and Shopping: Individuals, Consumerism, and Collective Action*.

Monterrosa, E. C., Frongillo, E. A., Drewnowski, A., de Pee, S. and Vandevijvere, S. (2020). Sociocultural influences on food choices and implications for sustainable healthy diets. *Food and Nutrition Bulletin* 41(2_suppl), 59S–73S. https://doi.org/10.1177/0379572120975874.

Nardi, V. A. M., Teixeira, R., Ladeira, W. J. and de Oliveira Santini, F. (2020). A meta-analytic review of food safety risk perception. *Food Control* 112, 107089. https://doi.org/10.1016/j.foodcont.2020.107089.

Nguyen, T. H., ben Taieb, S., Moritaka, M. and Fukuda, S. (2022). Implicit and explicit attitudes toward foods derived from genome editing and genetic modification technologies under different information treatments. *Journal of Food Products Marketing* 28(1), 1–30. https://doi.org/10.1080/10454446.2022.2037487.

Noussair, C., Robin, S. and Ruffieux, B. (2007). Measuring preferences for genetically modified food products. *Environmental Economics, Experimental Methods*, London, UK: Routledge. 344–365. https://doi.org/10.4324/9780203935361.

Okumus, B., Okumus, F. and McKercher, B. (2007). Incorporating local and international cuisines in the marketing of tourism destinations: The cases of Hong Kong and Turkey. *Tourism Management* 28(1), 253–261.

Onken, K. A., Bernard, J. C. and Pesek, J. D. (2011). Comparing willingness to pay for organic, natural, locally grown, and state marketing program promoted foods in the mid-Atlantic region. *Agricultural and Resource Economics Review* 40(1), 33–47. https://doi.org/10.1017/S1068280500004500.

Page, E. T. (2018). *Trends in Food Recalls: 2004–13*. Economic Information Bulletin 191.

Paul, J. and Rana, J. (2012). Consumer behavior and purchase intention for organic food. *Journal of Consumer Marketing* 29(6), 412–422. https://doi.org/10.1108/07363761211259223.

Pozo, V. F. and Schroeder, T. C. (2016). Evaluating the costs of meat and poultry recalls to food firms using stock returns. *Food Policy* 59, 66–77. https://doi.org/10.1016/j.foodpol.2015.12.007.

Pruitt, J. R., Melton, K. M. and Palma, M. A. (2021). Does physical activity influence consumer acceptance of gene edited food? *Sustainability* 13(14), 7759. https://doi.org/10.3390/su13147759.

© Burleigh Dodds Science Publishing Limited, 2024. All rights reserved.

Qaim, M. (2020). Role of new plant breeding technologies for food security and sustainable agricultural development. *Applied Economic Perspectives and Policy* 42(2), 129–150. https://doi.org/10.1002/aepp.13044.

Rana, J. and Paul, J. (2017). Consumer behavior and purchase intention for organic food: A review and research agenda. *Journal of Retailing and Consumer Services* 38(May), 157–165. https://doi.org/10.1016/j.jretconser.2017.06.004.

Röhr, A., Lüddecke, K., Drusch, S., Müller, M. J. and Alvensleben, R. v (2005). Food quality and safety--consumer perception and public health concern. *Food Control* 16(8), 649–655. https://doi.org/10.1016/j.foodcont.2004.06.001.

Rupp, N. G. (2004). The attributes of a costly recall: Evidence from the automotive industry. *Review of Industrial Organization* 25(1), 21–44. https://doi.org/10.1023/B:REIO.0000040514.22968.e1.

Saksena, M. J., Okrent, A. M., Anekwe, T. D., Cho, C., Dicken, C., Effland, A., Elitzak, H., Guthrie, J., Hamrick, K. S., Hyman, J., Jo, Y., Lin, B.-H., Mancino, L., McLaughlin, P. W., Rahkovsky, I., Ralston, K., Smith, T. A., Stewart, H., Todd, J. and Tuttle, C. (2018). *United States Department of Agriculture America's Eating Habits: Food Away from Home United States Department of Agriculture*. Available at: www.ers.usda.gov.

Sckokai, P., Veneziani, M., Moro, D. and Castellari, E. (2014). Consumer willingness to pay for food safety: The case of mycotoxins in milk. *Bio-Based and Applied Economics* 3(1), 63–81. https://doi.org/10.13128/BAE-12827.

Seys, S. A., Sampedro, F. and Hedberg, C. W. (2015). Assessment of Shiga toxin-producing *Escherichia coli* O157 illnesses prevented by recalls of beef products. *Foodborne Pathogens and Disease* 12(9), 800–805. https://doi.org/10.1089/fpd.2015.1968.

Shahriari, E., Torres, I. M., Zúñiga, M. A. and Yarlou, P. M. (2019). Values driving organic food purchase intention: A comparative analysis between a developing eastern country (Iran) and a developed Western country (US). *Journal of International Consumer Marketing* 31(4), 317–329. https://doi.org/10.1080/08961530.2018.1561345.

Shang, X. and Tonsor, G. T. (2017). Food safety recall effects across meat products and regions. *Food Policy* 69, 145–153. https://doi.org/10.1016/j.foodpol.2017.04.002.

Sharratt, L. and Chopra, T. (2019). GM contamination in Canada: The failure to contain living modified organisms. Available at: https://cban.ca/wp-content/uploads/GM-contamination-in-canada-2019.pdf.

Sikora, D. and Rzymski, P. (2021). Public acceptance of GM foods: A global perspective (1999–2019). In: *Policy Issues in Genetically Modified Crops*. Elsevier Inc. https://doi.org/10.1016/b978-0-12-820780-2.00013-3.

Sloman, S. and Fernbach, P. (2018). The Knowledge Illusion: The myth of individual thought and the power of collective wisdom. Pan Macmillan, 304 pages. ISBN 978-1509813087.

Sobal, J., Carole, A. and Bisogni, C. A. (2009). Constructing food choice decisions. *Annals of Behavioral Medicine* 38(suppl_1), s37–s46. https://doi.org/10.1007/s12160-009-9124-5.

Steenkamp, J.-B. E. M. (1989). Product quality: An investigation into the concept and how it is perceived by consumers. Agricultural University. Promotor(en): M.T.G. Meulenberg. Wageningen: Landbouwuniversiteit Wageningen - ISBN 9789023224150 – 281. Dissertation, internally prepared.

Steenkamp, J.-B. E. M. and van Trijp, H. C. M. (1996). Quality guidance: A consumer-based approach to food quality improvement using partial least squares. *European Review of Agricultural Economics* 23(2), 195–215. https://doi.org/10.1093/erae/23.2.195.

© Burleigh Dodds Science Publishing Limited, 2024. All rights reserved.

Steiner, R. (2011). *Agriculture Course: The Birth of the Biodynamic Method.*
Stroebele, N. and de Castro, J. M. (2004). Effect of ambience on food intake and food choice. *Nutrition* 20(9), 821–838. https://doi.org/10.1016/j.nut.2004.05.012.
Taylor, M., Klaiber, H. A. and Kuchler, F. (2016). Changes in U.S. consumer response to food safety recalls in the shadow of a BSE scare. *Food Policy* 62, 56–64. https://doi.org/10.1016/j.foodpol.2016.04.005.
Teixeira, S. F., Barbosa, B., Cunha, H. and Oliveira, Z. (2022). Exploring the antecedents of organic food purchase intention: An extension of the theory of planned behavior. *Sustainability* 14(1), 242. https://doi.org/10.3390/su14010242.
Thilmany, D., Bond, C. A. and Bond, J. K. (2008). Going local: Exploring consumer behavior and motivation for direct food purchases. *American Journal of Agricultural Economics* 90(5), 1303–1309.
Thilmany, D., Canales, E., Low, S. A. and Boys, K. (2021). Local food supply chain dynamics and resilience during COVID-19. *Applied Economic Perspectives and Policy* 43(1), 86–104. https://doi.org/10.1002/aepp.13121.
Thomas, M. and Feng, Y. (2021). Consumer risk perception and trusted sources of food safety information during the COVID-19 pandemic. *Food Control* 130. https://doi.org/10.1016/j.foodcont.2021.108279.
Tonsor, G. T., Schroeder, T. C. and Pennings, J. M. E. (2009). Factors impacting food safety risk perceptions. *Journal of Agricultural Economics* 60(3), 625–644. https://doi.org/10.1111/j.1477-9552.2009.00209.x.
Tully, K., Gedan, K., Epanchin-Niell, R., Strong, A., Bernhardt, E. S., BenDor, T., Mitchell, M., Kominoski, J., Jordan, T. E., Neubauer, S. C. and Weston, N. B. (2019). The invisible flood: The chemistry, ecology, and social implications of coastal saltwater intrusion. *BioScience* 69(5), 368–378. https://doi.org/10.1093/biosci/biz027.
van Bussel, L. M., Kuijsten, A., Mars, M. and van 't Veer, P. (2022). Consumers' perceptions on food-related sustainability: A systematic review. *Journal of Cleaner Production* 341, 130904. https://doi.org/10.1016/j.jclepro.2022.130904.
van der Lans, I. A., van Ittersum, K., De Cicco, A. and Loseby, M. (2001). The role of the region of origin and EU certificates of origin in consumer evaluation of food products. *European Review of Agricultural Economics* 28(4), 451–477. https://doi.org/10.1093/erae/28.4.451.
van Ruth, S. M., Luning, P. A., Silvis, I. C. J., Yang, Y. and Huisman, W. (2018). Differences in fraud vulnerability in various food supply chains and their tiers. *Food Control* 84, 375–381. https://doi.org/10.1016/j.foodcont.2017.08.020.
Wang, R., Zhong, D., Wu, S., Han, Y., Zheng, Y. and Tang, F., Ni, Z. and Liu, Y. (2021). The phytochemical profiles for walnuts (J. regia and J. sigillata) from China with protected geographical indications. *Food Science and Technology (Campinas)* 4 (Suppl. 2). https://doi.org/10.1590/fst.30320.
Wilcock, A. and Ball, B. (2014). Food safety: Consumer perceptions and practices. *Practical Food Safety, 9781118474,* 11–29. https://doi.org/10.1002/9781118474563.ch2.
Willer, H., Trávníček, J., Meier, C. and Schlatter, B. (2022). *World of Organic Agriculture Statistics and Emerging Trends 2022.* Frick Forschungsinstitut Für Biologischen Landbau Fibl.
Yang, J. and Goddard, E. (2011). Canadian consumer responses to BSE with heterogeneous risk perceptions and risk attitudes. *Canadian Journal of Agricultural Economics/Revue Canadienne d'Agroeconomie* 59(4), 493–518. https://doi.org/10.1111/j.1744-7976.2011.01225.x.

© Burleigh Dodds Science Publishing Limited, 2024. All rights reserved.

Yeung, R. M. W. and Morris, J. (2001). Food safety risk: Consumer perception and purchase behaviour. *British Food Journal* 103(3), 170–187. https://doi.org/10.1108/00070700110386728.

Yu, J. and Hooker, N. H. (2020). Assessing the impact of retailer disclosure on the effectiveness of U.S. meat and poultry recalls. *Journal of Food Protection* 83(12), 2095–2101. https://doi.org/10.4315/JFP-20-006.

Zellner, D., Geller, T., Lyons, S., Pyper, A. and Riaz, K. (2017). Ethnic congruence of music and food affects food selection but not liking. *Food Quality and Preference* 56, 126–129. https://doi.org/10.1016/j.foodqual.2016.10.004.

© Burleigh Dodds Science Publishing Limited, 2024. All rights reserved.

Chapter 9

Understanding consumer attitudes to organic food: using profile deviation analysis for consumer benchmarking

Tatiana Anisimova, Linnaeus University, Sweden; and Felix T. Mavondo, Monash University, Australia

1 Introduction

Organic food market growth, which remains inconsistent globally, constitutes a barrier to growth in sustainable consumption. European countries tend to account for the highest share of organic food sales as a percentage of their respective food markets. Denmark has the highest organic food sales share worldwide, with 12.1% in 2019, followed by Switzerland with a share of 10.4% (The Research Institute of Organic Agriculture FiBL, 2021). Although in Australia the organic food market share is slowly growing (Parkes, 2017), the growth currently sits at 2% (Prendergast, 2020).

The consumption of organic food has been gaining momentum in the wake of the coronavirus disease 2019 (COVID-19) pandemic. It has been reported that during the COVID-19 lockdown, consumption patterns shifted toward healthier food habits to maintain a good nutritional status and reduce health risks (Di Renzo et al., 2020; Sidor and Rzymski, 2020). Spanish consumers have shown healthier dietary habits during COVID-19 (Rodríguez-Pérez et al.,

http://dx.doi.org/10.19103/AS.2023.0129.20
© Burleigh Dodds Science Publishing Limited, 2024. All rights reserved.

2020). Li et al. (2021) found that food purchasing behaviour of Chinese females increased towards the higher consumption of sustainable and healthy diets during COVID19. According to The World of Organic Agriculture 2021 (FiBL), the year 2020 witnessed an accelerated growth of the organic market in many countries due to the pandemic. In the UK alone, Waitrose organic food sales increased by 13% during the lockdowns (Leathy, 2020).

Previous research has identified three central dimensions underlying an organic food consumer profile (Aertsens et al., 2009; Anisimova, 2016; Bonroy and Constantatos, 2008). Firstly, increasingly, hedonistic or 'me-centric' values matter to individuals when consuming organic foods (Schneider-Kamp, 2021). Secondly, consumers are conscious of how organic foods benefit their health (Peluso et al., 2021). Thirdly, consumer trust in the organic food industry remains a severe obstacle to organic consumption because organic food products are credence goods (Bonroy and Constantatos, 2008).

Our investigation is performed in a configuration-theoretic setting (Vorhies and Morgan, 2003; Kabadayi et al., 2007; Venkatraman and Prescott, 1990; Venkatraman, 1990). This involves viewing organic food consumers holistically using multiple dimensions to build an organic food consumer profile. Given the importance of consumer satisfaction and loyalty as key consumer performance indicators (de Chermatony et al., 1998; Hoeffler and Keller, 2002), we first identify the profile of the most loyal and most satisfied organic food consumers along the three key dimensions: consumer hedonism, healthism, and trust. A second step involved making comparisons between the ideal and non-ideal profiles to examine whether deviation from this ideal profile along the dimensions mentioned earlier results in a decrease in consumer satisfaction and loyalty. To enable the above steps, we applied a profile deviation (PD) analysis (Venkatraman, 1989, 1990; Venkatraman and Prescott, 1990).

The research question we seek to answer is as follows:

To what extent is deviation from an organic food consumer benchmark profile, consisting of healthism, hedonism, and trust, associated with a reduced consumer satisfaction and loyalty?

Our study is one of the few within the marketing and sustainability fields to opt for a configuration-theoretic and PD approach to explain consumer behavior (see Anisimova and Mavondo, 2010; Blome et al., 2014; Malhotra et al., 2013). We integrate the configuration theory and sustainable consumption literature and the PD analysis to provide novel theoretical and practical insights into key drivers of individual organic food satisfaction and loyalty. Thereby, we are aiming to advance research at the intersection of configuration and organic food consumption theories. Methodologically, the configurational design adopted in this study enriches conventional empirical approaches involving direct, linear relationships between predictor and outcome variables. Finally,

© Burleigh Dodds Science Publishing Limited, 2024. All rights reserved.

PD analysis is a diagnostic tool that can help practitioners to design more effective marketing communications.

2 Theoretical framework and hypothesis development

2.1 Configuration theory and profile deviation approaches

The central pillars of the configuration theory are the 'common alignments among many elements' and 'complex systems of interdependency brought about by central orchestrating themes' (Miller, 1996, p. 506). PD is used to operationalize this conceptual framework (Doty et al., 1993). It involves evaluating the degree of adherence to or alignment with a benchmark profile (Kabadayi et al., 2007; Venkatraman, 1989). Deviation from this profile implies negative and significant implications for performance outcomes (Venkatraman and Prescott, 1990). The PD approach requires three essential steps (Venkatraman and Prescott, 1990): (i) identifying the best performers; (ii) describing their profiles along the critical dimensions, which yields a benchmark profile; and (iii) testing the performance implications of deviations from such a benchmark profile. Specifying the ideal profile is labeled by Venkatraman and Prescott (1990, p. 6) as being the 'calibration sample' of best-performing entities. We argue that configuration theory can be instrumental for sustainable consumption research because organic food consumers can be categorized in terms of their performance regarding certain metrics (e.g. loyalty), including top performers (e.g. the most loyal organic food consumers), and those deviating from this performance benchmark.

2.2 Conceptualizing consumer-based performance indicators: satisfaction and loyalty

Satisfaction and loyalty are essential indicators of consumer-based performance along with traditional financial performance measures (Hoque, 2014; Kaplan and Norton, 1992; Oliver, 2014). Consumer satisfaction can be described as an emotional or cognitive response to a particular focus such as a product, expectations or a consumption experience at a particular time (Giese and Cote, 2000). Consumer loyalty has been defined as 'the attachment that a consumer has to a brand' (Aaker, 1991, p.39), denoting brand success (de Chermatony et al., 1998) and a competitive advantage (Watson et al., 2015). Consumer loyalty has been shown to improve brand performance and company sales (Chaudhuri and Holbrook, 2001; Yeung, and Ennew, 2000). Reichheld (1996) reported a positive relationship between consumer loyalty and their willingness to pay a price premium. Loyal consumers have also been seen to purchase their favorite brands more frequently and regularly and are less prone to switching to other brands (Yoo et al., 2000).

© Burleigh Dodds Science Publishing Limited, 2024. All rights reserved.

2.3 Configurations, alignment and misalignment in the context of consumer organic food choice

Consumer behavior is often conceptualized as being influenced by consumer attitudes toward that behavior. Attitudes have both cognitive and affective dimensions (Ajzen, 2001; Agarwal and Malhotra, 2005). This means there is a need to include both attitude components in behavior models in organic food choice models, which has often been overlooked in the previous research (Aertsens et al., 2009). In line with this multi-component view of attitude, we investigate the role of affective factors in the form of trustworthiness of organic food and hedonistic consumer motivations as well as cognitive factors embodied in consumer healthism on organic food consumption behavior (Aertsens et al., 2009; Thøgersen, 2009; Watson and Spence, 2007).

Previous research into cognitive attitudes in the organic food context has identified perceived health benefits as a major purchasing motive (Kriwy and Mecking, 2012; Pino et al., 2012; Aertsens et al., 2009). Health motives as triggers of organic food consumption constitute a cognitive-utilitarian consumption dimension because they reflect the consumers' evaluations of how useful the consumption of organic food is (Lee and Yun, 2015). Hedonistic motives, by contrast, are relatable to sensory or emotional experiences (Batra and Ahtola, 1991; Hirschman and Holbrook, 1982). Some researchers consider these self-centered attitudes even more influential than environmental or ethical concerns (Davis, 2013; Moser, 2016). Such motives have been found to influence consumer decision-making regarding organic food consumption (Fifita et al., 2020; Anisimova, 2016). Lee and Yun (2015) found a positive link between hedonic attitudes and intentions to buy organic food. Audebert *et al.* (2006) identified hedonism as an important driver of attitudes toward red meat. Similarly, Chryssohoidis and Krystallis (2005) report a significant link between the importance Greek consumers attach to 'fun and enjoyment' and the consumption of organic fruits and vegetables. However, there are some empirical studies that do not support the existence of a positive link between healthism and organic product consumption (Michaelidou and Hassan, 2008; Tarkiainen and Sundqvist, 2005), which warrants more research to understand the impact of hedonism on organic food consumption.

The organic food market is a complex market where consumer trust plays a key role (Hamzaoui et al., 2013; Sønderskov and Daugbjerg, 2011). This type of affective attitude is of paramount importance since organic foods possess credence attributes, with consumers having to trust producers, sellers or third parties for trustworthy information about the correctness of the attributes that organic foods promise to confer (Perrini et al., 2010). Previous research found a positive relationship between organic food trust and organic food consumption. Teng and Wang (2015) and Yin et al. (2010), for instance,

© Burleigh Dodds Science Publishing Limited, 2024. All rights reserved.

showed that trustworthy information about organic food stimulated consumer purchases of organic products. Castaldo et al. (2009) and Lassoued and Hobbs (2015) reported a positive link between consumer trust in food with credence attributes and consumer loyalty. If consumers lose trust in organic products, a credibility gap arises, which may have negative effects on their organic purchase intentions (Lee and Hwang, 2016). Previous studies have found the evidence that low consumer trust in organic food causes a decrease in organic food purchases (Aertsens et al., 2009; Lea and Worsley, 2005).

To sum up, we argue the following: (i) there is an ideal organic consumer profile in terms of healthism, hedonism and trust associated with a superior satisfaction and loyalty in organic food and (ii) organic food consumers deviating from the ideal consumer profile will result in lower levels of satisfaction and loyalty.

3 Hypothesis development

The above reasoning makes a case for the theoretical relevance to the organic food consumption of consumer healthism, hedonism and trust. Through the lens of configuration theory and PD analysis, these three key determinants of organic food choice constitute configurations that jointly and interdependently characterize the benchmark of organic food consumers. We define 'ideal' as those organic food consumers who associate organic products with healthism and experience hedonic feelings and display a high trust level toward organic foods. This calls for investigating the consistency of organic food consumers relative to this 'ideal' consumer profile. Assessing a fit requires measuring the deviations of other organic food consumers from the ideal consumer in terms of healthism, hedonism and trust. It is argued that if such a multivariate ideal profile exists for the most satisfied and loyal benchmark consumer, then any deviations from this ideal consumer profile should entail a negative and significant effect on satisfaction and loyalty. This logic is in line with recent studies in related fields. For example, Hult et al. (2007) report that the perception deviations of a firm's customers in terms of service, product, and eBusiness quality in relation to benchmark customers with the highest repurchase intentions, resulted in weaker repurchase intentions. Blome et al. (2014) found that a firm's degree of misalignment from the ideal profile in terms of sustainable supply chain collaboration is negatively associated with sustainable production.

While our focus is on analyzing PD in an organic food satisfaction and loyalty context from a holistic perspective, we first conducted a separate investigation of the disaggregated effects on the consumer satisfaction and loyalty of consumer deviations from the ideal consumer profile for each of the three dimensions – healthism, hedonism and trust. Based on the support of previous research as discussed previously, we postulate that organic food satisfaction

© Burleigh Dodds Science Publishing Limited, 2024. All rights reserved.

and loyalty of organic food consumers are negatively affected by a situation where the perceptions of these organic consumers, in terms of health benefits, are misaligned when compared with the corresponding perceived healthism of their benchmark counterparts. Hence, we formulated the following hypothesis:

H1. Deviation from the ideal profile in consumer perceptions of the health benefits of organic foods of the baseline consumers will lead to a negative effect on (a) consumer satisfaction and (b) loyalty.

Second, building on the premise that hedonism constitutes an important antecedent of organic food consumption, our second conjecture is that the satisfaction and loyalty of organic food consumers are negatively influenced by a misalignment between the hedonistic values of benchmark organic consumers relative to those of the baseline organic consumers. Therefore, our next hypothesis is as follows:

H2. Deviation from the ideal profile consumers for organic food-related hedonism of the baseline consumers will have a negative effect on (a) satisfaction and (b) loyalty.

Third, building on the importance of consumer trust for organic food consumption, we posit that organic food satisfaction and loyalty of organic food consumers are negatively related to a misalignment with their benchmark. Thus, we formulated our third hypothesis as follows:

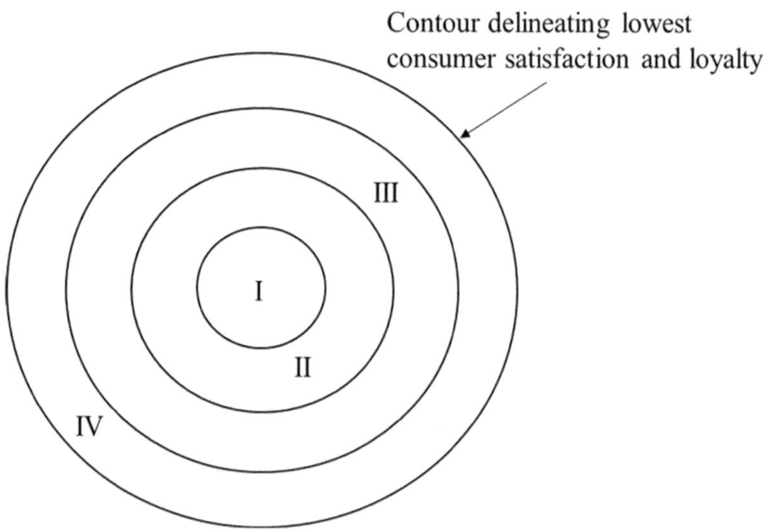

Figure 1 The systems approach to fit and profile deviation as concentric circles. Note: I = benchmark profile of most satisfied and most loyal organic food consumers in terms of healthism. hedonism and trust; II = somewhat dissatisfied or disloyal organic food consumers due to weak deviation from benchmark profile; III = dissatisfied or disloyal organic food consumers due to moderation deviation from benchmark profile; IV = strongly dissatisfied or disloyal organic food consumers due to strong deviation from benchmark profile.

© Burleigh Dodds Science Publishing Limited, 2024. All rights reserved.

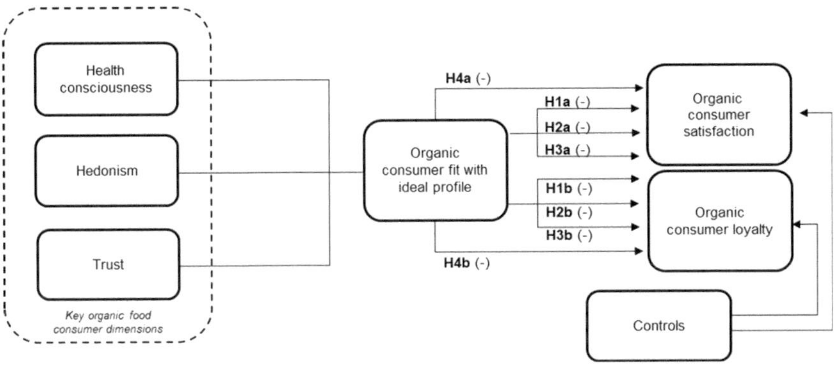

Figure 2 Conceptual framework of profile deviation for organic consumers.

H3. Deviation from the ideal profile on consumer trust of the baseline consumers will lead to a negative effect on (a) satisfaction and (b) loyalty.

Our final and aggregate hypothesis concerns the influence of healthism, hedonism and trust as interdependent configurations that jointly describe the archetype consumer of organic foods. In line with the systems approach to a fit, we examine the effects of satisfaction and loyalty of deviations of organic food consumers from the ideal and multivariate consumer profile consisting of healthism, hedonism, and trust. Our conjecture is that any deviations from this archetype multivariate consumer profile should entail negative effects regarding satisfaction and loyalty. Hence:

H4. Total deviation from the ideal profile of the baseline consumers will lead to a negative effect on (a) satisfaction and (b) loyalty.

We test our conceptual framework empirically by means of a PD analysis, which, as argued earlier represents a relevant empirical approach in this context. Figure 2 illustrates the conceptual framework of our study.

4 Method and data

4.1 Sample

The online structured questionnaire was used to collect the data. The online survey was chosen due to its' capacity to capture large sample sizes and a board geographic scope at a relatively low cost (Dillman et al., 2009). The respondents were recruited from a national research panel. A precondition for participation in the online survey was that respondents were at least 18 years and had bought organic foods in the past 6 months. Of the 1011 completed surveys, 82% of respondents had purchased organic food at least once in the previous 3 years and 44% had purchased organic food for more than three years. The sample was a demographically representative of the Australian adult

© Burleigh Dodds Science Publishing Limited, 2024. All rights reserved.

population of organic food consumers (see Anisimova, 2016 and Anisimova et al., 2019).

4.2 Measures and descriptive statistics

4.2.1 Dependent variables

Both dependent variables of consumer satisfaction and loyalty were measured on 7-point Likert scales. A seven-point Likert scale has been shown to be superior to a 5-point scale because it enhances reliability (Churchill, 1979) and is more closely related to a normal distribution. To enhance validity, a neutral response option was included (Churchill and Iacobucci, 2002). The satisfaction variable is a multi-item construct defined in line with the previous literature (Yi, 1991). The behavioral loyalty variable, too, was a multi-item construct adapted from Anisimova (2016), Anisimova et al (2019) and Yoo and Donthu (2001). For both constructs, the Cronbach's alpha exceed the acceptable level of 0.7 (Nunnally, 1978).

4.3 Profile deviation variables

We constructed aggregate and disaggregate PD variables for both satisfaction and loyalty. The calculations are based on the three key organic food-related perceptual dimensions: healthism, hedonism and trust. All three dimensions are multi-item constructs measured on 7-point Likert scales. The health consciousness construct is adapted from Chryssohoidis and Krystallis (2005); the consumer hedonism construct draws on Audebert et al. (2006); and the consumer trust construct builds on Chaudhuri and Holbrook (2001) and Kennedy et al. (2001). The Cronbach's alpha reliability scores for the PD dimensions exceed the acceptance level of 0.7.

Based on these three dimensions, we calculated both the aggregate and disaggregate PD variables. *Total PD* is the aggregated misalignment score of consumer satisfaction and loyalty study samples from the respective satisfaction and loyalty calibration sample mean in terms of healthism, hedonism and trust. This misalignment variable is calculated as follows:

$$Total\,profile\,deviation = \sqrt{\sum_{j=1}^{3}\left(x_{ij} - \overline{x}_{\text{ideal}}\right)^2}$$

where x_{ij} corresponds to the score for consumer *i* in the respective study sample along the *j*th dimension; $\overline{x}_{ideal}$ is the mean for the ideal profile (benchmark consumer healthism, hedonism, and trust) on the *j*th dimension, and three equates to the number of dimensions under consideration. This allows a

© Burleigh Dodds Science Publishing Limited, 2024. All rights reserved.

calculation of PD scores between the average organic food-related perceptions of benchmark consumers along the above three dimensions in each calibration sample and the corresponding perceptions of organic food consumers in each respective study baseline sample. Total PD corresponds to the aggregated total PD of a consumer in the organic food satisfaction and behavioral loyalty study samples from the respective satisfaction and loyalty calibration sample means in terms of the three individual dimensions considered for the PD (Anisimova and Mavondo, 2010). In addition, we calculated disaggregated PD scores for each of the three individual dimensions used in the total PD calculations (Anisimova and Mavondo, 2010). The disaggregated PD scores were computed as follows:

$$Profile\,deviation_j = \sqrt{\left(x_{ij} - \bar{x}_{\text{ideal}}\right)^2}$$

where *j* denotes the *j*th of our three PD dimensions. Hence, we constructed three PD variables based on the respective calibration and study samples used in the study: for healthism, hedonism and trust.

4.4 Control variables

We included several additional variables to control for the effects of drivers of organic food satisfaction and loyalty (Hansen et al., 2018; McFadden and Huffman, 2017; Tarkiainen and Sundqvist, 2009). *Gender, age and household size* measures were included. In addition, there were *annual household income* and *education level* measures. The education levels range from 1 ('no formal education') to 6 ('postgraduate qualification'). Finally, there was *organic food consumption persistence*, which was a measure of how many years the individual has been consuming organic food. The consumption duration ranges from 1 ('less than 1 year') to 5 ('more than 3 years').

4.5 Measurement evaluation

We used confirmatory factor analysis (CFA) to validate our measurement model with regard to our multi-item constructs. The CFA demonstrated a good model fit as follows: χ^2 (314) = 1438.796; χ^2/df = 4.58; a root mean squared error of approximation (RMSEA) = 0.060; fit index (CFI) = 0.953; RMR = 0.074; Tucker-Lewis index (TLI) = 0.947; and parsimony-adjusted measures (PCFI) = 0.852; AIC = 1566.796. Based on the CFA results, we also examined the convergent and discriminant validity of our measures. The factor composite reliability scores for all constructs exceeded 0.60 and factor loadings as well as the average variance explained are greater than 0.50 - which confirms convergent validity (Fornell and Larcker, 1981). Table 1 provides the evidence of the discriminant validity of our measures given that the average variances extracted are usually higher than the squared correlations among the constructs (Fornell and Larcker, 1981).

© Burleigh Dodds Science Publishing Limited, 2024. All rights reserved.

Table 1 Measures and measurement evaluation

Construct	Items	Standardized regression weight	Composite Reliability (CR)	Average Variance Extracted (AVE)
Satisfaction α = 0.90	'Overall, I am satisfied with the available range of organic foods'	0.677	0.90	0.71
	'I am satisfied with the current state of organic food certification'	0.841		
	'I am satisfied with the information available about organic foods'	0.921		
	'I am satisfied with the quality of organic food information'	0.925		
Behavioral loyalty α = 0.88	'I will continue purchasing organic foods' 'I would gladly buy more organic food if I could find it' 'I check to see if food products are organic before I purchase them' 'If organic and conventional food items were the same price, I would choose organic' 'If the "organic version" of the same product is more expensive than the "conventional" one, I would still choose organic' 'I would recommend purchasing organic food to a friend/relative' 'If I had to buy food today, I would buy certified organic food'	0.794 0.833 0.772 0.592 0.719 0.866 0.873	0.91	0.61
Health consciousness α = 0.86	'Organic food has health-promoting effects' 'Organic food helps me to live a healthy lifestyle' 'Organic food enhances my health' 'I consider myself very health-conscious'	0.863 0.954 0.936 0.450	0.89	0.68

(*Continued*)

© Burleigh Dodds Science Publishing Limited, 2024. All rights reserved.

Table 1 (*Continued*)

Construct	Items	Standardized regression weight	CR	AVE
Hedonism α = 0.86	'Purchasing organic food makes me feel good about myself' 'I eat organic food not because I have to but because I want to' 'I can treat myself by eating organic foods' 'I get pleasure from eating organic foods' 'The best time for me to enjoy organic foods is during meals with family or friends'	0.722 0.735 0.798 0.879 0.728	0.88	0.60
Trust α = 0.96	'I trust Australian institutions certifying organic foods' 'I trust Australian organic food manufacturers' 'I trust claims on organic food labels' 'I trust organic food products I purchase' 'I can rely on organic food products sold in Australia' 'I trust store personnel who sell organic foods' 'I trust a product that carries an organic label/an organic certificate'	0.876 0.908 0.894 0.908 0.896 0.787 0.885	0.95	0.77

4.6 Theory testing with profile deviation analysis

The first step consists of establishing calibration samples, that is, ideal organic food consumer profiles in terms of both consumer satisfaction and loyalty against which other organic consumers are benchmarked. To that end, for each performance variable (i.e. satisfaction and loyalty), we ranked the organic food consumers in terms of their satisfaction and loyalty, selecting the top 5% in each case for the calibration samples – one for satisfaction and one for loyalty.[1] These samples use the top 5% of organic food consumers to identify the ideal profiles in terms of the three key organic food consumption determinants: healthism, hedonism and trust. In the second step, we remove the bottom 5% along the

[1] PD studies usually select the highest performing 10% of observations in a dataset to calibrate ideal profiles (Venkatraman and Prescott 1990; Malhortra et al. 2013). We consider our 5% cutoff to be broadly in line with this approach.

© Burleigh Dodds Science Publishing Limited, 2024. All rights reserved.

Table 2 Discriminant validity of the measures

		1	2	3	4	5
1	Satisfaction	0.71				
2	Behavioral loyalty	0.08	0.61			
3	Health consciousness	0.06	0.50	0.68		
4	Hedonism	0.09	0.64	0.51	0.60	
5	Trust	0.35	0.34	0.25	0.29	0.78

Entries on the main diagonal are the average variances extracted; values below the main diagonal are the squared correlations.

satisfaction and loyalty scales to ensure our resulting analyses are unbiased for testing the misalignment propositions outlined in our conceptual framework (Venkatraman and Prescott, 1990).

We used regression modeling to test for PD. The PD procedure bears empirical advantages such as the possibility of including more variables without having to increase the sample size. Therefore, PD allows for a holistic identification of the benchmark organic food consumer. We estimated using the following regression model:

$$y_{ib} = \beta_0 + \beta_1 PD_i + \beta_2 c_{i1} + \ldots + \beta_n c_{in} + \varepsilon_i \quad (1)$$

In Equation (1), y_{ib} denotes the organic consumer i's score regarding behavioral outcome b, with b = 1 (satisfaction) or 2 (loyalty). PD_i is the aggregated PD score of consumer i in the study sample from the calibration sample mean in terms of the three key antecedents of organic food satisfaction and behavioral loyalty identified in the conceptual framework. For the calculation of this variable, we refer to Section 3.2. c_{in} is the consumer i's outcome for control variable n, and ε_i is the error term.

In addition, we estimated the following regression model involving the disaggregated PD variables:

$$y_{bi} = \beta_0 + \beta_1 PD_{i1} + PD_{i2} + PD_{i3} + \beta_2 c_{1i} + \ldots + \beta_n c_{ni} + \varepsilon_i \quad (2)$$

This model is identical to the model presented in Equation (1), except for the *PD* variables which are now disaggregated: PD_{i1} is the PD score regarding consumer health consciousness, PD_{i2} is the PD score regarding consumer hedonism and PD_{i3} is the PD score in terms of organic food trust. Tables 3–5 report descriptive statistics for the study sample (baseline profiles) for consumer satisfaction and loyalty. Table 6 presents PD results for organic food consumers.

© Burleigh Dodds Science Publishing Limited, 2024. All rights reserved.

Table 3 Descriptive statistics of the study (baseline) samples

	Organic food satisfaction				Organic food behavioral loyalty			
Variable	Mean	SD	Minimum	Maximum	Mean	SD	Minimum	Maximum
Organic food satisfaction	4.243	0.820	2.5	5.75				
Behavioral loyalty					4.872	0.818	3.286	6.429
Gender	0.482	0.500	0	1	0.489	0.500	0	1
Age[a]	7.764	3.269	2	16	7.813	3.239	2	16
Household size	2.159	0.720	1	4	2.154	0.711	1	4
Annual household income[b]	4.084	2.737	1	8	4.101	2.734	1	8
Education level[c]	4.019	1.272	1	6	4.012	1.280	1	6
Consumption persistence	3.453	1.573	1	5	3.529	1.544	1	5
Total profile deviation	1.894	1.037	0.139	6.959	2.508	1.200	0.245	7.389
PD-1: health consciousness	0.929	0.639	0.113	3.613	1.452	0.844	0.055	4.695
PD-2: hedonistic values	0.979	0.751	0.006	4.206	1.531	0.883	0.049	5.049
PD-3: trust in organic food	1.093	0.821	0.022	4.407	1.051	0.827	0.042	4.387

$N = 741$ for the organic food satisfaction study sample. $N = 779$ for the organic food behavioral loyalty sample.
[a] The mean value for age in both samples of approximately 7.7 means that the mean age group of sample participants is close to 45--49 (value label 8). The minimum value for age in both samples is 2, which refers to the age group 18–19. The maximum value for age in both samples is 16, which refers to the age group 85 and above.
[b] The mean value for annual household income in both samples of approximately 4 means that the mean annual household income in the samples is in the range 65 000–74 999 AUS$.
[c] The mean education level of 4 in both samples implies that individuals on average have Technical and Further Education (TAFE)/trade qualifications.

© Burleigh Dodds Science Publishing Limited, 2024. All rights reserved.

Table 4 Correlations of the consumer satisfaction study sample

		1	2	3	4	5	6	7	8	9	10	11
1	Organic food satisfaction	1										
2	Gender	-0.01	1									
3	Age	-0.02	-0.13*	1								
4	Household size	0.04	-0.01	-0.28*	1							
5	Household income	0.05	-0.06	-0.33*	0.28*	1						
6	Education level	0.01	-0.07*	-0.18*	-0.02	0.29*	1					
7	Consumption persistence	0.04	-0.03	0.12*	-0.06	0.02	0.12*	1				
8	Total profile deviation	-0.31*	-0.10*	0.06	-0.03	-0.06	-0.05	-0.20*	1			
9	PD-1: health consciousness	-0.16*	-0.07*	-0.02	0.00	-0.01	0.01	-0.17*	0.73*	1		
10	PD-2: hedonism	-0.11*	-0.09*	0.08*	-0.00	-0.03	-0.07	-0.20*	0.79*	0.51*	1	
11	PD-3: trust in organic food	-0.43*	-0.08*	0.09*	-0.04	-0.08*	-0.05	-0.11*	0.78*	0.38*	0.37*	1

N = 741.
* $p<0.05$.
** $p<0.01$.
† $p<0.1$.

© Burleigh Dodds Science Publishing Limited, 2024. All rights reserved.

Table 5 Pairwise correlations of the behavioral loyalty study sample

		1	2	3	4	5	6	7	8	9	10	11
1	Organic food loyalty	1										
2	Gender	0.14[a]	1									
3	Age	-0.014	-0.13[a]	1								
4	Household size	0.04	0.01	-0.28[a]	1							
5	Household income	0.05	-0.05	-0.33[a]	0.28[a]	1						
6	Education level	0.02	-0.06	-0.19[a]	-0.01	0.27[a]	1					
7	Consumption persistence	0.23[a]	-0.03	0.14[a]	-0.06	-0.00	0.13[a]	1				
8	Total profile deviation	-0.70[a]	-0.13[a]	0.09[a]	-0.08[a]	-0.07	-0.056	-0.17[a]	1			
9	PD-1: health consciousness	-0.59[a]	-0.10[a]	0.02	-0.06	-0.03	-0.01	-0.15[a]	0.81[a]	1		
10	PD-2: hedonism	-0.66[a]	-0.13[a]	0.13[a]	-0.06	-0.06	-0.07[a]	-0.17[a]	0.86[a]	0.60[a]	1	
11	PD-3: trust in organic food	-0.43[a]	-0.05	0.10[a]	-0.06	-0.07[a]	-0.04	-0.06	0.69[a]	0.37[a]	0.41[a]	1

N = 779.

© Burleigh Dodds Science Publishing Limited, 2024. All rights reserved.

Table 6 Profile of the most satisfied and most loyal organic food consumers

	Organic food satisfaction				Organic food behavioral loyalty			
Variable	Mean	SD	Minimum	Maximum	Mean	SD	Minimum	Maximum
Health consciousness	5.613	1.146	1.75	7	6.445	0.633	4.5	7
Hedonistic values	5.206	1.372	1	7	6.048	0.726	4.4	7
Trust in organic food	5.836	0.656	4	7	5.815	0.717	3.714	7

The table reports the average profile, in terms of health consciousness, hedonism and trust, of the study's two calibration samples, that is, the 5% most satisfied and loyal organic food consumers from the original organic food consumer sample. The cutoff value for the satisfaction calibration sample is 6 (i.e. 95% of the observations have an organic food satisfaction score below 6). The cutoff value for the behavioral loyalty calibration sample is 6.43 (i.e. 95% of the observations have an organic food behavioral loyalty score below 6.43).

© Burleigh Dodds Science Publishing Limited, 2024. All rights reserved.

5 Results and discussion

Table 7 reports estimation results for the six models (1–6) based on ordinary least square regressions. Models (2) and (3) present results for PD effects on consumer satisfaction with organic food, and Models 5 to 6 report findings for PD effects on consumer behavioral loyalty to organic food. Models 1 and 4 present regression results regarding only the effects of the control variables on the respective dependent variables. For all models, the variance inflation factors are well below the cut-off value of 10 which suggests that the multicollinearity issues are not a concern in the analysis (Neter et al., 1996).

Models 3 and 6 report results for our fine-grained analysis involving tests of the effects of disaggregated PD variables on organic food satisfaction (Model 3) and behavioral loyalty (Model 6). In relation to PD in terms of healthism, a negative effect on organic food satisfaction, as indicated by a statistically non-significant coefficient in Model 3, was not found. However, its effect on loyalty was negative and highly statistically significant ($\beta = -0.210$, $P < 0.001$ in Model 6). Similar to hedonism, the regression was not significant for consumer satisfaction Model 3 but highly statistically significant for loyalty Model 6, displaying a negative coefficient sign ($\beta = -0.377$, $P < 0.001$). In terms of consumer trust, the regression results are highly statistically significant in both consumer satisfaction (Model 3: $\beta = -0.488$, $P < 0.001$) and consumer behavioral loyalty (Model 6: $\beta = -0.137$, $P < 0.001$), with negative coefficient signs. Noteworthy here, is the fact that the effect size in the satisfaction model is more than three times that in the behavioral loyalty model. This somewhat contradicts the general trend that the behavioral loyalty model (Model 6) provides a better fit to the data than the satisfaction model (Model 3), as indicated by the variance as explained and which is more than twice as large (54% versus 19%).

Models 2 and 5 indicate that the aggregate total PD coefficients are negative and highly statistically significant for both satisfaction ($\beta = -0.349$, $P < 0.001$) and loyalty ($\beta = -0.598$, $P < 0.001$). The model variance as explained differs between both dependent variables, while it is 10% in the satisfaction case (Model 2), the behavioral loyalty model (specification 5) yields a significantly higher variance as explained, namely 51%. This is indicative of a better fit of the behavioral loyalty model. These findings are in line with Hult et al. (2007) who found that the closer the perceptions of a firm's customers in general match those of the firm's best customers (defined as those customers that have the highest repurchase intentions) on service, product and eBusiness qualities, the stronger the repurchase intentions of customers in general. Regarding control variable effects, the analysis, among other results, finds positive and highly statistically significant coefficients for *Gender* in Models 4–6. This, in line with previous findings, suggests that women's behavioral loyalty toward organic

© Burleigh Dodds Science Publishing Limited, 2024. All rights reserved.

Table 7 Profile deviation results for organic food consumers

	Satisfaction with organic food			Organic food behavioral loyalty		
	(1)	(2)	(3)	(4)	(5)	(6)
Gender	-0.016	-0.088	-0.083	0.240[c]	0.100[a]	0.093[a]
	(0.074)	(0.071)	(0.068)	(0.055)	(0.041)	(0.040)
Age	-0.002	0.004	0.007	-0.001	0.011	0.016[a]
	(0.012)	(0.012)	(0.011)	(0.009)	(0.007)	(0.007)
Household size	0.029	0.023	0.020	0.033	0.000	0.002
	(0.040)	(0.038)	(0.034)	(0.030)	(0.022)	(0.021)
Annual household income	0.049	0.037	0.026	0.040	0.024	0.026
	(0.042)	(0.040)	(0.038)	(0.030)	(0.022)	(0.021)
Education level	-0.009	-0.013	-0.013	-0.010	-0.020	-0.019
	(0.039)	(0.037)	(0.035)	(0.028)	(0.021)	(0.021)
Consumption persistence	0.042	-0.026	-0.005	0.200[c]	0.095[c]	0.088[c]
	(0.036)	(0.036)	(0.035)	(0.028)	(0.023)	(0.022)
Total profile deviation		-0.349[c]			-0.598[c]	
		(0.037)			(0.029)	
PD-1: health consciousness			-0.025			-0.210[c]
			(0.043)			(0.033)
PD-2: hedonistic values			0.062			-0.377[c]
			(0.045)			(0.033)
PD-3: trust in organic food			-0.488[c]			-0.137[c]
			(0.042)			(0.028)
R^2	0.01	0.10	0.19	0.08	0.51	0.54
N	741	741	741	779	779	779

The variables in the regressions are standardized with a mean of 0 and a standard deviation of 1.
[a] $P < 0.05$.
[b] $P < 0.01$.
[c] $P < 0.001$.
[d] $P < 0.1$.

food is higher than that of men (see e.g. Lockie et al., 2004; McFadden and Huffman, 2017).

6 Conclusion

Our PD analysis has revealed the significant negative consumer satisfaction and loyalty implications of deviating from the ideal organic consumer profile. We detected a specific consumer profile for healthism, hedonism and trust that is positively associated with higher satisfaction with and loyalty to organic foods. In the case of consumer satisfaction, PD in terms of trust is the only dimension with a significant negative effect. This suggests that improving consumer loyalty

© Burleigh Dodds Science Publishing Limited, 2024. All rights reserved.

is a more challenging and complex task because each of the three dimensions influences loyalty. Consumer satisfaction appears to be somewhat easier to manage since the disaggregated analysis (cf. Model 3 in Table 7) showed that it is consumer trust in organic food that accounts for the significance of the total PD result. Hence, from a theoretical perspective as well as from a managerial point of view, building trust in organic foods is critical to achieving improved consumer satisfaction.

7 Study contributions and implications

Applying our method has methodological advantages because of its holistic approach to defining ideal consumers on the basis of theoretically supported dimensions (Malhotra et al., 2013; Venkatraman, 1989; Venkatraman and Prescott, 1990). This study generates robust findings because the baseline sample was created by dropping not only the most loyal or satisfied (our 'calibration' sample) but also the least loyal or satisfied consumers to rule out the possibility of biased findings (Venkatraman and Prescott, 1990). PD has been rarely used by marketing scholars (Anisimova and Mavondo, 2010) and few previous studies in marketing using configuration theory and PD analysis, mainly focus on organizations as units of analysis (Kabadayi et al., 2007; Vorhies and Morgan, 2005). Our study is one of the few within the marketing field to opt for a configuration-theoretic and PD approach to explain individual consumer behavior (see Anisimova and Mavondo, 2010 and Malhotra et al., 2013 for exceptions).

To the best of our knowledge, this is the first study to integrate PD analysis in the context of organic food literature. This study provides novel theoretical insights into the negative relationship between (i) individual organic food satisfaction and (ii) loyalty and profile deviation. The three theoretically relevant dimensions: healthism, hedonism and trust are examined separately and then combined in a total ideal profile. The configuration-theoretic approach is guided by the holistic and systemic principles of inquiry; is typically based on profiles of ideal types and involves examining how a pattern of multiple independent variables is associated with a dependent variable rather than with how individual independent variables are associated with a dependent variable (Delery and Doty, 1996). Healthism, hedonism and trust could be argued to be only a few aspects, but they represent a set of 23 questions that could be used to build the ideal profile. This was avoided in the interest of parsimony and to reduce complexity by using summated variables.

Overall, the detected negative relationship between the aggregate PD variable and consumer satisfaction, as well as consumer loyalty provides evidence of the legitimacy of satisfaction and loyalty as key consumer performance indicators (de Chermatony et al., 1998; Hoeffler and Keller, 2002). In terms of satisfaction, the fact that the PD result seems to have been steered

© Burleigh Dodds Science Publishing Limited, 2024. All rights reserved.

by consumer trust suggests in turn that it is the attitudinal distance between organic food consumers and their archetype ideal counterparts in terms of trust in organic food that is particularly relevant for consumer satisfaction. To increase consumer trust, more certification of organic producers and enforcement of that status by industry or government bodies is needed. This is an important insight for brand equity in an organic food consumer context also because a decrease in consumer satisfaction risks reducing consumer loyalty (Lai et al., 2009; He et al., 2012).

Concerning the negative PD effects on consumer loyalty, our findings mean that deviating from the ideal consumer profile discussed previously may entail a reduced repeat purchasing behavior (LaBarbera and Mazursky, 1983), a reduced willingness to pay for that brand (Reichheld, 1996) as well as increased switching to other foods (Yoo et al., 2000). Such a reduced consumer attachment to organic foods, in turn, can have detrimental effects on producers' financial performances (Homburg and Gering, 2001; Chaudhuri and Holbrook, 2001). Overall, the results suggest that trust is critical to loyalty to organic food consumption, while all three dimensions are required to achieve consumer long-term loyalty. Operationally, our study and its findings can be considered robust because after identifying the most loyal or most satisfied customers, we respectively removed the lowest 5% for each model to reduce the possibility of biased findings. This has been alluded to in prior studies (Venkatraman and Prescott,1990) in PD studies but not previously implemented.

The managerial implications for retailers and manufacturers concern their resource reallocation to strengthen the dimensions of healthism, hedonism and trust in organic food marketing. Such targeted resources allocation is likely to be more advantageous in the form of customers continuing to purchase organic foods (Hult et al., 2007). More specifically, managers and practitioners can use PD as a diagnostic tool aimed at assisting practitioners in making more informed marketing-related decisions. Our results suggest that building trust is a key driver of satisfaction meaning that improved consumer satisfaction can in turn be expected to generate favorable spillover effects on consumer loyalty due to the previously mentioned positive direct link between these metrics.

The next set of priorities is to meet the hedonistic and healthiness needs of organic food consumers. It is suggested that using these insights to create holistic organic food marketing campaigns will ultimately achieve a convergence between the regular organic food consumer and their ideal counterparts in terms of consumer satisfaction and behavioral brand loyalty. To what extent this convergence will be accomplished may depend on the amount of financial and creative resources employed to increase consumer trust, healthism and hedonism levels in organic food consumption. Practitioners should work toward gathering like-minded individuals on social media platforms and enabling them to connect with each other as well as getting them to spread

© Burleigh Dodds Science Publishing Limited, 2024. All rights reserved.

the word, by creating a community culture and identity. The ultimate outcome of this process must be the achievement of a convergence - a conversion from misalignment to alignment - among organic food consumers along the key consumption dimensions and determinants identified in this study, i.e. healthism, hedonism, and trust. This is imperative in order to eliminate the negative performance implications in terms of satisfaction and loyalty of being misaligned from the benchmark organic food consumers - 'the organic food true believers' (Goulding et al., 2013; Pankraz, 2009). Furthermore, we contribute to sustainable consumption, which is in alignment with the UN Global Goals 2030.

7.1 Limitations and further research

One area of improvement could be the description of the ideal organic consumer. Future researchers could incorporate other relevant variables, which could permit better segmentation, targeting and positioning of organic products. Secondly, future researchers could extend this model by including other metrics including financial outcomes using a longitudinal study for additional insights.

8 Acknowledgement

The author gratefully acknowledges funding support from NASAA Organic, the Organic Federation of Australia and University of Canberra.

Funding: The research supporting this article was funded by the Organic Federation of Australia, University of Canberra and the National Association for Sustainable Agriculture, Australia (NASAA).

Declarations of conflict of interest: None.

9 Where to look for further information

The foundation profile configuration theory and profile deviation approaches theory behind this chapter is derived from seminar studies by:

- Venkatraman, N. (1989). The concept of fit in strategy research: toward verbal and statistical correspondence, *The Academy of Management Review* 14(3), 423–444.
- Venkatraman, N. (1990). Performance implications of strategic coalignment: a methodological perspective, *Journal of Management Studies* 27(1), 19–41.
- Venkatraman, N. and Prescott, J. E. (1990). Environment-strategy coalignment: an empirical test of its performance implications, *Strategic Management Journal* 11(1), 1–23.

© Burleigh Dodds Science Publishing Limited, 2024. All rights reserved.

Several reports about the current state of the Australian organic food industry are available online for further reading:

- https://austorganic.com/resources-and-research/publications/organic-market-report-2023/.
- https://www.agriculture.gov.au/sites/default/files/documents/organic-industry-data-collection-kg2-final-report-feb-2022.pdf.

10 References

Aaker, D. A. (1991). *Managing Brand Equity*, The Free Press, New York.

Aertsens, J., Verbeke, W., Mondelaers, K. and Van Huylenbroeck, G. (2009). Personal determinants of organic food consumption: a review, *British Food Journal* 121(9), 1140–1167.

Agarwal, J. and Malhotra, N. K. (2005). An integrated model of attitude and affect, *Journal of Business Research* 58(4), 483-493.

Ajzen, I. (2001). Nature and operation of attitudes, *Annual Review of Psychology* 52, 27-58.

Anisimova, T. (2016). Integrating multiple factors affecting consumer behavior toward organic foods: the role of healthism, hedonism, and trust in consumer purchase intentions of organic foods, *Journal of Food Products Marketing* 22(7), 809-823.

Anisimova, T. and Mavondo, F. T. (2010). The performance implications of company-salesperson corporate brand misalignment, *European Journal of Marketing* 44(6), 771–795.

Anisimova, T., Mavondo, F. and Weiss, J. (2019). Controlled and uncontrolled communication stimuli and organic food purchases: the mediating role of perceived communication clarity, perceived health benefits, and trust, *Journal of Marketing Communications* 25(2), 180–203.

Audebert, O., Deiss, V. and Rousset, S. (2006). Hedonism as a predictor of attitudes of young French women towards meat, *Appetite* 46(3), 239–247.

Batra, R. and Ahtola, O. T. (1991). Measuring the hedonic and utilitarian sources of consumer attitudes, *Marketing Letters* 2(2) , 159–170.

Blome, C., Paulraj, A. and Schuetz, K. (2014). Supply chain collaboration and sustainability: a profile deviation analysis, *International Journal of Operations and Production Management* 34(5), 639–663.

Bonroy, O. and Constantatos, C. (2008). On the use of labels in credence goods markets, *Journal of Regulatory Economics* 33(3), 237–252.

Castaldo, S., Perrini, F., Misani, N. and Tencati, A. (2009). The missing link between corporate social responsibility and consumer trust: the case of fair-trade products, *Journal of Business Ethics* 84(1), 1–15.

Chaudhuri, A. and Holbrook, M. B. (2001). The chain of effects from brand trust and brand affect to brand performance: the role of brand loyalty, *Journal of Marketing* 65(2), 81–93.

Chryssohoidis, G. M. and Krystallis, A. (2005). 'Organic consumers' personal values research: testing and validating the list of values (LOV) scale and implementing a value-based segmentation task, *Food Quality and Preference* 16(7), 585–599.

© Burleigh Dodds Science Publishing Limited, 2024. All rights reserved.

Churchill, G. (1979). *Marketing Research: Methodological Foundations*, Dryden Press, Orlando. 16(1), 64–73.
Churchill, G. and Iacobucci, D. (2002). *Marketing Research: Methodological Foundations* (8th edn.), Australia, Melbourne.
Cova, B. (2003). Analyzing and playing with "tribes which consume", *Finanza, Marketing and Produzione* 21(1), 66–89.
Cova, B. and Cova, V. (2002). Tribal marketing: the tribalization of society and its impact on the conduct of marketing, *European Journal of Marketing* 36(5/6), 595–620.
Cronbach, L. J. (1951). Cronbach alpha and the internal structure of tests, *Psychometrika* 16(3), 297–334.
Cronin, J. J. and Taylor, S. A. (1992). Measuring service quality: a re-examination and extension, *Journal of Marketing* 56(3), 55–68.
Davis, I. (2013). How (not) to market socially responsible products: a critical research evaluation, *Journal of Marketing Communications* 19(2), 136–150.
de Chematony, L., Dall'Olmo Riley, F. and Harris, F. (1998). Criteria to assess brand success, *Journal of Marketing Management* 14(7), 765–781.
Dess, G. G., Newport, S. and Rasheed, M. A. (1993). Configuration research in strategic management, *Journal of Management* 19, 775–795.
Dick, A. S. and Basu, K. (1994). Customer loyalty: towards an integrated framework, *Journal of the Academy of Marketing Science* 22(2), 99–113.
Dillman, D. A., Smyth, J. D. and Christian, L. M. (2009). *Internet, Mail and Mixed-Mode Surveys: The Tailored Design Method*, Wiley, Hoboken, NJ.
Di Renzo, L., Gualtieri, P., Pivari, F., Soldati, L., Attinà, A., Cinelli, G., Leggeri, C., Caparello, G., Barrea, L., Scerbo, F., Esposito, E. and De Lorenzo, A. (2020). Eating habits and lifestyle changes during COVID-19 lockdown: an Italian survey, *Journal of Translational Medicine* 18(1), 229.
Doty, D. H., Glick, W. H. and Huber, G. P. (1993). Fit, equifinality, and organizational effectiveness: a test of two configurational theories, *Academy of Management Journal* 36(6), 1196–1250.
Drazin, R. and Van de Ven, A. H. V. (1985). Alternative forms of fit in contingency theory, *Administrative Science Quarterly* 30(4), 514–539.
FiBL (2021). European organic market grew by double digits and organic area reached 13.5 million hectares in 2016, Available at: https://www.fibl.org/en/info-centre/news/european-organic-market-grew-to-euro-45-billion-in-2019.html Accessed 15 of March, 2022, Research Institute of Organic Agriculture.
Fifita, I. M., Seo, Y., Ko, E., Conroy, D. and Hong, D. (2020). Fashioning organics: wellbeing, sustainability, and status consumption practices, *Journal of Business Research* 117, 664–671.
Fornell, C. and Larcker, D. F. (1981). Structural equation models with unobservable variables and measurement error: Algebra and statistics, 382–388.
Giese, J. L. and Cote, J. A. (2000). Defining consumer satisfaction, *Academy of Marketing Science Review* 1(1), 1–22.
Goulding, C., Shankar, A. and Canniford, R. (2013). Learning to be tribal: facilitating the formation of consumer tribes, *European Journal of Marketing* 47(5/6), 813–832.
Hamzaoui-Essoussi, L., Sirieix, L. and Zahaf, M. (2013). Trust orientations in the organic food distribution channels: A comparative study of the Canadian and French markets, *Journal of retailing and consumer services* 20(3), 292–301.

© Burleigh Dodds Science Publishing Limited, 2024. All rights reserved.

Hansen, T., Sørensen, M. I. and Eriksen, M. R. (2018). How the interplay between consumer motivations and values influences organic food identity and behavior, *Food Policy* 74, 39–52.

He, H., Li, Y. and Harris, L. (2012). Social identity perspective on brand loyalty, *Journal of Business Research* 65(5), 648–657.

Hirschman, E. C. and Holbrook, M. B. (1982). Hedonic consumption: emerging concepts, methods, and propositions, *Journal of Marketing* 46(3), 92–101. doi:10.1177/002224298204600314.

Hoeffler, S. and Keller, K. L. (2002). Building brand equity through corporate societal marketing, *Journal of Public Policy and Marketing* 21(1), 78–89.

Homburg, C. and Gering, A. (2001). Personal characteristics as moderators of the relationship between consumer satisfaction and loyalty - an empirical analysis, *Psychology and Marketing* 18(1), 43–66.

Hoque, Z. (2014). 20 years of studies on the balanced scorecard: trends, accomplishments, gaps and opportunities for future research, *The British Accounting Review* 46(1), 33–59.

Hult, G. T. M., Boyer, K. K. and Ketchen Jr., D. J. (2007). Quality, operational logistics strategy, and repurchase intentions: a profile deviation analysis, *Journal of Business Logistics* 28(2), 105–132.

Kabadayi, S., Eyuboglu, N. and Thomas, G. P. (2007). The performance implications of designing multiple channels to fit with strategy and environment, *Journal of Marketing* 71(4), 195–211.

Kaplan, R. S. and Norton, D. P. (1992). The balanced scorecard–measures that drive performance, *Harvard Business Review* 70(1), 71–79.

Kennedy, M. S., Ferrell, L. K. and LeClair, D. T. (2001). Consumers' trust of salesperson and manufacturer: an empirical study, *Journal of Business Research* 51(1), 73–86.

Kriwy, P. and Mecking, R. A. (2012). Health and environmental consciousness, costs of behaviour and the purchase of organic food, *International Journal of Consumer Studies* 36(1), 30–37.

LaBarbera, P. A. and Mazursky, D. (1983). A longitudinal assessment of consumer satisfaction/ dissatisfaction: the dynamic aspect of the cognitive process, *Journal of Marketing Research* 20(4), 393–404.

Lai, F., Griffin, M. and Babin, B. J. (2009). How quality, value, image, and satisfaction create loyalty at a Chinese telecom, *Journal of Business Research* 62(10), 980–986.

Lassoued, R. J. and Hobbs, J. E. (2015). Consumer confidence in credence attributes: the role of brand trust, *Food Policy* 52, 99–107.

Lea, E. and Worsley, T. (2005). Australians' organic food beliefs, demographics, and values, *British Food Journal* 107(11), 855–869.

Leathy, E. (2020). Lockdown boost for organic produce. Available at: http://www.fruitnet.com/fpj/article/182772/lockdown-boost-for-organic-produce.

Lee, H.-J. and Hwang, J. (2016). The driving role of consumers' perceived credence attributes in organic food purchase decisions: a comparison of two groups of consumers, *Food Quality and Preference* 54, 141–151.

Lee, H.-J. and Yun, Z.-S. (2015). Consumers' perceptions of organic food attributes and cognitive and affective attitudes as determinants of their purchase intentions toward organic food, *Food Quality and Preference* 39, 259–267.

© Burleigh Dodds Science Publishing Limited, 2024. All rights reserved.

Liu, M. T., Wong, I . A., Tseng, T.-H., Chang, A. W. and Phau, I. (2017). Applying consumer-based brand equity in luxury hotel branding, *Journal of Business Research* 81, 192–202.

Lockie, S., Lyons, K., Lawrence, G. and Grice, J. (2004). Choosing organics: a path analysis of factors underlying the selection of organic food among Australian consumers, *Appetite* 43(2), 135–146.

Malhotra, N., Mavondo, F., Mukherjee, A. and Hooley, G. (2013). Service Quality of frontline employees: a profile deviation analysis, *Journal of Business Research* 66(9), 1338–1344.

McFadden, J. R. and Huffman, W. E. (2017). Willingness-to-pay for natural, organic, and conventional foods: the effects of information and meaningful labels, *Food Policy* 68, 214–232.

Michaelidou, N. and Hassan, L. M. (2008). The role of health consciousness, food safety concern and ethical identity on attitudes and intentions towards organic food, *International Journal of Consumer Studies* 32(2), 163–170.

Miller, D. (1996). Configurations revisited, *Strategic Management Journal* 17(7), 505–512.

Moser, A. K. (2016). Consumers' purchasing decisions regarding environmentally friendly products: an empirical analysis of German consumers, *Journal of Retailing and Consumer Services* 31, 389–397.

Neter, J., Kutner, M. H., Nachtsheim, C. J. and Wasserman, W. (1996). *Applied Linear Statistical Models* (4th edn.), Irwin Publishing, Boston, MA.

Nunnally, J. C. (1978). *Psychometric Theory*, McGraw-Hill Book Company, New York.

Oliver, R. L. (2014). *Satisfaction: A Behavioral Perspective on the Consumer*. McGraw-Hill Book Company, Boston, MA.

Parkes, B. (2017). Growing hunger for organics: Can Australia keep up with demand?, Available at: https://www.intheblack.com/articles/2017/09/01/organic-food-demand-australia/ Accessed Accessed 2 June 2022.

Peluso, A. M., Pichierri, M. and Pino, G. (2021). Age-related effects on environmentally sustainable purchases at the time of COVID-19: evidence from Italy, *Journal of Retailing and Consumer Services* 60, 102443.

Perrini, F., Castaldo, S., Misani, N. and Tencati, A. (2010). The impact of corporate social responsibility associations on trust in organic products marketed by mainstream retailers: a study of Italian consumers, *Business Strategy and the Environment* 19(8), 512–526.

Pino, G., Peluso, A. M. and Guido, G. (2012). Determinants of regular and occasional consumers' intentions to buy organic food, *Journal of Consumer Affairs* 46(1), 157–169.

Prendergast, J. (2020). Organic fruit and vegetables in demand as supermarkets move to reduce prices, Available at: https://www.abc.net.au/news/rural/2020-02-19/organic-fruit-and-vegetables-in-demand-push-for-lower-prices/11977032/. Accessed 3 June 2022.

Reichheld, F. F. (1996). *The Loyalty Effect*, Harvard Business School Press, Boston, MA.

Rodríguez-Pérez, C., Molina-Montes, E., Verardo, V., Artacho, R., García-Villanova, B., Guerra-Hernández, E. J. and Ruíz-López, M. D. (2020). Changes in dietary behaviours during the COVID-19 outbreak confinement in the Spanish COVIDiet study, *Nutrients* 12(6), 1–19.

© Burleigh Dodds Science Publishing Limited, 2024. All rights reserved.

Sidor, A. and Rzymski, P. (2020). Dietary choices and habits during COVID-19 lockdown: experience from Poland, *Nutrients* 12(6), 1-13.

Schneider-Kamp, A. (2021). Inclusion of the excluded: consumers' quest for hedonism in food consumption, *International Journal of Consumer Studies* 45(3), 320-334.

Sønderskov, K. M. and Daugbjerg, C. (2011). The state and consumer confidence in eco-labelling: organic labelling in Denmark, Sweden, The United Kingdom and The United States, *Agriculture and Human Values* 28(4), 507-517.

Tarkiainen, A. and Sundqvist, S. (2005). Subjective norms, attitudes and intentions of Finnish consumers in buying organic food, *British Food Journal* 107(11), 808-822.

Tarkiainen, A. and Sundqvist, S. (2009). Product involvement in organic food consumption: does ideology meet practice?, *Psychology and Marketing* 26(9), 844-863.

Teng, C. and Wang, Y. (2015). Decisional factors driving organic food consumption, *British Food Journal* 117(3), 1066-1081.

Thøgersen, J. (2009). Consumer decision-making with regard. *Traditional Food Production and Rural Sustainable Development: A European Challenge* 173.

Venkatraman, N. (1989). The concept of fit in strategy research: toward verbal and statistical correspondence, *The Academy of Management Review* 14(3), 423-444.

Venkatraman, N. (1990). Performance implications of strategic coalignment: a methodological perspective, *Journal of Management Studies* 27(1), 19-41.

Venkatraman, N. and Prescott, J. E. (1990). Environment-strategy coalignment: an empirical test of its performance implications, *Strategic Management Journal* 11(1), 1-23.

Vorhies, D. W. and Morgan, N. A. (2003). A configuration theory assessment of marketing organization fit with business strategy and Its Relationship with Marketing Performance, *Journal of Marketing* 67(1), 100-115.

Vorhies, D. W. and Morgan, N. A. (2005). Benchmarking marketing capabilities for sustainable competitive advantage, *Journal of Marketing* 69(1), 80-94.

Watson, L. and Spence, M. T. (2007). Causes and consequences of emotions on consumer behaviour - a review and integrative cognitive appraisal theory, *European Journal of Marketing* 41(5/6), 487-511.

Watson, G. F., Beck, J. T., Henderson, C. M. and Palmatier, R. W. (2015). Building, measuring, and profiting from customer loyalty, *Journal of the Academy of Marketing Science* 43(6), 790-825.

Yeung, M. C. H. and Ennew, C. T. (2000). From customer satisfaction to profitability, *Journal of Strategic Marketing* 8(4), 313-326.

Yi, Y. (1991). A critical review of consumer satisfaction. In: Zeithaml, V. A. (Ed.). *Review of Marketing* (vol. 1990), pp. 68-123, American Marketing Association, Chicago.

Yin, S., Wu, L., Du, L. and Chen, M. (2010). Consumers' purchase intention of organic food in China, *Journal of the Science of Food and Agriculture* 90(8), 1361-1367.

Yoo, B., Donthu, N. and Lee, S. (2000). An examination of selected marketing mix elements and brand equity, *Journal of the Academy of Marketing Science* 28(2), 195-211.

Yoo, B. and Donthu, N. (2001). Developing and validating a multidimensional consumer-based brand equity scale, *Journal of Business Research* 52(1), 1-14.

© Burleigh Dodds Science Publishing Limited, 2024. All rights reserved.

Chapter 10

Understanding consumer attitudes to environmental sustainability issues in agricultural and food production

Caroline Saunders, Tim Driver and Meike Guenther, Lincoln University, New Zealand

1 Introduction

With current concerns about the environment, especially climate change, consumers are increasingly interested in the environmental impacts of the food products they purchase and seek out products that have sustainability credentials that can be verified (Euromonitor, 2020, 2021; Peschel et al., 2016; Te Puna Whakaaronui, 2022). As a consequence, environmental labelling is growing, and in recent years, the number of certification schemes for environmental attributes in food products has increased significantly (Eco-Label Index, 2022; EC, 2022). However, the value that consumers place on environmental attributes in food products varies across different countries and products (Guenther et al., 2015; Tait et al., 2016). Understanding these differences in preferences and cultures is important for producers, food manufacturers, retailers, and policymakers to encourage consumers to make more sustainable and healthier choices.

There is a wide range of studies worldwide that have investigated consumer attitudes and preferences for different environmental attributes in food products employing different analytical approaches, e.g. choice experiments, auction mechanisms, cluster analysis, and regression analysis. This chapter commences with a description of different categories of product

http://dx.doi.org/10.19103/AS.2023.0129.22
© Burleigh Dodds Science Publishing Limited, 2024. All rights reserved.

attributes with a specific focus on environmental attributes in food products. Then ecolabelling is described in more detail. This is followed by an extensive literature review of studies that assessed consumer attitudes and preferences towards different environmental attributes in food across countries. The chapter finishes with a conclusion.

2 Product attributes

The attributes or characteristics of products can be distinguished between search, experience, and credence attributes, taking into account the various possibilities that consumers have for obtaining information about the attributes of products (Darby and Karni, 1973; Karl and Orwat, 2000; Nelson, 1970).

The concept of search and experience attributes of products is originally described by Philip Nelson (Nelson, 1970). Search attributes of a product imply that the consumer can identify the quality attributes of the product before the purchase by inspection (Karl and Orwat, 2000). For example, the weight of a product is a search attribute. In contrast, experience attributes of goods are attributes that can only be determined after purchase and during consumption of the product. An example for an experience attribute is the taste of a food product.

Darby and Karni (1973) added the concept of credence attribute to search and experience attributes. Credence attributes of goods imply that the impact and quality of the use or consumption cannot be identified, experienced, or inspected by consumers. Examples of credence attributes in food are food safety, environmental quality, climate-friendly, low carbon emissions, social responsibility, and many more. For credence attributes, there is an information asymmetry between the consumer and producer. The producer knows relevant information about the product that the consumer does not know. This information asymmetry exists particularly for the credence attributes of a product. However, this is less for search and experience attributes because as stated earlier these attributes can be experienced before or after purchase. A producer can inform consumers of a credence attribute of the product by providing information on a product label (Van Amstel et al., 2008).

3 Environmental attributes

Many examples of environmental attributes in food products exist. These can include generic environmental quality attributes (such as 'environmentally friendly') and specific attributes, such as organic and other production methods, reduced greenhouse gas (GHG) emissions and carbon footprinting, air quality, water quality and footprinting, waste reduction and suboptimal foods, and biodiversity and wildlife protection. It should be noted that some food product

© Burleigh Dodds Science Publishing Limited, 2024. All rights reserved.

attributes that international consumers have associated with environmental credentials, such as organic and local food, are covered elsewhere in this book and not included herein.

Many studies have examined consumer preferences for generic environmental credentials (e.g. 'environmentally friendly') in food products, assessing their relative importance against other basic, search, and credence attributes. Overall, these studies have shown that many international consumers prefer environmentally friendly goods and are willing to pay a premium for this attribute in the food products that they buy (Dalziel et al., 2018; Lagerkvist et al., 2014, 2017; Lai et al., 2018; Pomarici et al., 2018; Tait et al., 2018d; Ubilava et al., 2011; Viegas et al., 2014; Yang and Renwick, 2019). It is also important to point out that an 'environmentally friendly' attribute in food products can be seen as a proxy for other attributes by consumers. Numerous studies have identified a relationship between consumers' recognition of the generic environmental attributes of a product and other attributes, such as food safety and product quality. For example, Dalziel et al. (2018) found environmental condition to be an important factor of food safety in China, India, Indonesia, Japan, and the UK, suggesting that the condition of the environment in which food is produced/grown may influence consumers' safety perceptions. Similarly, Tait et al. (2020c) found a perceived positive relationship between environmental condition and the quality of a product, with other studies reinforcing this finding (Tait et al., 2018d; Tait et al., 2020c).

Consumers worldwide have shown a preference for product claims regarding specific environmental factors, such as air, soil, and water quality. These terms (air quality, soil quality, and water quality) most often refer to the protection of the physical/chemical properties of the air, soil, or water proximal to production sites (e.g. avoidance of contamination) but can also include efficiency in the factor's use (e.g. reduced water use). Both air and water quality have been frequently cited as an important aspect of environmental condition for international consumers (Guenther et al., 2015; Miller et al., 2017; Saunders et al., 2016). Studies examining consumer preferences and their willingness to pay (WTP) for improved air, soil, and water quality in food production have found that segments of consumers in international markets may be willing to pay a premium of up to 67% for these properties (Chen et al., 2018; Eldesouky et al., 2020; Giannoccaro et al., 2019; Krovetz et al., 2017; Pomarici et al., 2018; Saunders et al., 2016; Savchenko et al., 2019a, 2019b; Steiner et al., 2017; Tait et al., 2016, 2020f, 2020g; Ubilava et al., 2011; Viegas et al., 2014). Similar to GHG emissions, the efficiency of water use in food production can be communicated using a water footprint label. Many studies have examined consumer preferences for water footprint labels, indicating positive WTP for this attribute (Chen et al., 2018; Eldesouky et al., 2020; Giannoccaro et al., 2019; Krovetz et al., 2017; Pomarici et al., 2018; Savchenko et al., 2019a,b; Steiner

© Burleigh Dodds Science Publishing Limited, 2024. All rights reserved.

et al., 2017; Tait et al., 2016; Tait et al., 2020f,g). While relatively understudied, the literature has found that international consumers also prefer food products with associated soil quality credentials (Chen et al., 2018; Viegas et al., 2014).

Globally, consumers are also increasingly concerned about the GHG emissions associated with food production. Studies have examined consumer preferences for products with reduced GHG emissions in production, carbon zero, and those that carry a 'carbon footprint'. The carbon footprint of a product can be defined as the total or equivalent amount of carbon dioxide (CO_2) that is emitted in the production process, including the product's entire life cycle (Canavari and Coderoni, 2020). Studies have suggested that consumers value products that include carbon footprint labels that communicate information regarding GHG emissions in production, with some segments willing to pay a premium for such products (Canavari and Coderoni, 2020; Dudinskaya et al., 2021; Feucht and Zander, 2017; Li et al., 2016; Sonoda et al., 2018; Steiner et al., 2017; Tait et al., 2016, 2020f,g, 2021; Van Loo et al., 2014).

The provision of recyclable packaging and other food waste-saving actions are considered to be important by a range of international consumers in relation to food products. For example, Guenther et al. (2021) found that UK consumers rated recycling as the most important factor underpinning environmental condition, with Chinese and Japanese consumers also rating this factor highly. This is consistent with prior research, which has also found recycling to be among the most important factors underpinning environmental condition (Guenther et al., 2015; Saunders et al., 2013, 2015).

International consumers also value the inclusion of biodiversity and wildlife protection credentials in the food products that they purchase. While relatively under-researched, some studies have examined consumer preferences and WTP for food products that include biodiversity enhancement and wildlife protection credentials. This can include efforts to protect wildlife and minimise ecological damages within or proximal to food production sites, as well as the place values associated with wild food procurement. Studies have identified a range of premiums that consumers are willing to pay for the inclusion of biodiversity/wildlife protection credentials in food products, showing positive WTP for a range of biodiversity credentials in food products across the literature (Chang et al., 2019; Guenther et al., 2015, 2019; Foti et al., 2019; Tait et al., 2016, 2018c, 2020f; Uchida et al., 2014).

4 Ecolabelling

For products promoting environmental attributes in food, the role of verification including labelling is of significant importance as these credence attributes cannot be identified by the consumers before or after consumption of the product. Labelling is a common method of communicating certain credence

© Burleigh Dodds Science Publishing Limited, 2024. All rights reserved.

attributes to influence consumer's decision-making and buying behaviour. Specifically, product ecolabels provide information about the environmental attributes of a product to consumers and encourage them to make sustainable purchase decisions (Siraj et al., 2022; Yang and Renwick, 2019). Companies use ecolabels for their products to gain consumers' trust and to reduce the information asymmetry associated with the sustainability of the product marketed (Jørgensen et al., 2022; Lee et al., 2020). While sustainability claims generally include more than just environmental credentials and can include social, economic and cultural attributes, this chapter concentrates on environmental attributes.

Environmental labelling may be mandatory, such as hazard warnings (e.g. pesticide hazards) but predominantly environmental labelling is voluntary. Ecolabels can be categorised into 'single attribute' and 'multiple attribute' ecolabels. This means the information on the label can involve one environmental issue such as recyclability or it can relate to numerous environmental aspects of the product life cycle. Most of the environmental labelling schemes cover multiple environmental issues such as different kinds of pollution or amounts of waste. The underlying environmental criteria are, in particular, quantitative threshold or limit values for product-related environmental damage or other qualitative product requirements that occur through a product's life cycle (Karl and Orwat, 2000).

Ecolabels vary in their stringency. While some ecolabels are based on independent third-party certification with an assessment of compliance, other labels are self-declarations made by companies themselves or certificates awarded by private institutes and bodies, which assessed specific features of the product (mostly single attribute labels such as recyclability) or just indicating that the product was 'environmentally friendly' on a broad scale and in a simplistic way (Iraldo et al., 2020).

Consumers' use of ecolabels is growing. This is a consequence of the arising awareness of food consumers on environmental issues in food production. According to the Eco Label Index (2022), the largest global directory of ecolabels, there were 455 ecolabels in 25 industry sectors in 199 countries in 2022. For example, in the last decade, the number of products and services carrying the EU Ecolabel (Euroflower) has increased three-fold between 2010 and 2022 (from 21 000 goods and services in 2010 to 89 000 goods and services across 24 product groups in 2022) (EC, 2022). Other examples of international ecolabels are the Blauer Engel (the first environmental label founded in Germany in 1978) Nordic Swan, UK Red Tractor, the U.S. Green Seal, and many more. The majority of ecolabels are founded in developed countries. However, emerging economies such as India, Korea, or Thailand have also started to develop ecolabel schemes (Iraldo et al., 2020).

© Burleigh Dodds Science Publishing Limited, 2024. All rights reserved.

5 Literature review

The values, attitudes, perceptions, and preferences of consumers regarding and their WTP for environmental attributes of the food products that they purchase have been investigated in studies worldwide. This section will provide a literature review of studies that have assessed consumer preferences and their WTP for environmental attributes in food and preferences, including certification schemes.

5.1 Cross-country studies

Cross-country studies have shown that consumer preferences and attitudes towards these attributes vary across countries. Consumers in different markets have demonstrated different associations of specific factors with environmental condition, with some factors emerging as more important than others (Guenther et al., 2015, 2021; Miller et al., 2017; Saunders et al., 2013, 2015). For example, consumers in developing countries have shown greater preference in the form of WTP values for a range of environmental attributes relative to developed markets (Saunders et al., 2013; Guenther et al., 2015; Miller et al., 2017; Tait et al., 2016). However, these studies have also shown similarities in consumer preferences, especially for basic attributes (e.g. quality, price, and taste) and provenance attributes (e.g. locally produced); these attributes are generally valued higher than environmental attributes in most markets (Akaichi et al., 2020; Dudinskaya et al., 2021; Feucht and Zander, 2017). Examining environmental attributes in isolation, and given the aforementioned heterogeneity between markets, it could be inferred from the cross-country studies examined here that consumers have the highest preference for reduced GHG emissions, followed by water quality and biodiversity (Akaichi et al., 2020; Dudinskaya et al., 2021; Feucht and Zander, 2017; Saunders et al., 2013; Tait et al., 2016).

Illustrating this, Yang and Renwick (2019) conducted a meta-analysis of consumers' WTP for the credence attributes of livestock products, including environmental attributes. The authors conducted a systematic literature review and applied a meta-regression analysis in an effort to introduce some generality to WTP studies. A total of 566 WTP estimates from 94 studies were initially identified. As shown in Table 1, the environment-friendly attribute was found to have the lowest associated price premiums by a significant margin. The authors also created two subsamples within the data to separate red meat from dairy. The applied regression model highlighted that the environment-friendly attribute valued the second-least by consumers, next to traceability. For dairy products, food safety was associated with the highest premiums, with environment friendly and grass-based the lowest (Yang and Renwick, 2019).

© Burleigh Dodds Science Publishing Limited, 2024. All rights reserved.

Table 1 WTP values for credence attributes (price premium %)

Attribute	Whole sample model	Red meat model	Dairy model
Environment friendly	24.1%	18.9%	25%
Animal welfare	31.9%	19.3%	31%
Organic	35.8%	31.37%	28.5%
Hormone/antibiotic free	32.2%	24%	34.3%
Grass based	24.9%	22.3%	25.1%
Food safety	29.9%	23%	39.2%
Protected designations of origin/protected geographical indications (PGIs)	24.7%	22.4%	25.7%
Country of origin/region of origin	29.8%	22.5%	29.9%
Traceability	20.1%	17.7%	26.1%
Mixed attributes	25.7%	19.2%	25.8%

Source: Yang and Renwick (2019).

A series of research work conducted by the Agribusiness and Economics Research Unit (AERU) at Lincoln University, New Zealand, has investigated the comparative preferences of consumers in a number of international markets (Guenther et al., 2015, 2021; Saunders et al., 2013, 2015; Tait et al., 2016, 2018a–d, 2020a–i, 2021, 2022). In a recent cross-country study, Guenther et al. (2021) surveyed consumers in China, Japan, and the UK in the years 2019, 2020, and 2021 about the importance of factors underpinning the attribute environmental condition in relation to food supply. Results showed that water quality was the most important factor underpinning environmental condition for China and Japan across all survey years, while recycling was the most important factor associated with environmental condition for UK consumers across all survey years. Air quality was also an important factor underpinning environmental condition in all countries, as were factors of biodiversity and wildlife protection, including protecting coastal and sea life and protecting endangered animals and plants. In China, an important factor for environmental condition was organic production - however, this factor was not as important in Japan and the UK, where it was rated among the least important factors of environmental condition. The factors, wilderness and GHG emissions, were also not important factors for China and UK participants (Guenther et al., 2021). The ranking of the importance of these factors was consistent over the three survey years for all countries (Guenther et al., 2021). This study reinforced previous work by the AERU, which showed similar findings to the previous one (Guenther et al., 2015; Miller et al., 2017; Saunders et al., 2013, 2016; Tait et al., 2016).

Tait et al. (2016) conducted a cross-country analysis of lamb consumer preferences and WTP for environmental attributes (GHG emissions minimisation, water management, and biodiversity enhancement) against other attributes in

© Burleigh Dodds Science Publishing Limited, 2024. All rights reserved.

developed (UK) and developing economies (China and India). Results reported in Table 2 show that food safety, followed by animal welfare, appeared to be the most valued attribute, with premiums of between 9% and 49% for certified products. Of the environmental attributes, GHG minimisation certification was valued the most highly (by a thin margin) across all countries. A key difference was that Indian respondents indicated much higher WTP for environmental attributes compared with UK and Chinese consumers (Tait et al., 2016). This is consistent with previous work indicating that Chinese and Indian consumers were willing to pay higher premiums for the environmental attributes of water pollution minimisation, GHG minimisation, and improved biodiversity relative to UK consumers (Saunders et al., 2013).

Similarly, Feucht and Zander (2017) examined European consumers' (France, Germany, Italy, Norway, Spain, and the UK) WTP for 'climate-friendly' milk products (products with lower carbon footprints), including products that displayed two types of CO_2 labels, as well as the product claims 'climate friendliness', local production, and the EU organic label. Results showed that the highest WTP was for local and organic production relative to carbon labelling or 'climate friendly' claims across all countries (see Table 3) (Feucht and Zander, 2017).

This is in line with Akaichi et al. (2020) who examined and compared Spanish and UK beef mince consumer preferences and WTP for a range of credence attributes, including the associated level of GHG emissions in production, as well as combinations of attributes (such as low fat and low GHG emissions). As shown in Table 4, it was found that consumers in both countries were willing to pay a premium for the inclusion of low GHG emissions and moderate GHG emission labels in beef mince products. The UK participants showed a much higher preference for the low GHG emissions label relative to the moderate GHG emissions label (+133%), as well as relative to Spanish

Table 2 WTP for lamb attributes – China, India, UK (n = 2067 total; China n = 686, India n = 695, UK n = 686)

		WTP (%)		
		China	India	UK
Food safety (vs. not certified)	Certified	34%	49%	15%
Farm animal welfare (vs. not certified)	Certified	9%	29%	18%
Water management (vs. not certified)	Certified	7%	21%	6%
GHG minimisation (vs. not certified)	Certified	8%	28%	6%
Biodiversity enhancement (vs. not certified)	Certified	5%	26%	4%
Country of origin (vs. no label)	Domestic	−27%	-	5%
	Foreign	-	13%	-5%

Source: Tait et al. (2016).

© Burleigh Dodds Science Publishing Limited, 2024. All rights reserved.

Table 3 Willingness to pay for the environmental attributes of milk products – European countries (€per 1 litre UHT milk product)

	France (n = 1000)	Germany (n = 1001)	Italy (n = 1003)	Norway (n = 1001)	Spain (n = 1002)	UK (n = 1000)
CO_2 label 1	17%	26%	33%	13%	24%	17%
CO_2 label 2	5%	6%	12%	0%	19%	10%
'Climate friendly'	9%	10%	19%	8%	25%	7%
Local	29%	40%	37%	24%	25%	25%
Organic	18%	20%	32%	13%	27%	15%

Source: Feucht and Zander (2017).

Table 4 Marginal WTP (mean) for credence attributes of beef mince products – Spain and UK [local currency – Spain (€); UK (£)]

	Marginal WTP (local currency – €)	
	Spain	UK
Low fat (3 g per 100 g serving of beef mince)	€1.94	€3.41
Moderate fat (12 g per 100 g serving of beef mince)	€1.57	€2.23
Local (origin)	€1.61	€1.54
Rest of the country/national (origin)	€1.37	€1.33
Organic	€1.09	€1.02
Low GHG emissions (5.9 kg of CO_2e per 500 g beef mince)	€0.96	€2.05
Moderate GHG emissions (19.1 kg of CO_2e per 500 g of beef mince)	€1.24	€0.88
Low fat and local	€0.24	€0.25
Low fat and organic	€0.29	-€0.38
Low fat and low GHG	€0.53	-€1.09
Moderate fat and local	€0.44	€0.00
Moderate fat and organic	€0.22	€0.00
Moderate fat and low GHG	€0.29	-€0.98

Source: Akaichi et al. (2020).

consumers (+114%). However, the total premium that UK participants were willing to pay for the moderate GHG emissions label was 41% lower than the price premium that Spanish participants were willing to pay for the same label (Akaichi et al., 2020).

Also examining the carbon footprint of a range of meat products, Dudinskaya et al. (2021) conducted a large-scale WTP study for red meat (beef, lamb, and goat) product attributes across seven countries (Finland, France, Greece, Italy, Spain, Turkey, and the UK). Results showed positive WTP estimates for the carbon footprinting of meat products in all countries (excluding Italy),

© Burleigh Dodds Science Publishing Limited, 2024. All rights reserved.

Table 5 Estimated WTP for red meat label attributes for seven European countries – Finland (FI) (*n* = 413), France (FR) (*n* = 414), Germany (GR) (*n* = 400), Italy (IT) (*n* = 417), Spain (ES) (*n* = 417), Turkey (TR) (*n* = 391), United Kingdom (UK) (*n* = 414) (local currency)

	FI (€)	FR (€)	GR (€)	IT (€)	ES (€)	TR (TRY)	UK (£)
Halal	−€0.693	−€2.041	€0.183	−€0.542	−€1.335	13.230	−£0.713
National origin	€2.277	€3.737	€2.299	€3.052	€2.584	11.070	£0.433
EU origin	€0.636	€1.695	€0.082	€0.557	€1.068	−0.993	£0.143
Protected geographical indications/protected designations of origin	€0.035	€0.357	€0.973	€0.815	€0.472	6.857	£0.302
Carbon footprint	€0.330	€0.495	€0.412	−€0.032	€0.516	3.853	£0.047
Organic	€0.839	€2.058	€1.265	€0.657	€0.463	4.458	£0.491
Low fat	€0.330	€1.134	€0.181	€0.554	€0.357	0.242	£0.137
High protein	−€0.332	−€0.147	−€0.147	€0.183	−€0.150	−3.048	−£0.136
Ready to cook	€0.310	−€0.705	−€0.816	−€0.200	−€1.300	−1.646	−£0.285

Source: Dudinskaya et al. (2021).

© Burleigh Dodds Science Publishing Limited, 2024. All rights reserved.

with the highest overall values indicated for products of national origin, followed by organic (Table 5).

Guenther et al. (2012) examined Japanese and UK consumer knowledge of sustainability issues and preferences for sustainability labels in relation to food products – specifically recyclability, environmentally friendly sourcing, eco-friendly packaging, and GHG emissions. The research found that consumers in both markets exhibited some knowledge of and concern regarding a number of environmental issues, particularly carbon labelling, signalling a desire for the provision of carbon labels on the food products they bought. However, the authors also found that consumers had low awareness and knowledge of the processes and terms underpinning these claims (Guenther et al., 2012). This theme is consistent in the literature, as Schaufele and Hamm (2017) found in their review of international WTP literature for sustainability credentials in wine products. Specifically, the authors found that consumers across different countries showed a WTP a premium for wine products with associated sustainable production methods. However, consumers' perceptions and knowledge of sustainable wine production processes were low across a range of countries, with consumers often associating sustainability labels with quality cues in wine products (Schaufele and Hamm, 2017).

5.2 Europe

In addition to cross-country analysis, studies have assessed consumer preferences for environmental credence attributes in food products in individual European countries. For example, Tait et al. (2020i) investigated UK consumer preferences and WTP for the credence attributes of lamb products, including environmental attributes, with latent class analysis highlighting three distinct lamb consumer groups. These included Group 1 (more likely to be male, from urban location, slightly higher income, higher overall purchase frequency, and less concerned about UK origin), Group 2 [pay higher prices overall, interested in organic production, strongest preferences for various country of origin (COOs)], and Group 3 (more likely to purchase NZ lamb, and the broadest set of preferences overall). As shown in Table 6, consumers in all groups were willing to pay a premium for environmental attributes, particularly water quality protection. However, non-environmental attributes were generally valued higher by the respondents, with provenance attributes (e.g. COO) valued the highest by participants in all consumer groups (Tait et al., 2020i). In a later survey, Tait et al. (2022) included additional environmental attributes (carbon neutral and biodiversity protection). Results showed positive WTP for all environmental attributes across consumer groups. This included Group 1 (slightly higher income, more likely to be from an urban location, highest preference for a range of attributes, highest purchase frequency and

© Burleigh Dodds Science Publishing Limited, 2024. All rights reserved.

Table 6 WTP for lamb attributes (average % of product price) – UK

	WTP (average % of product price)		
	Group 1 (20% of consumers)	Group 2 (20% of consumers)	Group 3 (60% of consumers)
Water quality protection		8%	6%
Organic			14%
Enhanced animal welfare	14%	8%	14%
No GM feed	13%	10%	15%
100% grass fed	21%	25%	34%
100% pasture raised	12%	18%	22%
No added antibiotics		20%	24%
No added growth hormones	18%	21%	24%
Produced in New Zealand	46%	21%	16%
Produced on Māori farms		34%	34%
Produced in Scotland	28%	34%	42%
Produced in Wales	37%	46%	47%
Produced in England	41%	52%	48%

Source: Tait et al. (2020i). Note: In this table, WTP was included only if the attribute was statistically significant.

prices paid for lamb products), Group 2 (price and safety conscious – highest preference for low price and food safety), and Group 3 (ambivalent – lowest importance placed on a range of factors, including social responsibility, use of additives, organic production, and genetically modified). While all three groups indicated a positive WTP for water quality protection, only Group 2 showed WTP for carbon neutral (13% premium), and only Group 3 showed WTP for biodiversity protection (16% premium). This reinforces the prior study, again showing the highest WTP values for product provenance over other credence attributes (Tait et al., 2022).

Macdiarmid et al. (2021) conducted a series of choice experiments examining UK consumer preferences and WTP for trade-offs between the attributes healthiness, carbon footprint, and meat content in ready meals. In particular, within the choice experiment, participants selected one of three carbon footprint options – a red footprint (associated with the lowest GHG emissions reductions but the highest meat content), an orange footprint (associated with both moderate GHG emissions and meat content), and a green footprint (associated with the highest GHG emissions reductions but the least

© Burleigh Dodds Science Publishing Limited, 2024. All rights reserved.

meat content). Results showed that consumers were willing to pay a premium for the inclusion of a carbon footprint but were less willing to pay this premium when it was revealed that this was associated with less meat content per meal. This suggests that, in this instance, UK consumers may make trade-offs between the basic (e.g. meat content) and environmental credence attributes (e.g. lower carbon footprint) of the food products that they buy (Macdiarmid et al., 2021).

Considering verified environmental claims, Lagerkvist et al. (2014) examined Swedish consumers' WTP for COO and ethical attributes, including environmental attributes, in the presence or absence of a price attribute. The WTP results for the part of the sample that were shown product prices are outlined in Table 7. Consumers were willing to pay a premium for all attributes, including verified environmental impact of livestock production. However, the environmental attribute had one of the lowest associated premiums, with participants willing to pay an average 10% premium for this attribute (Lagerkvist et al., 2014). The findings of this research have also been corroborated by a later study that showed highly similar valuations of these attributes (Lagerkvist et al., 2017).

Also focussing on the effects of product price, Van Loo et al. (2014) surveyed Belgian consumers regarding their WTP for a range of attributes by combining different environmental and ethical attributes in a choice experiment for chicken breast products. Results presented in Table 8 showed that consumers with higher incomes were generally willing to pay higher premiums for these attributes relative to those with lower incomes. In addition, across both income groups,

Table 7 WTP for beef attributes - Sweden (n = 1070; n = 630 - no-price sample; n = 440 - price sample)

Attribute	WTP Premium (%)
Origin information (vs. zone of origin inside or outside EU)	43%
Animal-specific reference code (vs. not present)	6%
Traceability to specific slaughterhouse (vs. not present)	12%
Traceability to group or specific animal (vs. not present)	11%
Traceability to the specific breeder (vs. not present)	12%
Verified animal welfare for livestock production (vs. not present)	16%
Organic production (vs. not present)	14%
Verified environmental impact of livestock production (vs. not present)	10%
Verified health impact from consumption of beef	8%
Verified social responsibility for livestock production (vs. not present)	10%
Information about medication use (vs. not present)	16%
Type of animal feed (vs. not present)	7%

Source: Lagerkvist et al. (2014).

© Burleigh Dodds Science Publishing Limited, 2024. All rights reserved.

Table 8 WTP for chicken breast attributes - Belgium (n = 359)

Attribute		Low income Premium (%)	High income Premium (%)
Organic logo	Biogarantie logo (Belgium)	23%	34%
	EU Organic logo	12%	18%
Animal welfare (label vs. none)	European animal welfare label	26%	39%
Free range claims (vs. none)	Free range	43%	64%
	Traditional free range	50%	74%
	Free range - total freedom	63%	93%
Carbon footprint label	20% CO_2 reduction 5.6 kg CO_2e compared to 7 kg CO_2	18%	27%
	30% CO_2 reduction 4.9 kg CO_2e compared to 7 kg CO_2	24%	36%

Source: Van Loo et al. (2014)

the most valued attributes were free range, followed by animal welfare, 30% reduction in CO_2 emissions, and the Belgian organic logo (Van Loo et al., 2014).

Regarding ecolabelling, Pomarici et al. (2018) examined young Italian consumers' WTP for ecolabelled wine products with associated water efficiency using a non-hypothetical experimental auction mechanism. The authors found a range of positive premiums for the water-saving-labelled wine products, with participants placing average bids of €4.51 for the front-labelled product, €4.32 for the back-labelled product, and €4.16 for the conventional unlabelled product. The authors also found that participants' wine consumption frequency, trust in label claims, and environmentally friendly attitudes positively influenced WTP for water-saving labelled wine products (Pomarici et al., 2018).

Similarly, Boccia et al. (2019) conducted a series of choice experiments examining Italian consumers' WTP for the brand and corporate environmental and social responsibility programme participation in relation to ready-meal products. Results indicated that participants are willing to pay a €2.46 premium for products with recognisable brand names that also participated in the previously discussed programmes. In addition, participants were willing to pay a €1.53 premium for products participating in environmentally friendly social responsibility programmes, while they were only willing to pay a €0.19 premium for only social responsibility programme participation, suggesting an overall preference for the environmental attribute (Boccia et al., 2019). This is consistent with other studies, such as Paci et al. (2018) who examined Italian consumers' WTP for the inclusion of environmental and health attributes in fresh fish burger products. The authors found that consumers were willing to pay up to an additional €0.57 for the 'environment' attribute and €0.37 for the

© Burleigh Dodds Science Publishing Limited, 2024. All rights reserved.

'health' attribute, suggesting a higher valuation of the environmental attribute in this case (Paci et al., 2018).

The provision of negative information regarding a food product's environmental impacts can also influence consumers' WTP for products. This was demonstrated by Koistinen et al. (2013), who examined Finnish consumer preferences and WTP for the physical, production, and carbon footprint attributes of minced meat (beef and pork) products. Taking a latent class approach, the authors classified participants into six distinct consumer groups based on their attitudes and preferences. Specifically for the valuation of the carbon footprint of products, participants were asked to indicate their WTP for beef and pork products before and after revealing information regarding their carbon footprints. Results shown in Table 9 indicate that participants were generally less willing to pay a higher price for beef products after carbon footprint information was revealed but were generally willing to pay a higher price for pork products - however, this price impact was relatively low. This suggests that the popularity of beef products decreased due to their negative environmental impact relative to pork products in this instance, with members of the method of production-conscious class willing to pay the highest premium for pork over beef mince products (Koistinen et al., 2013).

In line with this, studies have shown that the information presented to consumers can influence their perceptions of the quality and their intention to purchase suboptimal food products (i.e. 'ugly' fruit and vegetables that would otherwise be wasted) (Pfeiffer et al., 2021; Savchenko et al., 2019a,b; van Giesen and de Hooge, 2019). For example, van Giesen and de Hooge (2019) found that consumers presented with information about the sustainability and authenticity of suboptimal food products were more likely to choose these items, increasing their purchase intention for and quality perceptions of suboptimal products.

5.3 North America

Studies have also assessed consumer preferences for environmental credence attributes in food products in North America, most predominantly the USA. For example, Li et al. (2016) examined US consumers' household WTP for a programme aimed at reducing GHG emissions associated with beef production. This found four consumer segments based on their willingness to support a programme certifying 'carbon-friendly' beef products - 'does not support', 'supports but will not pay more', 'supports and will pay more', and 'willing to pay specific premium for certified beef'. For the latter two segments combined, results indicated that participants in these segments would be willing to pay an average US$306 per year to support this programme (51.6% of their average annual total beef product spend). Across all segments, including those that would not support this programme, the average annual WTP was valued at

© Burleigh Dodds Science Publishing Limited, 2024. All rights reserved.

Table 9 WTP for beef and pork mince products after carbon footprint information revealed, by consumer group – Finland (% of usual price paid)

	Class 1: Price conscious	Class 2: Fat content conscious	Class 3: Ideological but passive	Class 4: Content with conventional	Class 5: Beef preferring	Class 6: Method of production conscious
Class size	23%	23%	17%	14%	12%	11%
WTP for beef mince	−0.9%	−3.1%	−1.7%	−2.1%	−0.2%	−2.9%
WTP for pork mince	1.0%	5.1%	4.8%	−0.9%	1.0%	13.8%

Source: Koistinen et al. (2013).

© Burleigh Dodds Science Publishing Limited, 2024. All rights reserved.

US$64 (just over 10% of all participants' average annual total beef product spend). Average WTP was also shown to be higher for participants that donated to environmental organisations (Li et al., 2016).

In a similar study, Krovetz et al. (2017) examined Californian consumers' WTP for food products with higher water efficiency in production (avocados, almonds, lettuce, and tomatoes) in the context of historic drought conditions. Results identified WTP for water efficiency of approximately US$0.12 per gallon of water saved on average, which exceeds the typical cost of water for producers during drought years (between US$0.3 and US$0.5 per gallon). The authors also found that providing participants with information on Californian drought conditions in the context of food production was associated with higher WTP values for low water footprint food products. Findings were also heterogeneous depending on participants' stated level of environmental concern, as well as a number of demographic factors. WTP estimates per gallon of water saved were also examined for different products – approximately US$0.054 per gallon for avocadoes, US$0.093 per gallon for almonds, US$0.48 per gallon for tomatoes, and US$0.0013 per gallon for lettuce (Krovetz et al., 2017).

Also examining production practices, Chen et al. (2018) examined US consumers' WTP for sustainability claims in fresh strawberries, including reduced pesticide and fertiliser use and reduced negative impacts on ecosystem services (including air, soil, and water quality). Participants indicated the highest WTP for reduced pesticide use (US$3.07/lb), followed by reduced impacts on water quality (US$2.95/lb). However, reported WTP values for all environmental attributes were very similar, ranging from US$2.92/lb to US$2.95/lb (Chen et al., 2018).

Likewise, Tait et al. (2018b) examined Californian consumer preferences and WTP for beef attributes, including environmentally sustainable, across three beef products (ground beef, top sirloin steak, and ribeye steak). Results showed that WTP estimates were generally higher for ribeye steak, followed by top sirloin steak, and ground beef. Californian consumers indicated positive WTP for the environmentally sustainable attribute (7% for ground beef, 3% for top sirloin steak, and 4% for ribeye steak) – however, this was one of the lowest positive WTP values for of all attributes, valued just above traceability (Tait et al., 2018b).

Also in California, Tait et al. (2018c) examined consumer preferences and WTP for New Zealand wine product credence attributes, including a range of environmental management attributes (biodiversity, water, by-products, energy, and GHG management). Results showed that Californian consumers were willing to pay a premium for all attributes. Of the environmental management attributes, water management had the highest premium (18%), followed by by-products management (16%), GHG management (12%), energy management (11%), and biodiversity management (9%). Overall, as

© Burleigh Dodds Science Publishing Limited, 2024. All rights reserved.

with prior studies, specific appellation had the highest associated premiums (USA followed by New Zealand), followed by organic attributes (100% organic followed by made with organic grapes) (Tait et al., 2018c).

In a different study, Tait et al. (2020f) examined New York consumers' WTP for the attributes of sauvignon blanc wine, including a series of environmental management attributes. The researchers identified three distinct consumer groups, each with WTP for attributes. These included Group 1 (younger, more likely to have children and be educated, higher wine purchase, consumption and engagement, greater use of technology for wine information and purchasing), Group 2 (more likely to live in an urban area, more likely to buy wine at winery tasting rooms), and Group 3 (older, rank NZ wine the highest of all groups, lowest level of experience, engagement and trust in wine sustainability claims, more likely to buy wine at a liquor store). As shown in Table 11, there was a diversity of WTP for these attributes, with consumers in Group 2 generally willing to pay the highest premiums for attributes, including environmental attributes. For the non-appellation attributes, participants in Group 3 showed a higher WTP for biodiversity management and GHG management, as well as the lowest overall WTP for the appellation attributes – this suggests the presence of a more environmentally concerned consumer segment in the New York wine market (Tait et al., 2020f). The study was repeated in Texas, with similar results reported

Table 10 WTP for pork chop attributes – USA (n = 839*; brand CEs n = 642, non-brand CEs n = 197)

	Brands	Choices with brand Premium (%)	Choices without brands Premium (%)
Third-party certified antibiotic-free production (vs. no certification)	Hormel	22%	18%
	Tyson	10%	
	Store brand	18%	
	No brand	28%	
Third-party certified environment-friendly production: water and air quality (vs. no certification)	Hormel	22%	7%
	Tyson	7%	
	Store brand	4%	
	No brand	9%	
Third-party certified animal welfare in the production process (vs. no certification)	Hormel	17%	12%
	Tyson	12%	
	Store brand	5%	
	No brand	19%	
	Store brand	16%	
	No brand	11%	

Source: Ubilava et al. (2011).

© Burleigh Dodds Science Publishing Limited, 2024. All rights reserved.

Table 11 WTP for wine (sauvignon blanc) credence attributes – New York, USA (n = 495)

	Consumer groups		
Attribute	Group 1 (47%)	Group 2 (22%)	Group 3 (31%)
Biodiversity management	9%		24%
Energy management	9%		
Pest and disease management	13%	16%	15%
Social responsibility	12%	6%	
GHG management	14%	8%	21%
100% organic	14%	21%	
Critic rating (per point >80)		2%	2%

Source: Tait et al. (2020f). Note: In this table, WTP was included only if the attribute was statistically significant.

Table 12 WTP for wine (sauvignon blanc) credence attributes – Texas, USA

	Consumer groups		
Attribute	Group 1 (40%)	Group 2 (25%)	Group 3 (35%)
Biodiversity management	7%	14%	16%
Water management			12%
By-product management	11%		
Energy management	5%		
Pest and disease management	12%		14%
Social responsibility	13%	14%	7%
GHG management	11%	17%	21%
Made with organic grapes			
100% organic	13%	14%	
Critic rating (per point >80)		3%	4%

Source: Tait et al. (2020g). Note: In this table, WTP was included only if the attribute was statistically significant.

across three consumer groups. These groups showed similar characteristics to the previous study and included Group 1 (younger, more likely to have children, had higher wine purchase, consumption and engagement, and greater use of technology for wine information and purchasing), Group 2 (older, ranked NZ wine higher, had the lowest level of experience, engagement and trust in wine sustainability claims, and were more likely to buy wine at discount stores), and Group 3 (higher daily internet access, more likely to buy wine at grocery stores). As shown in Table 12, consumers in Group 3 were willing to pay the highest premiums for most credence attributes in this study. For the non-appellation attributes, participants in Group 3 showed the higher WTP for most environmental management attributes, including GHG management (21%

© Burleigh Dodds Science Publishing Limited, 2024. All rights reserved.

premium), biodiversity management (16% premium), and water management (12% premium). Similarly high WTP values were shown for GHG management across all three consumer groups, with a range of positive WTP values shown for biodiversity management. Overall, as with prior studies, the appellation attributes had the highest premiums of all attributes, particularly New Zealand wines, including those produced by Māori enterprise (Tait et al., 2020g).

A food product's brand may influence consumers' WTP for its environmental attributes, as studied by Ubilava et al. (2011). The authors compared US consumers' WTP for the certification of credence attributes for branded and non-branded pork products. In a split-sample, some choice experiments included a brand (Hormel, Tyson, Store brand, and no brand), while the others did not. Attributes included antibiotic use, animal welfare, and environmental friendliness in the production process. As shown in Table 10, WTP results ranged from 4% to 28% for certified antibiotic-free, environmentally friendly, and animal welfare attributes. The study also reported a greater variation in WTP for the non-branded product, suggesting increased uncertainty when no brand information is provided (Ubilava et al., 2011).

Turning to ecolabels, Van Loo et al. (2015) investigated the consumer preferences for sustainability certification of coffee products, including Fair Trade (FT), Rainforest Alliance, USDA Organic, and carbon footprint. Three consumer segments were identified based on a cluster analysis: indifferent, sustainability and price-conscious, and price-oriented consumers. Results showed that respondents, on average, were willing to pay the most for USDA Organic certified coffee (16%), with the highest premium from the sustainability and price-conscious group (19%). The results also showed that visual attention to attributes is related to preferences for attributes, whereby taking more time and placing greater attention on a particular attribute is related to higher WTP (Van Loo et al., 2015).

Prior information regarding food products may also influence US consumer WTP for their environmental attributes. In this sense, Ortega et al. (2014) explored consumer WTP for imported seafood products for which past food contamination and adulteration incidents may have impacted consumer preferences for Chinese tilapia. Two surveys were conducted (for shrimp and Chinese tilapia products) with 335 respondents each. The corresponding CEs included a variety of credence attributes, including the environmental attribute eco-friendly production, as well as the effect of these products having a US, China, or Thailand COO. Results showed that consumers were willing to pay the highest premium for enhanced food safety (188% premium for domestic shrimp, a 41% premium for shrimp from China, and a 46% premium for shrimp from Thailand). A similar relationship was found for no antibiotic use and eco-friendly production, which were both associated with a higher WTP for the US product by US consumers. Specifically, US consumers were

© Burleigh Dodds Science Publishing Limited, 2024. All rights reserved.

willing to pay a 60% premium for eco-friendly domestic shrimp production and a 53% premium for eco-friendly domestic tilapia production (Ortega et al., 2014).

5.4 Asia

Studies have also assessed consumer preferences for environmental credence attributes in food products across Asian countries, most predominantly China. For example, Tait et al. (2020b) examined Chinese (Beijing) consumers' WTP for beef tenderloin product attributes. The choice experiment highlighted three distinct beef tenderloin consumer groups in Beijing representing 71%, 17%, and 12% of those surveyed, respectively. These included Group 1 (higher beef cut purchasing, spend the most domestically and at hypermarkets, value greater product variety), Group 2 (highest beef consumption, more likely to purchase and rank NZ beef highly, most engaged online), and Group 3 (lower income, lowest overall beef cut purchasing, less likely to think that beef production is important for the economy and that environmental impacts are well-managed, lowest online and technology engagement). Table 13 shows the results of the WTP analysis by consumer group. Of the environmental attributes, carbon neutral was the most preferred by Beijing consumers across most consumer groups (70% premium in Group 2, 21% premium in Group 1), followed by water

Table 13 WTP for beef tenderloin attributes – Beijing, China (average % of product price)

	Consumer groups		
Attributes	Group 1 (71%)	Group 2 (17%)	Group 3 (12%)
Organic	15%		30%
Enhanced animal welfare	8%		
GMO free	19%		
Carbon neutral	21%	70%	
Biodiversity enhancement	24%		
Water quality protection		42%	
Feedlot raised	94%	74%	
100% pasture raised	17%	37%	
No added antibiotics			
No added hormones	16%	8%	20%
Social responsibility		30%	
Traceability		13%	20%
100% grass fed	111%		8%
Grain fed	8%		

Source: Tait et al. (2020b). Note: In this table, WTP was included only if the attribute was statistically significant.

© Burleigh Dodds Science Publishing Limited, 2024. All rights reserved.

quality protection (42% premium in Group 2) and biodiversity enhancement (24% premium in Group 1). As with prior studies, provenance attributes were associated with the highest premiums, as well as production practices such as feedlot raised and 100% grass fed (Tait et al., 2020b).

Similarly, Tait et al. (2020c) examined Chinese (Beijing) consumers' WTP for the attributes of ultraheat-treated (UHT) milk products using the same latent class approach as the previous study. This found three groups – Group 1 (highest UHT milk purchase frequency, less likely to consider the health impacts of dairy consumption, highly engaged online), Group 2 (lowest overall dairy product purchasing), and Group 3 (higher income, highest NZ milk purchase frequency, and ranking, concerned with health and environmental sustainability). Table 14 shows participants' WTP for a range of UHT milk product attributes, including the environmental attributes carbon neutral, biodiversity enhancement, and water quality protection. These environmental attributes were among the highest valued attributes overall, particularly biodiversity enhancement (56% premium in Group 3 and 35% premium in Group 1) and carbon neutral (50% premium in Group 2) (Tait et al., 2020c). This is consistent with previous studies that have shown a higher preference for low-carbon food products among specific consumer groups in the Chinese market (Chuanmin et al., 2014; Yang et al., 2021; Zhong and Chen, 2019).

In an earlier study, Tait et al. (2018d) explored Chinese (Shanghai) consumers' WTP for the credence attributes of New Zealand yogurt products.

Table 14 WTP for UHT milk attributes – Beijing, China (average % of product price)

	Consumer groups		
Attributes	Group 1 (38%)	Group 2 (31%)	Group 3 (32%)
Enhanced animal welfare		12%	27%
Organic		44%	18%
Increased protein		12%	11%
Increased calcium		32%	23%
Care for workers		15%	10%
Contribute to local communities	33%		23%
Support for farmers		15%	
Carbon neutral		50%	
Biodiversity enhancement	35%		56%
Water quality protection	13%		
100% pasture raised		24%	
Feedlot raised	16%		
100% grass fed	12%		
Grain fed	15%	16%	14%

Source: Tait et al. (2020c). Note: In this table, WTP was included only if the attribute was statistically significant.

© Burleigh Dodds Science Publishing Limited, 2024. All rights reserved.

Results showed that, for non-COO attributes, environmental sustainability was the third-highest valued attribute (47% premium), preceded by enhanced food safety (51% premium) and organic (51% premium) (Tait et al., 2018d). Also in Shanghai, Tait et al. (2018a) examined consumers' WTP for the attributes of New Zealand kiwifruit. Results showed that, for the environmental attributes, water use and pollution minimisation in production were valued the highest (45%), followed by waste minimisation (40%) and GHG emissions minimisation (32%). It should be noted that the environmental attributes were among the lowest valued of all attributes in this study, with GHG emissions valued second lowest (Tait et al., 2018a). Conversely, Lai et al. (2018) used a series of choice experiments to determine Chinese consumers' (Beijing and Shanghai) WTP for attributes of pork products, including environmental, food safety, animal welfare standards, and provenance attributes, showing that environmental standards were the third-highest valued attribute in both cities, behind food safety and Chinese origin pork products (Lai et al., 2018).

For Japan, Tait et al. (2020e) examined consumers' WTP for a range of kiwifruit attributes. As with previous studies, the authors conducted a WTP analysis via a series of choice experiments, grouping participants into three distinct consumer groups. These included Group 1 (younger, greater focus on personal health, value for money and taste), Group 2 (higher income, more educated, environmental sustainability and social responsibility most important, highest weekly kiwifruit purchase frequency, usually pay the highest prices for kiwifruit), and Group 3 (older, more likely to be from rural location, lowest prices paid for all varieties, more likely to purchase and rank NZ kiwifruit highly). Table 15 shows the results of the WTP analysis by consumer group.

Table 15 WTP for kiwifruit attributes – Japan (average % of product price – large gold kiwifruit)

	Consumer groups		
Attributes	Group 1 (27%)	Group 2 (43%)	Group 3 (30%)
Sweet taste	46%		51%
Balance of acidic and sweet	96%		49%
Carbon neutral		155%	
Water quality protection	67%		10%
Organic	25%	32%	22%
Enhanced food safety	27%		8%
Social responsibility		44%	9%
Large size		88%	18%
Increased fibre			16%
Increased Vitamin C		74%	15%

Source: Tait et al. (2020e). Note: In this table, WTP was included only if the attribute was statistically significant.

© Burleigh Dodds Science Publishing Limited, 2024. All rights reserved.

Of the environmental attributes, participants in Group 2 were willing to pay a significant premium for carbon-neutral kiwifruit (155% premium), among the highest premiums in this study. This is consistent with Motoshita et al. (2015), who found that consumers in Japan indicated that they would adjust their purchase behaviour when given information about how their food choice affects GHG emissions. In the Tait et al. (2020e) study, participants in Group 1 and Group 3 were also willing to pay a premium for the inclusion of the water quality protection attribute, signalling WTP of 67% and 10%, respectively. Across all attributes, and as seen in prior studies, the provenance attributes carried the highest WTP, with Japanese consumers indicating a strong preference for domestically grown kiwifruit, followed by those grown in New Zealand and Italy (Tait et al., 2020e).

Similarly, Sonoda et al. (2018) examined Japanese consumers' marginal WTP for a range of beef product attributes, including the generic environmentally friendly attribute alongside COO and animal welfare. The authors used a latent class analysis approach, grouping consumers into several groups – label conscious (11% of sample), domestic preferring (31.6%), price conscious (21.4%), animal welfare preferring (23.1%), and not interested in production (12.9%). Table 16 shows that consumers in a range of classes were willing to pay a premium for the environmentally friendly attribute, with this attribute preferred over the animal welfare attribute by consumers in 3 out of 5 groups. As with previous studies, the authors also found a strong domestic preference among Japanese consumers over most other attributes, with the highest premium being indicated for Japanese beef by consumers in Class 2 (Sonoda et al., 2018).

In the case of biodiversity and wildlife protection attributes, Chang et al. (2019) examined Japanese consumer support and WTP for fish-friendly rice production with associated biodiversity credentials. Specifically, the researchers presented consumers with two different types of information, one that provided information on multifunctionality in agricultural production (e.g. rice terraces that support fish life) and the other one that provided information on environmentally friendly agricultural products. Results showed that consumers' WTP significantly increased when presented with either form of information relative to the control group (who saw neither information source), suggesting a WTP a premium for products with associated ecosystem and wildlife protection credentials (Chang et al., 2019).

Similarly, Uchida et al. (2014) examined Japanese consumer preferences for salmon, taking into account two-way interactions motivated by consumer valuations of different product attributes in relation to ecolabel characteristics. The study included a split-sample choice experiment across three types of information effects regarding fisheries (specifically overfishing and the decline of fish-stock): (1) minimal information without the source of the claim; (2) Food

© Burleigh Dodds Science Publishing Limited, 2024. All rights reserved.

Table 16 WTP for beef attributes – Japan [local currency – Yen (¥) per 100 g beef]

	Consumer groups				
Attributes	Class 1: Label conscious (11.0%)	Class 2: Domestic preferring (31.6%)	Class 3: Price conscious (21.4%)	Class 4: Animal welfare preferred (23.1%)	Class 5: Not interested in production (12.9%)
Animal welfare	¥596	¥331	¥45	¥726	¥25
Environmentally friendly	¥574	¥571	¥54	¥505	¥71
Origin: Japan	¥557	¥1817	¥51	¥369	¥441
Origin: USA	−¥193	−¥79	−¥94	−¥219	¥91
No choice	¥1453	¥136	−¥455	−¥358	−¥1342

Source: Sonoda et al. (2018).

© Burleigh Dodds Science Publishing Limited, 2024. All rights reserved.

and Agriculture Organization (FAO) based information with charts and graphics; and (3) scientific information accompanied by a diagram. Hence, instead of using a conventional approach of 'no information' vs. 'some information', the authors applied minimum information as the baseline. Results showed that Japanese consumers were willing to pay a 27% premium for domestic fish compared to imported fish, with a similar premium found for the ecolabel. Overall, the interaction effects revealed that the value of ecolabels increased value for the wild product, in particular for the domestic product. Compared to the baseline, results showed that added information increased the value of the ecolabel, although only marginally, when the FAO or science-based information was considered to be credible and interesting (Uchida et al., 2014).

6 Conclusion

There is an increasing expectation from consumers that products have sustainability credentials and labels communicating these attributes. These sustainability criteria cover a whole range of attributes from environmental to social and ethical dimensions. The purpose of this chapter was to assess consumer attitudes and preferences for different environmental attributes in food products.

A wide range of studies worldwide have investigated consumer attitudes and preferences for different environmental attributes in food products. These studies have shown that consumer attitudes and preferences and their WTP for environmental attributes vary across international markets and across products. However, the literature review revealed some similarities between countries. For example, across all regions and markets, consumers have generally indicated a higher preference and WTP for the basic and provenance attributes over the environmental attributes of food products. In particular, this generally includes the highest premiums for food safety, and local or domestic food production in most regions, with mixed preferences shown by Chinese consumers.

WTP values were varied for a range of specific environmental attributes, with generic environmental protection, as well as reduced GHG emissions, water quality protection, and biodiversity protection showing relatively low WTP In addition, across all markets and regions, consumer preferences and WTP for the environmental attributes of food products can be positively influenced by the information treatments applied to products. Specifically, ranges of WTP values observed across these studies are shown in Table 17.

Consumers in international markets have exhibited different associations of specific factors with environmental condition, with some factors emerging as more important than others, such as air quality, water quality, and wildlife protection. Some Asian market segments have indicated significantly higher WTP for environmental attributes relative to other segments, with some willing to pay

© Burleigh Dodds Science Publishing Limited, 2024. All rights reserved.

Table 17 Average and range of WTP values observed in this chapter

Attribute type	Average WTP (% of product price)	Range of WTP values (% of product price)
Generic environmental credentials (e.g. environmentally friendly)	12%	3-25%
Water protection	20%	6-67%
GHG emissions reduction	23%	5-155%
Biodiversity/wildlife protection	19%	4-56%
Waste management/reduction	22%	11-40%

significantly higher premiums for environmental attributes, and higher premiums relative to European and North American markets. This is consistent with previous findings that consumers in Asian markets rate the importance of environmental condition and indicate much higher relative WTP for environmental attributes than European markets. While provenance attributes (e.g. COO) were generally valued higher than all other attributes across the reviewed Asian studies, they were much lower in the examined European and North American studies.

This chapter has shown that, while highly heterogeneous, international consumers exhibit a positive preference and WTP for environmental attributes in food products. However, relative to other food product attributes, such as basic and provenance attributes, environmental attributes are valued relatively low across a range of countries. This information is useful for food product market positioning and particularly for international exporting firms. Research into consumer WTP for the environmental attributes of food products is ongoing and will be required over time as trends change to allow stakeholders to effectively target evolving market segments.

7 Where to look for further information

The following articles provide a good overview of the subject:

- Guenther, M., Saunders, C., Dalziel, P., Rutherford, P. and Driver, T. 2015. Maximising Export Returns. Consumer attitudes towards attributes of food and beverages in developed and emerging export markets relevant to New Zealand. AERU Research Report 336, November 2015. Agribusiness and Economics Research Unit (AERU), Lincoln University: Lincoln, New Zealand.
- Tait, P., Driver, T. and Saunders, C. 2020a. Consumer willingness to pay for environmental attributes - results from AERU research. AERU Client Report prepared for MfE and MBIE, March 2020 Agribusiness and Research Unit (AERU), Lincoln University: Lincoln, New Zealand.

© Burleigh Dodds Science Publishing Limited, 2024. All rights reserved.

- Yang, W. and Renwick, A. 2019. Consumer Willingness to Pay Price Premiums for Credence Attributes of Livestock Products – A Meta-Analysis. *Journal of Agricultural Economics* 70 (3); 618–639.

The following website provides information on consumer preferences and WTP for food product attributes:

- Agribusiness and Economic Research Unit (AERU), 2022. Accessible here: https://www.aeru.co.nz/.

8 References

Akaichi, F., Giha, C. R., Glenk, K. and Gil, J. M. 2020. How consumers in the UK and Spain value the coexistence of the claims low fat, local, organic and low greenhouse gas emissions. *Nutrients* 12(1).

Boccia, F., Manzo, R. M. and Covino, D. 2019. Consumer behaviour and corporate social responsibility: an evaluation by a choice experiment. *Corporate Social Responsibility and Environmental Management* 2018; 1-9.

Canavari, M. and Coderoni, S. 2020. Consumer stated preferences for dairy products with carbon footprint labels in Italy. *Agricultural and Food Economics* 8(1); 4.

Chang, T., Takahashi, D. and Shobayashi, M. 2019. 'Consumers' recognition of multifunctionality in agriculture and price premiums for environmentally friendly agricultural products: evidence from a survey experiment. *Food Ethics* 2(2–3); 111–125.

Chen, X., Gao, Z., Swisher, M., House, L. and Zhao, X. 2018. Eco-labeling in the fresh produce market: not all environmentally friendly labels are equally valued. *Ecological Economics* 154; 201–210.

Chuanmin, S., Xiaomin, Y., Yukun, Z., Chuanxi, S. and Penghui, D. 2014. *Consumer Behaviour on Low-Carbon Agri-Food Purchase: A Carbon Labelling Experimental Study in China*, Agricultural Economics Communications, Czech Republic (vol. 60): 133–146.

Dalziel, P., Saunders, C., Tait, P., Saunders, J., Miller, S., Guenther, M., Rutherford, P. and Driver, T. 2018. Rewarding responsible innovation when consumers are distant from producers: evidence from New Zealand. *International Food and Agribusiness Management Review* 21(4); 487–504.

Darby, M. R. and Karni, E. 1973. Free competition and the optimal amount of fraud. *Journal of Law and Economics* 16(1); 67–88.

Dudinskaya, E. C., Naspetti, S., Arsenos, G., Caramelle-Holtz, E., Latvala, T., Martin-Collado, D., Orsini, S., and Zanoli, R. 2021. European Consumers' willingness to pay for red meat labelling attributes. *Animals* 11; 556.

Eco label index 2022. *Ecolabel Directory*. Accessed 6th June 2022 Available at: https://www.ecolabelindex.com/.

Eldesouky, A., Mesias, F. J. and Escribano, M. 2020. Perception of Spanish consumers towards environmentally friendly labelling in food. *International Journal of Consumer Studies* 44(1); 64–76.

© Burleigh Dodds Science Publishing Limited, 2024. All rights reserved.

Euromonitor International 2020. Top 10 global consumer trends 2020. Accessed 18th July 2022 Available at: https://www.euromonitor.com/top-10-global-consumer-trends-2020/report.

Euromonitor International 2021. Top 10 global consumer trends 2021. Accessed 18th July 2022 Available at: https://go.euromonitor.com/white-paper-EC-2021-Top-10-Global-Consumer-Trends.html.

Feucht, Y. and Zander, K. 2017. 'Consumers' willingness to pay for climate-friendly food in European countries'. *International Journal of Food System Dynamics, Proceedings in System Dynamics and Innovation in Food Networks* 2017.

Foti, V. T., Scuderi, A., Stella, G. and Timpanaro, G. 2019. Consumer purchasing behaviour for "biodiversity-friendly" vegetable products: increasing importance of informal relationships. *Agricultural Economics*, 65(9); 404–414.

Giannoccaro, G., Carlucci, D., Sardaro, R., Roselli, L. and De Gennaro, B. C. 2019. Assessing consumer preferences for organic vs eco-labelled olive oils. *Org. Agr.* 9; 483–494.

Guenther, M., Saunders, C. M. and Tait, P. R. 2012. Carbon labelling and consumer attitudes. *Carbon Management* 3(5); 445–455.

Guenther, M., Saunders, C., Dalziel, P., Rutherford, P. and Driver, T. 2015. Maximising export returns. Consumer attitudes towards attributes of food and beverages in developed and emerging export markets relevant to New Zealand. AERU Research Report 336, *November 2015*, Agribusiness and Economics Research Unit (AERU), Lincoln University, Lincoln, New Zealand.

Guenther, M., Saunders, C., Driver, T., Dalziel, P., Tait, P. and Rutherford, P. 2021. Consumer attitudes towards food attributes: a cross-country comparison. *European Union Studies Association Asia Pacific (EUSAAP) Conference: New Directions, New Leadership in a Post-Covid Environment, 28–29 June 2021, Melbourne, Australia*.

Iraldo, F., Griesshammer, R. and Kahlenborn, W. 2020. The future of ecolabels. *The International Journal of Life Cycle Assessment* 25(5); 833–839.

Jørgensen, S., Pedersen, L. J. T. and Skard, S. 2022. How going green builds trusting beliefs. *Business Strategy and the Environment* 31(1); 297–311.

Karl, H. and Orwat, C. 2000. Economic aspects of environmental labeling. In: Folmer, H. and Tietenberg, T. (Eds). *The International Yearbook of Environmental and Resource Economics, 1999/2000*, Edward Elgar Publishing, Cheltenham.

Koistinen, L., Pouta, E., Heikkila, J., Forsman-Hugg, S., Kotro, J., Makela, J. and Niva, M. 2013. The impact of fat content, production methods and carbon footprint information on consumer preferences for minced meat. *Food Quality and Preference* 29(2); 126–136.

Krovetz, H., Taylor, R. and Villas-Boas, S. 2017. Willingness to pay for low water footprint choices during drought. National Bureau of Economic Research Working Paper 23495, *June 2017*.

Lagerkvist, C. J., Berthelsen, T., Sundstrom, K. and Johansson, H. 2014. Country of origin or EU/non-EU labelling of beef? Comparing structural reliability and validity of discrete choice experiments for measurement of consumer preferences for origin and extrinsic quality cues. *Food Quality and Preference* 34; 50–61.

Lagerkvist, C. J., Hess, S. and Johansson, H. 2017. How much credence does it take? Evidence on the trade-off between country-of-origin information and credence attributes for beef from a choice experiment in Sweden. *Foods* 6(10); 84.

© Burleigh Dodds Science Publishing Limited, 2024. All rights reserved.

Lai, J., Wang, H. H., Ortega, D. L. and Widmar, N. J. O. 2018. Factoring Chinese consumers' risk perceptions into their willingness to pay for pork safety, environmental stewardship, and animal welfare. *Food Control* 85; 423–431.

Lee, E. J., Bae, J. and Kim, K. H. 2020. The effect of environmental cues on the purchase intention of sustainable products. *Journal of Business Research* 120; 425–433.

Li, X., Jensen, K. L., Clark, C. D. and Lambert, D. M. 2016. Consumer willingness to pay for beef growing using climate friendly production practices. *Food Policy* 64; 93–106.

Macdiarmid, J. I., Cerroni, S., Kalentakis, D. and Reynolds, C. 2021. How important is healthiness, carbon footprint and meat content when purchasing a ready meal? Evidence from a non-hypothetical discrete choice experiment. *Journal of Cleaner Production* 282; 124510.

Miller, S., Driver, T., Saunders, C., Tait, P. and Rutherford, P. 2017. New Zealand food and beverage consumer preferences for product attributes and alternative retailers, and in-market use of digital media and smart technology. AERU Research Report No. 342, May 2017, Agribusiness and Economics Research Unit (AERU), Lincoln University, Lincoln, New Zealand.

Motoshita, M., Sakagami, M., Kudoh, Y., Tahara, K. and Inaba, A. 2015. Potential impacts of information disclosure designed to motivate Japanese consumers to reduce carbon dioxide emissions on choice of shopping method for daily foods and drinks. *Journal of Cleaner Production* 101; 205–214.

Nelson, P. 1970. Information and consumer behavior. *Journal of Political Economy* 78(2); 311–329.

Ortega, D. L., Wang, H. H. and Widmar, N. J. O. 2014. Aquaculture imports from Asia: an analysis of U.S. consumer demand for select food quality attributes. *Agricultural Economics* 45(5); 625–634.

Peschel, A. O., Grebitus, C., Steiner, B. and Veeman, M. 2016. How does consumer knowledge affect environmentally sustainable choices? Evidence from a cross-country latent class analysis of food labels. *Appetite* 106; 78–91.

Pfeiffer, B. E., Sundar, A. and Deval, H. 2021. Not too ugly to be tasty: guiding consumer food inference for the greater good. *Food Quality and Preference* 92; 104218.

Pomarici, E., Asioli, D., Vecchio, R. and Naes, T. 2018. Young consumers' preferences for water-saving wines: an experimental study. *Wine Economics and Policy* 7(1); 65–76.

Saunders, C., Guenther, M., Tait, P. and Saunders, J. 2013. Consumer attitudes and willingness to pay for attributes of food, in particular from New Zealand. *Contributed Paper Prepared for Presentation at the 57th AARES Annual Conference*, Sydney, New South Wales, *5-8th February, 2013*.

Saunders, C., Guenther, M., Driver, T., Tait, P., Dalziel, P. and Rutherford, P. 2015. Consumer attitudes to New Zealand food product attributes and technology use in key international markets. AERU Research Report No. 333*, May 2015*, Agribusiness and Economics Research Unit (AERU), Lincoln University, Lincoln, New Zealand.

Saunders, C., Guenther, M., Saunders, J., Dalziel, P. and Rutherford, P. 2016. Consumer preferences for attributes in food and beverages in developed and emerging export markets and their impacts. Paper prepared for presentation at EAAE Seminar 159: China-EU Agricultural Trade Relations in a Global Context, 13-15 October 2016, Nanjing, China.

Savchenko, O. M., Kecinski, M., Li, T. and Messer, K. D. 2019a. Reclaimed water and food production: cautionary tales from consumer research. *Environmental Research* 170; 320–331.

© Burleigh Dodds Science Publishing Limited, 2024. All rights reserved.

Savchenko, O. M., Li, T., Kecinski, M. and Messer, K. D. 2019b. Does food processing mitigate consumers' concerns about crops grown with recycled water? *Food Policy* 88; 101748.

Schaufele, I. and Hamm, U. 2017. 'Consumers' perceptions and willingness-to-pay for wine with sustainability characteristics: a review. *Journal of Cleaner Production* 147; 379-394.

Siraj, A., Taneja, S., Zhu,Y., Jiang, H., Luthra, S. and Kumar, A. 2022. Hey, did you see that label? It's sustainable!: understanding the role of sustainable labelling in shaping sustainable purchase behaviour for sustainable development. *Business Strategy and the Environment* 31(7); 2820-2838.

Sonoda, Y., Oishi, K., Chomei, Y. and Hirooka, H. 2018. How do human values influence the beef preferences of consumer segments regarding animal welfare and environmentally friendly production? *Meat Science* 146; 75-86.

Steiner, B. E., Peschel, A. O. and Grebitus, C. 2017. Multi-product category choices labeled for ecological footprints: exploring psychographics and evolved psychological biases for characterizing latent consumer classes. *Ecological Economics* 140; 251-264.

Tait, P. R., Saunders, C. M., Guenther, M. and Rutherford, P. 2016. Emerging versus developed economy consumer willingness to pay for environmentally sustainable food production: a choice experiment approach comparing Indian, Chinese and United Kingdom lamb consumers. *Journal of Cleaner Production* 124; 65-72.

Tait, P., Rutherford, P., Driver, T., Li, X., Saunders, C., Dalziel, P. and Guenther, M. 2018a. Consumer Insights and Willingness to Pay for Attributes: Kiwifruit in Shanghai, China. AERU Research Report No. 346, June 2018, Agribusiness and Economics Research Unit (AERU), Lincoln University, Lincoln, New Zealand.

Tait, P., Rutherford, P., Driver, T., Li, X., Saunders, C., Dalziel, P. and Guenther, M. 2018b. Consumer Insights and Willingness to Pay for Attributes: New Zealand Beef Products in California, USA. AERU Research Report No. 348*, June 2018*, Agribusiness and Economics Research Unit (AERU), Lincoln University, Lincoln, New Zealand.

Tait, P., Rutherford, P., Driver, T., Li, X., Saunders, C., Dalziel, P. and Guenther, M. 2018c. Consumer Insights and Willingness to Pay for Attributes: New Zealand Wine in California, USA. AERU Research Report No. 349*, June 2018*, Agribusiness and Economics Research Unit (AERU), Lincoln University, Lincoln, New Zealand.

Tait, P., Rutherford, P., Driver, T., Li, X., Saunders, C., Dalziel, P. and Guenther, M. 2018d. Consumer Insights and Willingness to Pay for Attributes: New Zealand Yogurt Products in Shanghai, China. AERU Research Report No. 347*, June 2018*, Agribusiness and Economics Research Unit (AERU), Lincoln University, Lincoln, New Zealand.

Tait, P., Driver, T. and Saunders, C. 2020a. Consumer willingness to pay for environmental attributes - results from AERU research. AERU Client Report Prepared for MfE and MBIE, March 2020. Agribusiness and Research Unit (AERU), Lincoln University, Lincoln, New Zealand.

Tait, P., Saunders, C., Dalziel, P., Rutherford, P., Driver, T. and Guenther, M. 2020b. Beijing beef consumer consumption behaviours and product preferences: a Latent Class Analysis. AERU Research Report No. 360*, November 2020*. Agribusiness and Research Unit (AERU), Lincoln University, Lincoln, New Zealand.

Tait, P., Saunders, C., Dalziel, P., Rutherford, P., Driver, T. and Guenther, M. 2020c. Beijing U H T Milk consumer consumption behaviours and product preferences: a Latent

© Burleigh Dodds Science Publishing Limited, 2024. All rights reserved.

Class Analysis. AERU Research Report No. 363, *November 2020*. Agribusiness and Research Unit (AERU), Lincoln University, Lincoln, New Zealand.

Tait, P., Saunders, C., Dalziel, P., Rutherford, P., Driver, T. and Guenther, M. 2020e. Japanese Kiwifruit consumer consumption behaviours and product preferences: a Latent Class Analysis. AERU Research Report No. 361, *November 2020*. Agribusiness and Research Unit (AERU), Lincoln University, Lincoln, New Zealand.

Tait, P., Saunders, C., Dalziel, P., Rutherford, P., Driver, T. and Guenther, M. 2020f. New York Sauvignon blanc wine consumer consumption behaviours and product preferences: a Latent Class Analysis. AERU Research Report No. 364, *November 2020*. Agribusiness and Research Unit (AERU), Lincoln University, Lincoln, New Zealand.

Tait, P., Saunders, C., Dalziel, P., Rutherford, P., Driver, T. and Guenther, M. 2020g. Texas Sauvignon blanc consumer consumption behaviours and product preferences: a Latent Class Analysis. AERU Research Report No. 365, *November 2020*. Agribusiness and Research Unit (AERU), Lincoln University, Lincoln, New Zealand.

Tait, P., Saunders, C., Dalziel, P., Rutherford, P., Driver, T. and Guenther, M. 2020h. United Arab Emirates beef consumer consumption behaviours and product preferences: a Latent Class Analysis. AERU Research Report No. 359, *November 2020*. Agribusiness and Research Unit (AERU), Lincoln University, Lincoln, New Zealand.

Tait, P., Saunders, C., Dalziel, P., Rutherford, P., Driver, T. and Guenther, M. 2020i. United Kingdom lamb consumer consumption behaviours and product preferences: a Latent Class Analysis. AERU Research Report No. 362, *November 2020*. Agribusiness and Research Unit (AERU), Lincoln University, Lincoln, New Zealand.

Tait, P., Saunders, C., Dalziel, P., Rutherford, P., Driver, T. and Guenther, M. 2021. California apple consumer consumption behaviours and product preferences: a Latent Class Analysis AERU Research Report No. 366, *March 2021*. Agribusiness and Research Unit (AERU), Lincoln University, Lincoln, New Zealand.

Tait, P., Saunders, C., Dalziel, P., Rutherford, P., Driver, T. and Guenther, M. 2022. United Kingdom lamb consumer consumption behaviours and product preferences: a Latent Class Analysis (2020). AERU Research Report No. 371, *March 2022*. Agribusiness and Research Unit (AERU), Lincoln University, Lincoln, New Zealand.

Ubilava, D., Foster, K. A., Lusk, J. L. and Nilsson, T. 2011. Differences in consumer preferences when facing branded versus non-branded choices. *Journal of Consumer Behaviour* 10(2); 61–70.

Uchida, H., Onozaka, Y., Morita, T. and Managi, S. 2014. Demand for ecolabeled seafood in the Japanese market: a conjoint analysis of the impact of information and interaction with other labels. *Food Policy* 44; 68–76.

Van Amstel, M., Driessen, P. and Glasbergen, P. 2008. Eco-labeling and information asymmetry: a comparison of five eco-labels in the Netherlands. *Journal of Cleaner Production* 16(3); 263–276.

Van Giesen, R. I. and de Hooge, I. E. 2019. Too ugly, but I love its shape: Reducing food waste of suboptimal products with authenticity (and sustainability) positioning. *Food Quality and Preference* 75; 249–259.

Van Loo, E. J., Caputo, V., Nayga Jr., R. M. and Verbeke, W. 2014. 'Consumers' valuation of sustainability labels on meat. *Food Policy* 49; 137–150.

Van Loo, E. J., Caputo, V., Nayga Jr., R. M., Seo, H.-S., Zhang, B. and Verbeke, W. 2015. Sustainability labels on coffee: consumer preferences, willingness-to-pay and visual attention to attributes. *Ecological Economics* 118; 215–225.

© Burleigh Dodds Science Publishing Limited, 2024. All rights reserved.

Viegas, I., Nunes, L. C., Madureira, L., Fontes, M. A. and Santos, J. L. 2014. 'Beef Credence Attributes: implications of Substitution Effects on Consumers' WTP. *Journal of Agricultural Economics* 65(3); 600–615.

Whakaaronui, T. Puna 2022. Well_NZ - Reframing New Zealand's Food Sector Opportunities. Te Puna Whakaaronui. Accessed 19th July 2022 Available at: https://fitforabetterworld.org.nz/assets/Te-Puna-Whakaaronui-publications/Reframing-New-Zealands-Food-Sector-Opportunities.pdf.

Yang, W. and Renwick, A. 2019. Consumer willingness to pay price premiums for credence attributes of livestock products – a meta-analysis. *Journal of Agricultural Economics* 70(3); 618–639.

Yang, X., Chen, Q., Xu, Z., Zheng, Q., Zhao, R., Yang, H., Ruan, C., Han, F. and Chen, Q. 2021. 'Consumers' preferences for health-related and low-carbon attributes of rice: a choice experiment. *Journal of Cleaner Production* 295; 126443.

Zhong, S. and Chen, J. 2019. How environmental beliefs affect consumer willingness to pay for the greenness premium of low-carbon agricultural products in China: theoretical model and survey-based evidence. *Sustainability* 11(3); 592.

© Burleigh Dodds Science Publishing Limited, 2024. All rights reserved.

Index

© Burleigh Dodds Science Publishing Limited, 2024. All rights reserved.

© Burleigh Dodds Science Publishing Limited, 2024. All rights reserved.

© Burleigh Dodds Science Publishing Limited, 2024. All rights reserved.

© Burleigh Dodds Science Publishing Limited, 2024. All rights reserved.

© Burleigh Dodds Science Publishing Limited, 2024. All rights reserved.